Using Real–Time Data and AI for Thrust Manufacturing

D. Satishkumar
Nehru Institute of Technology, India

M. Sivaraja
Nehru Institute of Technology, India

A volume in the Advances in
Computational Intelligence and
Robotics (ACIR) Book Series

Published in the United States of America by
IGI Global
Engineering Science Reference (an imprint of IGI Global)
701 E. Chocolate Avenue
Hershey PA, USA 17033
Tel: 717-533-8845
Fax: 717-533-8661
E-mail: cust@igi-global.com
Web site: http://www.igi-global.com

Library of Congress Cataloging-in-Publication Data

CIP Data in progress

This book is published in the IGI Global book series Advances in Computational Intelligence and Robotics (ACIR) (ISSN: 2327-0411; eISSN: 2327-042X)

British Cataloguing in Publication Data
A Cataloguing in Publication record for this book is available from the British Library.

All work contributed to this book is new, previously-unpublished material.
The views expressed in this book are those of the authors, but not necessarily of the publisher.

For electronic access to this publication, please contact: eresources@igi-global.com.

Advances in Computational Intelligence and Robotics (ACIR) Book Series

ISSN:2327-0411
EISSN:2327-042X

Editor-in-Chief: Ivan Giannoccaro, University of Salento, Italy

MISSION

While intelligence is traditionally a term applied to humans and human cognition, technology has progressed in such a way to allow for the development of intelligent systems able to simulate many human traits. With this new era of simulated and artificial intelligence, much research is needed in order to continue to advance the field and also to evaluate the ethical and societal concerns of the existence of artificial life and machine learning.

The **Advances in Computational Intelligence and Robotics (ACIR) Book Series** encourages scholarly discourse on all topics pertaining to evolutionary computing, artificial life, computational intelligence, machine learning, and robotics. ACIR presents the latest research being conducted on diverse topics in intelligence technologies with the goal of advancing knowledge and applications in this rapidly evolving field.

COVERAGE

- Cyborgs
- Fuzzy Systems
- Evolutionary Computing
- Artificial Intelligence
- Brain Simulation
- Adaptive and Complex Systems
- Natural Language Processing
- Heuristics
- Computational Intelligence
- Pattern Recognition

IGI Global is currently accepting manuscripts for publication within this series. To submit a proposal for a volume in this series, please contact our Acquisition Editors at Acquisitions@igi-global.com or visit: http://www.igi-global.com/publish/.

Titles in this Series

For a list of additional titles in this series, please visit:
http://www.igi-global.com/book-series/advances-computational-intelligence-robotics/73674

Predicting Natural Disasters With AI and Machine Learning
D. Satishkumar (Nehru Institute of Technology, India) and M. Sivaraja (Nehru Institute of Technology, India)
Engineering Science Reference • copyright 2024 • 340pp • H/C (ISBN: 9798369322802) • US $315.00 (our price)

Impact of AI on Advancing Women's Safety
Sivaram Ponnusamy (Sandip University, Nashik, India) Vibha Bora (G.H. Raisoni College of Engineering, Nagpur, India) Prema M. Daigavane (G.H. Raisoni College of Engineering, Nagpur, India) and Sampada S. Wazalwar (G.H. Raisoni College of Engineering, Nagpur, India)
Engineering Science Reference • copyright 2024 • 315pp • H/C (ISBN: 9798369326794) • US $315.00 (our price)

Empowering Low-Resource Languages With NLP Solutions
Partha Pakray (National Institute of Technology, Silchar, India) Pankaj Dadure (University of Petroleum and Energy Studies, India) and Sivaji Bandyopadhyay (Jadavpur University, India)
Engineering Science Reference • copyright 2024 • 330pp • H/C (ISBN: 9798369307281) • US $300.00 (our price)

AIoT and Smart Sensing Technologies for Smart Devices
Fadi Al-Turjman (AI and Robotics Institute, Near East University, Nicosia, Turkey & Faculty of Engineering, University of Kyrenia, Kyrenia, Turkey)
Engineering Science Reference • copyright 2024 • 250pp • H/C (ISBN: 9798369307861) • US $300.00 (our price)

Industrial Applications of Big Data, AI, and Blockchain
Mahmoud El Samad (Lebanese International University, Lebanon) Ghalia Nassreddine (Rafik Hariri University, Lebanon) Hani El-Chaarani (Beirut Arab University, Lebanon) and Sam El Nemar (AZM University, Lebanon)

For an entire list of titles in this series, please visit:
http://www.igi-global.com/book-series/advances-computational-intelligence-robotics/73674

701 East Chocolate Avenue, Hershey, PA 17033, USA
Tel: 717-533-8845 x100 • Fax: 717-533-8661
E-Mail: cust@igi-global.com • www.igi-global.com

Editorial Advisory Board

List of Reviewers

Jose Anand, *KCG college of Technology, India*
Paul Arokiadass Jerald. M., *Periyar Arts college, India*
R. Chitra, *Karunya University, India*
Nirmala Devi, *KCG college of Technology, India*
Deepa Jeyan, *Nehru College of Engineering and Research Center, India*
D. Kaleswaran, *Rathinam Technical Campus, India*
T. Kavitha, *AMC Engineering College, India*
P. K. Manoj Kumar, *Nehru Arts and Science college, India*
Sudheer Sankara Marar, *Nehru College of Engineering and Research Center, India*
D. Marshiana, *Sathyabama University, India*
K. Lakshmi Narayanan, *Francis Xavier Engineering College, India*
S. Padur Nisha, *Nehru Institute of Technology, India*
Lakshmana Pandiyan, *Puducherry Technological University, India*
C. V. Priya, *Muthoot Institute of Technology and Science, India*
Vineetha K. R., *Nehru College of Engineering and Research Center, India*
G. T. Rajan, *Sathyabama University, India*
A. Rameshbabu, *Sathyabama University, India*
Srinivasan Selvaganapathy, *Nokia Bell Labs CTO, India*
T. Senthilkumar, *GRT Institute of Technology and Science, India*
Pooja Singh, *Amity University, Noida, India,*
Sivaranjani, *Kobgu Engineering College, India*
L. Sumathi, *Government College of Technology, India*
Sundarsingh, *Sathyabama University, India*
S. Suresh, *PA college of Engineering and Technology, India*
S. Venkatalakshmi, *Sri Krishna College of Engineering and Technology, India*
S. A. Yuvaraja, *GRT Institute of Technology and Science, India*

Table of Contents

Detailed Table of Contents

Chapter 1
    A. Gobinath, Velammal College of Engineering and Technology, India
    Manjula Devi, Velammal College of Engineering and Technology, India
    P. Rajeswari, Velammal College of Engineering and Technology, India
    A. Srinivasan, Velammal College of Engineering and Technology, India
    Pavithra Devi, Velammal College of Engineering and Technology, India

The merger of artificial intelligence (AI) and robotics constitutes a paradigm shift in the field of automation, resulting in AI-enabled robotics. This abstract provides a succinct summary of the essential characteristics and consequences of artificial intelligence-enabled robotics. AI-enabled robotics enhances the capabilities of robotic systems by utilizing advanced machine learning algorithms and computational intelligence. AI and robotics collaboration gives robots the ability to detect, learn, adapt, and make intelligent judgments, resulting in unparalleled levels of autonomy and efficiency. Computer vision, natural language processing, and machine learning algorithms are key components of AI-enabled robotics, allowing robots to perceive and respond to complex stimuli in real time. This combination of AI and robotics not only allows for the automation of repetitive tasks, but also allows for the execution of complex and context-sensitive processes across multiple domains.

Chapter 2
    P. Selvakumar, Nehru Institute of Technology, India
    Sivaraja Muthusamy, Nehru Institute of Technology, India
    D. Satishkumar, Nehru Institute of Technology, India
    P. Vigneshkumar, KGiSL Institute of Technology, India
    C. Selvamurugan, Dhaanish Ahmed Institute of Technology, India
    P. Satheesh Kumar, Dr. N.G.P. Institute of Technology, India

The field of artificial intelligence has a long history and is always developing. It focuses on intelligent agents, which are devices composed of sensors that sense their environment and act in a way that maximizes the chance of achieving a goal. In this essay, the authors discuss the foundations of modern AI along with several examples of its use. Artificial intelligence approaches, including machine learning, deep learning, and predictive analysis, have been the subject of recent research with the goal of enhancing human learning, planning, reasoning, and initiative. Artificial intelligence driven literature review applications that can help save time and effort, enhance your comprehension and writing abilities, and generate high caliber research. Give them a try and discover how they might greatly enhance your literature review procedure. On the other hand, they can be an invaluable resource for assisting in locating pertinent research publications, recognizing key ideas, and monitoring the evolution of research over time.

Chapter 3

Dhinakaran Damodaran, Vel Tech Rangarajan Dr. Sagunthala R&D Institute of Science and Technology, India
Selvaraj Damodaran, Panimalar Engineering College, India
M. Thiyagarajan, Vel Tech Rangarajan Dr. Sagunthala R&D Institute of Science and Technology, India
L. Srinivasan, Dr. N.G.P. Institute of Technology, India

This chapter delves into the dynamic integration of artificial intelligence (AI) and machine learning (ML) in redefining trust in the digital age. Traditional trust-building methods encounter transformation through the capabilities of AI and ML, presenting both challenges and opportunities. The exploration encompasses ethical considerations in trustworthy algorithms, transparency in automated decision systems, and the crucial role of data in risk prediction. It delves into the reshaping of security and privacy paradigms, addressing bias, and fostering collaboration between humans and machines in trust manufacturing. The chapter also investigates the role of emotional intelligence in AI-enabled trust. In presenting a comprehensive overview, it contributes to the discourse on potential future scenarios where AI and ML revolutionize the establishment and scaling of trust in an interconnected digital landscape.

Dive deep into the many ways AI is revolutionizing thrust manufacturing in the aerospace sector with this in-depth study piece. Many parts of production are being rethought by artificial intelligence technology, from design optimization to real-time monitoring. Precision, efficiency, and creativity in the production of thrust systems are showcased in this study, which investigates the revolutionary influence of AI on predictive maintenance, quality control, supply chain management, and human-machine collaboration.

This chapter explores the transformative impact of integrating real-time data and artificial intelligence (AI) in the field of thrust manufacturing, particularly within the aerospace and automotive industries. As manufacturing processes evolve, the synergy between real-time data and AI advancements emerges as a catalyst for unparalleled efficiency, precision, and innovation. The chapter examines the foundational role of real-time data in providing a granular view of operations, complemented by the sophisticated capabilities of AI—from automation to adaptive intelligence. Through case studies, the document showcases successful applications of this synergy in optimizing production, predictive maintenance, and quality control. Despite the promise, challenges such as data security and workforce upskilling are acknowledged. The chapter concludes by envisioning a future where the convergence of real-time data and AI defines the landscape of intelligent thrust manufacturing, presenting opportunities for smart factories and adaptive supply chains.

    Renugadevi Ramalingam, R.M.K. Engineering College, India

    Malathi Murugesan, E.G.S. Pillay Engineering College, India

    S. Vaishnavi, Manipal Institute of Technology, India

    N. Priyanka, Vellore Institute of Technology, India

    S. Nalini, SRM Institute of Science and Technology, India

    T. Chandrasekar, Kalasalingam Academy of Research and Education, India

Artificial intelligence (AI) has the capacity to revolutionize the manufacturing sector. Positive effects include things like more output, lower costs, better quality, and less downtime. Large factories are just one group of people who can take advantage of this technology. It is important for many smaller firms to understand how simple it is to obtain high-quality, affordable AI solutions. AI has a wide range of potential applications in manufacturing. It enhances defect identification by automatically classifying faults in a variety of industrial products using sophisticated image processing techniques. Artificial intelligence has various potential applications in manufacturing since industrial IoT and smart factories generate enormous amounts of data every day. To better analyze data and make choices, manufacturers are increasingly using artificial intelligence solutions like deep learning neural networks and machine learning (ML). One common use of artificial intelligence in manufacturing is predictive maintenance.

    Geetha Manoharan, SR University, India

    Sunitha Purushottam Ashtikar, SR University, India

    M. Nivedha, Robert Gordon University, UK

AI is revolutionising industry with unprecedented efficiency and innovation. AI's promise has transformed large manufacturing organizations. This chapter covers manufacturing AI applications from optimization to process automation. Image and video recognition, prescriptive modeling, smart automation, advanced simulation, complex analytics, and more employ AI. Machine learning and deep learning apply AI to manufacturing using neural networks and algorithms. AI improves computer vision and image identification for quality control, improving product inspections. AI helps supply chain management estimate demand, optimize stocks, improve logistics, and distribute procedures. Strong cyber security and staff upskilling are needed to protect sensitive production data and smoothly move to AI-driven processes. Manufacturing's widespread AI deployment raises ethical, legal, and social challenges that researchers, industry stakeholders, and policymakers must solve. AI-powered manufacturers have various obstacles in maximizing efficiency, creativity, and long-term success.

Chapter 8

    Atharva Paymode, VIT Bhopal University, India
    Janhvi Shukla, VIT Bhopal University, India
    D. Lakshmi, VIT Bhopal University, India

The impact of AI on Industry 4.0 is discussed in this chapter, with a focus on how it works best with IoT to create autonomous factories. It is emphasized how important AI is for predictive maintenance, downtime reduction, and utilizing ML and deep learning for increased productivity. This study looks at how NLP and machine vision may revolutionize document processing and work in tandem with intelligent document processing to remove administrative bottlenecks. In addition to technical details, the chapter explores wider ramifications, including proactive field service and customized solutions. Supply chain optimization, cost reduction, and value extraction from datasets all depend on the integration of AI and data analytics. Designed with experts and policymakers in mind, this succinct story offers insightful information about AI's contribution to the evolution of manufacturing to a global audience ready to understand the rapid changes taking place.

Chapter 9

    P. Vijayakumar, Nehru Institute of Technology (Autonomous),
        Coimbatore, India
    S. Satheesh Kumar, Nehru Institute of Technology (Autonomous),
        Coimbatore, India
    B. S. Navaneeth, Nehru Institute of Technology (Autonomous),
        Coimbatore, India
    R. Anand, Nehru Institute of Technology (Autonomous), Coimbatore,
        India

Industry 4.0, the fourth industrial revolution, brought about a revolutionary shift in manufacturing by utilizing digital technology to improve productivity and communication. This chapter examines the deep integration of artificial intelligence (AI) technologies inside the industrial landscape as we approach Industry 5.0. Industry 5.0 represents a turn toward human-machine harmony, with artificial intelligence (AI) emerging as the key to coordinating this mutually beneficial partnership. It examines the transition from automation to cooperation, emphasizing AI technologies' role in enabling this peaceful cohabitation. This chapter explains how enhancing human capabilities with AI technology can empower workers. To ensure that the workforce continues to be a key factor in Industry 5.0, measures such as upskilling initiatives and the development of new positions that use AI are investigated. This chapter thoroughly reviews the relationship between AI and Industry 5.0, laying the basis for future research and discussion in the domains of ethics, industry, and technology.

This chapter discusses the convergence of the Fifth Industrial Revolution (Industry 5.0) with artificial intelligence (AI) technologies, emphasizing the revolutionary influence on industries, economies, and societal paradigms. Industry 5.0 is the most recent stage in the progression of industrial revolutions, and it is distinguished by the incorporation of smart technology, networking, and human-machine collaboration. The seamless integration of physical and digital systems lies at the heart of this revolution, resulting in a comprehensive and interconnected industrial environment. Artificial intelligence technology emerges as a key enabler of Industry 5.0, enabling intelligent automation, predictive analytics, and decision-making skills. This abstract investigates the deep implications of artificial intelligence in industrial environments, such as increased efficiency, optimal resource usage, and the development of adaptive, self-learning systems.

In the era of Industry 5.0, the integration of artificial intelligence (AI) transforms manufacturing, reshaping operational efficiency, maintenance practices, and production dynamics. This chapter delves into the core of smart factories and intelligent manufacturing, exploring the synergy of AI, machine learning, and real-time data analytics that underpins Industry 5.0. From shop floors to supply chains, AI optimization drives data-driven decision-making, creating a precision manufacturing landscape for maximum efficiency. The chapter illuminates how AI revolutionizes equipment management through proactive predictive maintenance, reducing downtime and enhancing sustainability. AI's impact extends to manufacturing flexibility, where smart factories adapt seamlessly to dynamic market demands, ensuring adaptability and agility. Through case studies and industry applications, this chapter unveils a future where AI and manufacturing converge, defining a transformative era of precision, adaptability, and efficiency in Industry 5.0.

> *R. Renugadevi, R.M.K. Engineering College, India*
> *J. Shobana, SRM Institute of Science and Technology, India*
> *K. Arthi, SRM Institute of Science and Teechnology, India*
> *Kalpana A. V., SRM Institute of Science and Technology, India*
> *D. Satishkumar, Nehru Institute of Technology, India*
> *M. Sivaraja, Nehru Institute of Technology, India*

Artificial intelligence (AI) is a system endowed with the capability to perceive its surroundings and execute actions aimed at maximizing the probability of accomplishing its objectives. It possesses the capacity to interpret and analyze data in a manner that facilitates learning and adaptation over time. Generative AI pertains to artificial intelligence models specifically designed for the creation of fresh content, spanning written text, audio, images, or videos. Its applications are diverse, ranging from generating stories mimicking a particular author's style to producing realistic images of non-existent individuals, composing music in the manner of renowned composers, or translating textual descriptions into video clips.

> *Siva Kumar A., SRM Institute of Science and Technology, India*
> *G. Indra, R.M.K. College of Engineering and Technology, India*
> *Umamageswaran Jambulingam, SRM Institute of Science and Technology, India*
> *Ramyadevi K., R.M.K. Engineering College, India*
> *Praveen Kumar B., GITAM School of Technology, India*
> *Kalpana A. V., SRM Institute of Science and Technology, India*

In recent decades, technological advancements have reshaped industries, particularly in communication, consumer electronics, and medical electronics. The rise of the industrial internet of things (IIoT) has been transformative, revolutionizing industrial processes. Accurate node localization within IIoT-enabled systems is crucial for operational efficiency and informed decision-making. This chapter introduces the 3D adaptive stochastic control algorithm (ASCA), designed explicitly for IIoT environments with machine learning capabilities, including convolutional neural networks. The algorithm leverages received signal strength indicator data and sophisticated machine learning algorithms to accurately determine node positions in a 3D industrial space, offering a cost-effective and energy-efficient alternative. Comparative evaluations show the superiority of the 3D ASCA algorithm, with an average localization error ranging from 0.37 to 0.7, surpassing benchmarks set by other algorithms.

Preface

Welcome to this comprehensive edited reference book, *Real-Time Data and AI for Thrust Manufacturing*, meticulously curated and edited by D. Satishkumar and M. Sivaraja. In an era where the fusion of technology and manufacturing is reshaping industries globally, this compilation explores the transformative impact of Artificial Intelligence (AI) on Trust Manufacturing.

The manufacturing landscape has evolved significantly, driven by the imperative need for informed decision-making. This book delves into the critical role of AI in achieving precision, efficiency, and seamless operations throughout the manufacturing process. From Industry 4.0 and the Internet of Things (IoT) to intelligent document processing and predictive maintenance, each chapter is a gateway to understanding the profound changes AI is bringing to the manufacturing sector.

The opening chapters lay the groundwork by examining the broader context of AI in manufacturing, emphasizing the Industry 4.0 paradigm and the pivotal role of machine learning (ML), deep learning, natural language processing, and machine vision. The authors underscore the significance of real-time data, presenting compelling statistics on the vast amounts of data generated within the manufacturing sector.

The subsequent chapters dissect specific applications of AI in manufacturing, ranging from intelligent document processing and communication between vendors and manufacturers to real-time monitoring, predictive analytics, and AI-enabled robotics. The diverse topics covered offer a panoramic view of how AI is optimizing operations, enhancing productivity, and driving innovation across various manufacturing sub-sectors.

This book is tailored for a wide-ranging audience, including professionals, policy-makers, researchers, academicians, and industry experts. It aspires to provide valuable insights into the societal and community benefits arising from AI in manufacturing, as well as its implications for resource utilization and daily life.

The listed chapters serve as an intellectual journey, exploring the present and envisioning the future of AI in Trust Manufacturing. As the manufacturing industry embraces AI-powered tools, real-time data analytics, and innovative solutions, this reference book stands as a beacon, guiding readers through the intricate nexus of technology and manufacturing.

We extend our gratitude to the contributors who have shared their expertise and experiences, making this book a valuable resource for anyone seeking a deeper understanding of the dynamic intersection between AI and manufacturing.

ORGANIZATION OF THE BOOK

Chapter 1: AI-Enabled Robotics

Exploring the synergy between artificial intelligence and robotics, this chapter emphasizes how AI-enabled robotics enhances autonomy and efficiency in automation. Key components such as computer vision, natural language processing, and machine learning algorithms empower robots to adapt intelligently to complex stimuli. The chapter showcases the transformative impact of AI in automating repetitive tasks and executing context-sensitive processes across various domains.

Chapter 2: AI-Powered Tools

This chapter delves into the foundations of modern AI, discussing various approaches, including machine learning, deep learning, and predictive analysis. It explores applications of AI-driven literature review tools, emphasizing their potential to enhance comprehension, writing abilities, and research quality. The chapter serves as a guide for leveraging AI-powered tools to streamline literature review processes.

Chapter 3: Forging Trust: A Futuristic Exploration of AI and ML in Trust Manufacturing

This chapter explores the integration of AI and ML in redefining trust-building methods. It delves into ethical considerations, transparency in automated decision systems, and the role of emotional intelligence in AI-enabled trust manufacturing. The chapter contributes to the discourse on potential future scenarios where AI and ML revolutionize trust establishment in a connected digital landscape.

Chapter 4: Applications of Artificial Intelligence in Thrust Manufacturing Enhancing Precision and Efficiency

Diving into the aerospace sector, this chapter explores how AI revolutionizes thrust manufacturing. From design optimization to real-time monitoring, the authors showcase AI's impact on predictive maintenance, quality control, and supply chain management. The chapter highlights the role of AI in reshaping human-machine collaboration and ensuring precision in manufacturing processes.

Chapter 5: Revolutionizing Thrust Manufacturing - The Synergy of Real-Time Data and AI Advancements

This chapter explores the transformative impact of integrating real-time data and artificial intelligence (AI) in thrust manufacturing within the aerospace and automotive industries. The authors examine the foundational role of real-time data, complemented by AI advancements, showcasing successful applications in optimizing production, predictive maintenance, and quality control. Challenges such as data security and workforce upskilling are acknowledged.

Chapter 6: The Usage of Artificial Intelligence in Manufacturing Industries - A Real-Time Application

Highlighting the potential of AI to revolutionize the manufacturing sector, this chapter discusses applications such as defect identification, predictive maintenance, and data analytics. The research emphasizes the simplicity of obtaining high-quality, affordable AI solutions and the broad range of applications in manufacturing, including image processing techniques for defect identification and predictive maintenance.

Chapter 7: Harnessing the Power of Artificial Intelligence in Reinventing Manufacturing Sector

Addressing the revolution brought by AI in the manufacturing sector, this chapter covers applications from optimization to process automation. It discusses the use of AI in image and video recognition, prescriptive modeling, and analytics, emphasizing the impact on quality control, supply chain management, and workforce upskilling.

Chapter 8: Revolutionizing the Manufacturing Sector - A Comprehensive Analysis of AI's Impact on Industry 4.0

Focusing on the impact of AI on Industry 4.0, this chapter explores how AI works in tandem with IoT to create autonomous factories. It discusses the contributions of AI to predictive maintenance, downtime reduction, and document processing. The chapter offers insights into wider ramifications, including proactive field service and customized solutions, shaping the evolution of manufacturing.

Chapter 9: Industrial Revolution (Industry 5.0) and Artificial Intelligence Technology

Examining the transition from Industry 4.0 to Industry 5.0, this chapter emphasizes the deep integration of AI technologies. It explores the shift towards human-machine harmony, investigating measures such as upskilling initiatives and the development of new positions that use AI. The chapter reviews the relationship between AI and Industry 5.0, laying the foundation for future research in ethics, industry, and technology.

Chapter 10: Industrial Revolution (Industry 5.0) and Artificial Intelligence Technology: Unleashing the Power of Industry 5.0

This chapter discusses the convergence of Industry 5.0 with AI technologies, emphasizing the revolutionary influence on industries, economies, and societal paradigms. It explores the seamless integration of physical and digital systems, resulting in a comprehensive and interconnected industrial environment. The chapter investigates the implications of AI in Industry 5.0, including increased efficiency, optimal resource usage, and the development of adaptive, self-learning systems.

Chapter 11: Precision Paradigm: AI Infused Evolution of Manufacturing in Industry 5.0

In the era of Industry 5.0, this chapter explores the integration of AI, machine learning, and real-time data analytics in smart factories and intelligent manufacturing. It highlights AI's role in driving data-driven decision-making, proactive predictive maintenance, and manufacturing flexibility. The chapter envisions a future where AI defines a transformative era of precision, adaptability, and efficiency in Industry 5.0.

Chapter 12: Using Real-Time Data and AI - Real-Time Applications of Artificial Intelligence Technology in Daily Operations

This chapter introduces artificial intelligence (AI) as a system capable of perceiving and analyzing data to maximize goal achievement. It explores generative AI, designed for creating fresh content across various mediums. The applications range from generating stories to producing realistic images, composing music, and translating textual descriptions into video clips, showcasing the diverse possibilities of AI technology in daily operations.

Chapter 13: 3D Localization in the Era of IIoT by Integrating Machine Learning with 3D Adaptive Stochastic Control Algorithm

This chapter explores the transformative role of the 3D Adaptive Stochastic Control Algorithm (ASCA) in the Industrial Internet of Things (IIoT). The algorithm, infused with machine learning capabilities, utilizes Received Signal Strength Indicator data to accurately locate nodes in a 3D industrial space. Comparative evaluations demonstrate the superiority of the 3D ASCA algorithm, positioning it as a cost-effective and energy-efficient solution for precise node localization within IIoT-enabled systems.

IN SUMMARY

As we reach the culmination of this meticulously curated edited reference book, it is with great satisfaction and anticipation that we reflect on the diverse range of chapters contributed by esteemed researchers and experts in the field. "Advancements in Artificial Intelligence and Manufacturing Technologies: A Comprehensive Reference" transcends disciplinary boundaries, offering a panoramic view of the transformative impact that artificial intelligence (AI) and manufacturing technologies have on various industries.

The chapters within this compendium collectively represent a tapestry of innovative ideas, cutting-edge research, and practical applications that redefine the landscape of industrial processes. From the precision of 3D localization in IIoT to the nuanced forecasting capabilities of Random Forest Regression, and from the fusion of AI and robotics to the critical role of AI in cybersecurity, each chapter unravels a distinct facet of the symbiotic relationship between AI and manufacturing technologies.

The contributors, hailing from diverse backgrounds and expertise, have delved into realms such as thrust manufacturing in aerospace, the ethical considerations in AI-enabled trust building, and the seamless integration of AI in the transition from Industry 4.0 to Industry 5.0. The compendium offers a holistic exploration of AI's multifaceted role in optimizing processes, enhancing efficiency, and shaping the future of manufacturing across various sectors.

We find ourselves at the intersection of AI's evolution and its impact on manufacturing, witnessing the dawn of Industry 5.0 where human-machine collaboration, data-driven decision-making, and adaptive intelligence converge. The chapters not only highlight the possibilities and promises but also address challenges, such as data security, workforce upskilling, and ethical considerations, providing a well-rounded perspective on the journey ahead.

As editors, we extend our gratitude to the authors who have dedicated their expertise, time, and insights to enrich this compendium. The collaborative effort has resulted in a reference book that is both informative and forward-looking, serving as a valuable resource for researchers, practitioners, and policymakers navigating the dynamic intersection of AI and manufacturing technologies.

In conclusion, "Advancements in Artificial Intelligence and Manufacturing Technologies: A Comprehensive Reference" stands as a testament to the dynamic synergy between AI and manufacturing, offering a comprehensive guide that transcends disciplinary boundaries, fosters interdisciplinary dialogue, and paves the way for continued exploration and innovation in the ever-evolving landscape of technology and industry. We trust that this reference book will inspire further research and catalyze advancements that propel us into a future where intelligent manufacturing becomes synonymous with efficiency, precision, and sustainability.

D. Satishkumar
Nehru Institute of Technology, India

M. Sivaraja
Nehru Institute of Technology, India

Acknowledgment

The editors would like to acknowledge the help of all the people involved in this project and, more specifically, to the authors and reviewers that took part in the review process. Without their support, this book would not have become a reality.

First, the editors would like to thank each one of the authors for their contributions. Our sincere gratitude goes to the chapter's authors who contributed their time and expertise to this book.

Second, the editors wish to acknowledge the valuable contributions of the reviewers regarding the improvement of quality, coherence, and content presentation of chapters. Most of the authors also served as referees; we highly appreciate their double task.

Third the editors would like to express gratitude towards members of Nehru Institute of Technology, Coimbatore, India for their kind co-operation and encouragement which help me in completion of this project.

However, it would not have been possible without the kind support and help of many individuals and organizations. I would like to extend my sincere thanks to all of them.

D. Satishkumar
Nehru Institute of Technology, India

M. Sivaraja
Nehru Institute of Technology, India

Chapter 1
AI–Enabled Robotics

A. Gobinath
Velammal College of Engineering and Technology, India

Manjula Devi
Velammal College of Engineering and Technology, India

P. Rajeswari
Velammal College of Engineering and Technology, India

A. Srinivasan
Velammal College of Engineering and Technology, India

Pavithra Devi
Velammal College of Engineering and Technology, India

ABSTRACT

The merger of artificial intelligence (AI) and robotics constitutes a paradigm shift in the field of automation, resulting in AI-enabled robotics. This abstract provides a succinct summary of the essential characteristics and consequences of artificial intelligence-enabled robotics. AI-enabled robotics enhances the capabilities of robotic systems by utilizing advanced machine learning algorithms and computational intelligence. AI and robotics collaboration gives robots the ability to detect, learn, adapt, and make intelligent judgments, resulting in unparalleled levels of autonomy and efficiency. Computer vision, natural language processing, and machine learning algorithms are key components of AI-enabled robotics, allowing robots to perceive and respond to complex stimuli in real time. This combination of AI and robotics not only allows for the automation of repetitive tasks, but also allows for the execution of complex and context-sensitive processes across multiple domains.

DOI: 10.4018/979-8-3693-2615-2.ch001

INTRODUCTION

Artificial intelligence (AI) stands at the forefront of technological evolution, encompassing a broad spectrum of disciplines within computer science. At its core, AI seeks to emulate human intelligence in machines, enabling them to perform tasks that traditionally required human cognitive abilities. The field is characterized by various sub-disciplines, with machine learning, natural language processing, and computer vision being integral components. Machine learning involves the development of algorithms that enable machines to learn from data, adapt, and improve performance over time. Natural language processing focuses on the interaction between computers and human language, allowing machines to understand, interpret, and generate human-like text. Meanwhile, computer vision empowers machines to interpret and make decisions based on visual data, expanding their capabilities to recognize patterns and objects (J. Howard, 2019).

The practical applications of AI are vast and ever-expanding. Speech recognition systems, driven by AI algorithms, have become integral components of virtual assistants like Siri and Alexa. Image and pattern recognition, another facet of AI, is evident in technologies such as facial recognition and object identification in images (Arinez et.al, 2020). AI is a driving force in the development of autonomous vehicles, contributing to navigation, obstacle detection, and decision-making processes. Healthcare benefits from AI through medical diagnosis, drug discovery, and personalized medicine, where machine learning algorithms analyze vast datasets to identify patterns and provide insights(Andras et.al, 2020).

However, the proliferation of AI is not without its challenges. Ethical concerns have risen to the forefront, encompassing issues such as privacy, bias in algorithms, and the potential for job displacement. The explainability of AI systems, particularly in deep learning, poses a significant challenge, as these algorithms are often perceived as "black boxes" due to their complex decision-making processes. Ensuring the safety and security of AI systems, especially in critical applications like healthcare and autonomous vehicles, is a paramount consideration (Q. Bai et.al, 2020).

On the other hand, Robotics, as a distinct discipline, focuses on the design, construction, and operation of robots. A robot is a programmable or virtual agent equipped with sensors for perception and actuators for interaction with the environment. The field encompasses diverse applications, ranging from industrial automation and healthcare to space exploration and agriculture. Industrial robots, for example, have revolutionized manufacturing processes by performing tasks such as assembly, welding, and material handling with precision and efficiency. In healthcare, robots are employed in surgical procedures, rehabilitation, and assistance for individuals with disabilities. The use of robotic systems in space exploration,

Figure 1. Architecture of AI and robotics communications

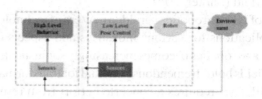

including rovers and unmanned spacecraft, allows for the exploration and study of distant planets and celestial bodies (Lee et.al, 2019).

Despite the progress, challenges persist in the field of robotics. The complexity of tasks, especially those requiring manipulation of diverse objects or navigation in unpredictable environments, remains a significant hurdle. Ensuring interoperability between different robotic systems is crucial, particularly in scenarios where diverse robots need to collaborate seamlessly. Cost considerations also pose challenges, limiting the widespread adoption of robotic systems, particularly for small and medium-sized enterprises (Jiang, 2020).

The merging of AI and robotics represents a paradigm shift in technological capabilities. The integration of AI into robotics enhances the intelligence and adaptability of robotic systems. Machine learning algorithms play a vital role in enabling robots to learn from data and improve their performance, making them more adaptable to unpredictable tasks and environments. Computer vision allows robots to perceive and understand their surroundings, enabling tasks such as object recognition and navigation (Panesar et.al, 2019).

Applications of AI in robotics are diverse and impactful. Robotic Process Automation (RPA) leverages AI in industrial robotics, automating tasks in manufacturing processes where robots can adapt to variations in production requirements. Humanoid robots, equipped with AI, can navigate complex environments, recognize faces, and engage in conversations, making them suitable for applications in customer service and assistance. Collaborative Robotics (Cobots) showcase the synergy between AI and robotics, as these robots are designed to work alongside human workers, enhancing efficiency and safety (Le Nguyen et.al, 2019). Autonomous vehicles, including self-driving cars and drones, heavily rely on AI algorithms for perceiving the environment, making decisions, and navigating safely.

However, challenges persist in the realm of AI robotics. Ensuring the safety and reliability of AI-powered robots, particularly in scenarios where they interact closely with humans, remains a paramount concern. The explainability of AI systems in robotics, necessary for understanding their decision-making processes, is a challenge that requires ongoing research and development. Interdisciplinary collaboration between AI researchers and roboticists is crucial for addressing these challenges and advancing the field (Sarker, 2021).

The integration of AI into robotics heralds a new era of technological advancement with profound implications for various industries and aspects of daily life. As AI continues to evolve, and robotics becomes increasingly sophisticated, the collaboration between these two fields holds tremendous potential for transformative breakthroughs. Navigating the challenges requires a concerted effort from researchers, engineers, and policymakers to ensure the responsible and ethical development of AI robotics, shaping a future where intelligent machines work harmoniously with humans for the betterment of society (S. He, L.G. Leanse, Y. Feng, 2021).

Significance of Their Intersection

The intersection of Artificial Intelligence (AI) and Robotics is a convergence that holds profound significance, shaping the landscape of technology and its impact on various aspects of society. This amalgamation of intelligent algorithms with physical systems has ushered in a new era of capabilities, extending far beyond what each field could achieve independently (Linaza et.al, 2021).

Firstly, the synergy between AI and Robotics enhances automation in ways that were once deemed futuristic. The integration of intelligent algorithms into robotic systems enables automation that is adaptive and responsive. Traditional automation often relied on pre-programmed instructions for repetitive tasks, but AI empowers robots to learn from data, make decisions, and adapt to dynamic environments (Belk, 2021). This adaptive automation is particularly crucial in industries such as manufacturing, where tasks can vary, and conditions are subject to change. The result is increased efficiency, precision, and the ability to handle complex tasks that previously required human intervention (Elallid et.al, 2022).

Moreover, the intersection of AI and Robotics has led to the development of robots with enhanced cognitive abilities, opening up new possibilities for human-robot collaboration. These intelligent robots can understand and respond to natural language, recognize patterns, and even learn from human behavior. This cognitive synergy is transforming industries such as customer service, where humanoid robots equipped with AI are capable of interacting with customers in a more intuitive and personalized manner. The collaborative potential extends beyond customer service to various domains, including healthcare, manufacturing, and education, where robots can work alongside humans as capable and adaptable partners (Ning et.al, 2021).

The healthcare sector, in particular, benefits significantly from the integration of AI and Robotics. Surgical robots, guided by AI algorithms, can perform intricate procedures with precision and accuracy, minimizing the invasiveness of surgeries and reducing recovery times. Additionally, AI-powered robotic systems aid in rehabilitation, assisting individuals in regaining mobility and independence (El-Shamouty et.al, 2019). The ability of these systems to continuously learn and adapt to individual patient needs makes them valuable assets in personalized healthcare (Lammie et.al, 2020).

The intersection of AI and Robotics is also driving advancements in autonomous systems, leading to the development of intelligent vehicles and drones. Autonomous vehicles, powered by AI algorithms, can perceive their surroundings, make decisions in real-time, and navigate complex environments. This has implications not only for the automotive industry but also for transportation, logistics, and urban planning. Drones equipped with AI can be deployed for various applications, from monitoring agricultural fields to delivering medical supplies in remote areas. The autonomy and intelligence embedded in these systems contribute to increased efficiency, safety, and innovation (Cooke, 2020).

Furthermore, the collaborative potential of AI and Robotics extends to the exploration of hazardous or unreachable environments. Robots equipped with AI can be deployed for search and rescue operations in disaster-stricken areas, where human intervention may be challenging or dangerous. Similarly, in space exploration, AI-powered robotic systems can navigate and conduct experiments in environments where human presence is limited. The fusion of AI intelligence with robotic capabilities expands the scope of exploration and contributes to our understanding of the universe. Despite the transformative potential, the intersection of AI and Robotics also raises ethical considerations and societal implications. As intelligent machines become more integrated into our daily lives, questions regarding job displacement, privacy, and the ethical use of AI-powered robotics emerge. Striking a balance between technological advancement and ethical considerations is crucial for ensuring that the benefits of this intersection are realized without compromising human well-being.

The intersection of AI and Robotics represents a paradigm shift in technological capabilities, with far-reaching implications for various industries and societal domains. This synergy enhances automation, introduces cognitive abilities to robots, and fosters collaboration between humans and machines. From manufacturing to healthcare, autonomous systems to exploration, the impact of this intersection is shaping the future of technology. However, responsible development and ethical considerations must accompany these advancements to ensure a harmonious integration that benefits society as a whole. As AI and Robotics continue to evolve, their intersection stands as a testament to the potential for innovation and positive transformation in the way we live and work.

Figure 2. Concepts behind AI in robotics

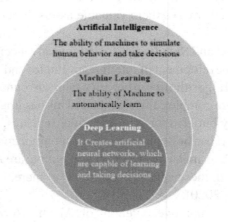

Fundamental Concepts Behind AI in Robotics

The integration of Artificial Intelligence (AI) into robotics represents a groundbreaking synergy that has propelled the capabilities of robotic systems to new heights. At the heart of this convergence lie fundamental concepts that form the backbone of intelligent robotic applications. Understanding these concepts is crucial for unraveling the intricacies of how AI empowers robots to perceive, learn, and adapt, ultimately reshaping the landscape of automation and autonomy.

1.   Machine Learning in Robotics:

One of the foundational pillars of AI in robotics is machine learning (ML). Machine learning algorithms empower robots to learn from data, recognize patterns, and make decisions without explicit programming. In the realm of robotics, this translates to the ability of machines to adapt and improve their performance based on experience. Subtopics under machine learning in robotics include (Soori, 2017):

Supervised Learning: In supervised learning, robots are trained on labeled datasets, allowing them to learn associations between input data and desired outputs. This is applied in tasks like object recognition and path planning.

Unsupervised Learning: Unsupervised learning enables robots to identify patterns and relationships within data without explicit guidance. This is particularly useful in scenarios where the robot encounters unknown or unpredictable environments.

Reinforcement Learning: This paradigm involves training robots through a system of rewards and punishments, allowing them to learn optimal behaviors by trial and error. Applications range from game-playing robots to adaptive control systems.

2.   Computer Vision:

Computer vision is a cornerstone of AI in robotics, enabling machines to interpret and understand visual information from the environment. In robotic applications, computer vision allows robots to perceive objects, navigate spaces, and interact with their surroundings. Subtopics within computer vision include:

Object Recognition: Robots equipped with computer vision can identify and classify objects in their field of view, a critical capability for tasks such as pick-and-place in manufacturing or even assisting the visually impaired.

Image Segmentation: This involves dividing an image into segments to understand the spatial layout of objects. It is valuable in tasks where precise identification of object boundaries is essential.

Visual SLAM (Simultaneous Localization and Mapping): Visual SLAM allows robots to create maps of their environment in real-time while simultaneously determining their own location within that space. This is crucial for autonomous navigation.

3.   Natural Language Processing (NLP) in Human-Robot Interaction:

Natural Language Processing is the interface that enables effective communication between humans and robots. This concept allows robots to understand, interpret, and generate human-like text or speech, facilitating seamless interactions. Subtopics in NLP for robotics include:

Speech Recognition: Robots equipped with NLP can understand and respond to spoken commands, making them more intuitive to interact with. This is especially relevant in applications like virtual assistants and customer service robots (Asmael, 202).

Language Understanding: NLP enables robots to comprehend the nuances of human language, allowing them to respond appropriately to queries and instructions. This is vital for collaborative tasks in which clear communication is essential.

Sentiment Analysis: Robots with NLP capabilities can discern the emotional tone in human communication, enhancing their ability to respond appropriately to different emotional contexts.

4. Sensor Fusion and Perception:

Robotic systems rely on a multitude of sensors to perceive and understand their environment. Sensor fusion involves integrating data from different sensors to create a comprehensive representation of the surroundings. Subtopics in sensor fusion include:

Lidar and Radar Integration: Combining data from lidar and radar sensors allows robots to perceive both the shape and motion of objects in their environment, contributing to enhanced situational awareness.

Inertial Measurement Units (IMUs): IMUs provide information about the robot's acceleration and orientation. Sensor fusion with IMU data is crucial for tasks requiring precise positioning and movement control.

Tactile Sensors: Integration of tactile sensors enables robots to sense touch and pressure, adding a level of dexterity and safety in interactions with objects and humans.

5. Decision-Making Algorithms:

AI in robotics extends to decision-making algorithms that enable robots to choose optimal actions based on their perception and learned knowledge. Subtopics within decision-making include:

Path Planning: Decision algorithms in robotics involve planning the most efficient path from one point to another, considering obstacles and constraints. This is essential for navigation in dynamic environments.

Reactive vs. Deliberative Systems: Reactive systems allow robots to respond quickly to immediate stimuli, while deliberative systems involve more strategic, long-term planning. The balance between these approaches depends on the specific requirements of the task.

Adaptive Control: Adaptive control algorithms enable robots to adjust their behavior in real-time, accommodating changes in the environment or the robot's internal dynamics.

In the dynamic field of AI in robotics, these fundamental concepts lay the groundwork for the development of intelligent and adaptive robotic systems. As technology advances, the synergy between AI and robotics will continue to evolve, pushing the boundaries of what is achievable in automation, autonomy, and human-robot collaboration. A deep understanding of these concepts is essential for researchers, engineers, and enthusiasts alike as they navigate the exciting and transformative intersections of AI and robotics.

CURRENT STATE OF AI ROBOTICS

1. Industrial Automation:

In the realm of industrial automation, the convergence of AI and Robotics has led to a paradigm shift in manufacturing processes. AI algorithms, particularly those rooted in machine learning, empower robots to adapt to changing production requirements. These robots can learn from historical data, adjusting their actions based on variables such as product variations and equipment conditions. Predictive maintenance, enabled by AI, ensures that robots can preemptively identify potential malfunctions, reducing downtime and enhancing overall efficiency. The implementation of collaborative robots (cobots) working alongside human workers is becoming more prevalent, illustrating the seamless integration of AI-driven robotics into diverse manufacturing environments.

2. Healthcare Applications:

AI-driven Robotics in healthcare is revolutionizing patient care, diagnostics, and surgical procedures. Surgical robots, guided by AI algorithms, offer unprecedented precision, reducing the invasiveness of surgeries and accelerating patient recovery. Machine learning models analyze vast datasets to identify patterns in medical imaging, aiding in early disease detection. The development of AI-powered robotic exoskeletons enhances rehabilitation processes, assisting individuals with mobility challenges. Telepresence robots equipped with AI enable remote healthcare consultations, extending medical services to remote or underserved areas.

3. Autonomous Vehicles:

Autonomous vehicles represent a pinnacle achievement in AI-driven Robotics. The current state of AI in autonomous vehicles involves the integration of advanced computer vision, machine learning, and sensor technologies. These vehicles leverage real-time data processing to perceive their surroundings, make decisions, and navigate diverse environments. Companies in the automotive sector are investing heavily in AI research to enhance the safety and reliability of self-driving cars. The collaboration between AI and Robotics in this domain extends beyond passenger vehicles to include drones, delivery robots, and unmanned aerial vehicles, transforming transportation and logistics.

4. Humanoid Robots and Human-Robot Interaction:

Humanoid robots equipped with AI technologies are becoming increasingly sophisticated, mimicking human-like movements and interactions. Advances in natural language processing enable these robots to understand and respond to human speech, facilitating intuitive communication. Emotionally intelligent AI algorithms enable robots to recognize and respond to human emotions, making them suitable for roles in customer service and companion robotics. Human-robot interaction research focuses on enhancing the ability of robots to interpret non-verbal cues, fostering more natural and meaningful exchanges between machines and humans.

5. Space Exploration:

AI-powered robotics plays a pivotal role in space exploration, enabling autonomous navigation and scientific exploration in extraterrestrial environments. Rovers equipped with AI algorithms can adapt to the challenges of unpredictable terrains, autonomously planning routes and conducting experiments. Autonomous spacecraft use AI for trajectory planning, collision avoidance, and on-the-fly decision-making. The integration of AI in space missions extends beyond robotic exploration to include autonomous satellite systems that can dynamically respond to changing mission objectives, paving the way for more agile and adaptive space technologies.

6. Service and Assistance Robots:

Service and assistance robots equipped with AI technologies are finding applications in diverse settings. Delivery robots, equipped with computer vision and navigation algorithms, navigate through urban environments to deliver packages. Cleaning robots use AI to adapt to different floor surfaces and efficiently perform cleaning tasks. These robots showcase adaptability and autonomy, making them valuable assets in environments where human intervention may be impractical or risky. The ongoing development of AI-driven service robots is expanding their role in sectors such as hospitality, retail, and logistics.

7. Challenges and Considerations:

Despite the remarkable progress, challenges in the integration of AI in Robotics persist. Ethical considerations regarding the responsible use of AI technologies, especially in critical applications like healthcare and autonomous vehicles, require careful attention. Privacy concerns arise in scenarios where robots equipped with AI collect and process sensitive information. Job displacement due to increased automation raises societal questions about workforce dynamics. The explainability of AI decision-making processes remains a challenge, particularly in applications where transparency is crucial, such as medical diagnoses and legal frameworks surrounding autonomous vehicles.

8.   Research and Innovation Trends:

Current trends in AI-driven Robotics research focus on addressing existing challenges and pushing the boundaries of technological capabilities. Explainable AI (XAI) research aims to enhance the transparency of AI decision-making processes, providing insights into how algorithms arrive at specific conclusions. Swarm robotics, inspired by the collective behavior of social insects, explores the coordination of multiple robots to achieve complex tasks collectively. Bio-inspired robotics draws inspiration from nature to create robots with enhanced adaptability and resilience. The interdisciplinary collaboration between AI, robotics, and other fields such as materials science and neuroscience is driving innovative trends.

9.   Collaboration with Edge Computing:

The integration of AI in Robotics increasingly relies on edge computing to process data closer to the source, reducing latency and enhancing real-time decision-making. Edge AI enables robots to operate with greater autonomy and responsiveness, making them more effective in dynamic and fast-paced environments. The collaboration between AI and edge computing extends beyond traditional robotic applications to include scenarios such as the Internet of Things (IoT), where distributed intelligence enhances the capabilities of interconnected devices.

10.   Future Outlook:

Looking ahead, the future of AI in Robotics holds exciting prospects. Continued advancements in AI algorithms, coupled with the development of more powerful hardware, will likely result in robots with enhanced cognitive capabilities. The integration of AI in robotics will extend to new domains, including agriculture, environmental monitoring, and disaster response. The collaborative potential between humans and robots will be further explored, with an emphasis on creating symbiotic relationships that leverage the strengths of both. As these technologies mature, ethical considerations, regulations, and ongoing research will play crucial roles in shaping a future where AI-driven Robotics contributes positively to society.

ADVANCEMENTS IN AI ROBOTICS

Recent advancements in AI robotics have propelled the field to new heights, fostering innovation across various domains. Notable breakthroughs include developments in swarm robotics, human-robot collaboration, and the pursuit of explainable AI. These advancements not only showcase the increasing capabilities of intelligent machines but also address critical challenges, paving the way for more responsible and effective integration of AI in robotics.

Swarm Robotics:

One of the cutting-edge areas in AI robotics is swarm robotics, drawing inspiration from the collective behavior of social insects. Recent research has made significant strides in understanding and implementing swarm intelligence in robotic systems. Swarm robots operate collaboratively, coordinating their actions to achieve complex tasks collectively. This approach offers advantages in terms of adaptability, fault tolerance, and scalability. Applications range from search and rescue missions, where a swarm of robots can explore large areas more efficiently than individual units, to environmental monitoring and surveillance. The decentralized nature of swarm robotics allows for flexibility in adapting to dynamic environments and addressing challenges that may arise during missions.

Human-Robot Collaboration:

Advancements in human-robot collaboration focus on creating synergistic partnerships between humans and robots, enabling them to work together seamlessly. This is particularly relevant in industrial settings, where collaborative robots, or cobots, are designed to operate alongside human workers. Recent developments in this area emphasize safety, adaptability, and ease of interaction. AI algorithms play a crucial role in enhancing the capabilities of cobots, allowing them to understand and respond to human gestures, collaborate in shared workspaces, and adapt their actions based on real-time feedback. The goal is to create a harmonious and efficient collaboration where the strengths of both humans and robots are leveraged to optimize productivity and task execution.

Explainable AI (XAI):

Explainable AI has emerged as a pivotal focus area in AI robotics, addressing the challenge of transparency in complex decision-making processes. Recent advancements aim to make AI algorithms more interpretable and understandable, especially in applications where accountability and trust are paramount. XAI techniques provide insights into how AI systems arrive at specific decisions, making them more accessible to end-users and facilitating human oversight. In robotics, this is crucial for applications such as autonomous vehicles and medical robots, where the ability to explain actions and decisions is essential for safety and ethical considerations. Ongoing research in XAI explores diverse approaches, including rule-based systems, interpretable machine learning models, and visualizations, to enhance the explainability of AI in robotics.

Integrating Learning and Adaptation:

Recent advancements in AI robotics underscore the integration of learning and adaptation capabilities in robotic systems. Machine learning algorithms, particularly reinforcement learning, enable robots to learn from experience, adapt to changing environments, and optimize their performance over time. This is evident in applications such as robotic grasping, where a robot learns to manipulate objects with different shapes and textures through trial and error. The ability of robots to adapt to unforeseen circumstances, unforeseen obstacles, or changes in tasks enhances their versatility and applicability across diverse scenarios. Additionally, continual learning frameworks enable robots to acquire new skills and knowledge throughout their operational lifespan, contributing to long-term autonomy and efficiency.

Enhanced Sensory Perception:

Advancements in sensor technologies have significantly enhanced the sensory perception of AI robots. Robots are now equipped with more sophisticated sensors, such as lidar, radar, and advanced cameras, enabling them to perceive their surroundings with greater accuracy and detail. This heightened sensory perception is instrumental in tasks like navigation, object recognition, and environmental monitoring. The integration of AI algorithms with these advanced sensors allows robots to interpret complex data in real-time, facilitating more informed decision-making. This is particularly valuable in applications such as autonomous vehicles, where precise perception of the environment is critical for safe navigation.

Recent advancements in AI robotics are characterized by breakthroughs in swarm robotics, human-robot collaboration, explainable AI, learning and adaptation, and enhanced sensory perception. These developments collectively contribute to the evolution of intelligent machines with increased autonomy, adaptability, and transparency. As researchers and engineers continue to push the boundaries of what is possible, the future holds promises of even more sophisticated AI robotics applications, reshaping industries, improving human-robot interactions, and addressing societal challenges in increasingly impactful ways.

THE FUTURE OF AI ROBOTICS: A TRANSFORMATIVE LANDSCAPE

The trajectory of AI robotics suggests a future where machines with advanced cognitive capabilities redefine our interaction with technology. As learning algorithms become more sophisticated, robots will adapt dynamically to diverse environments, demonstrating an unprecedented level of autonomy. This evolution in intelligence will not only broaden the spectrum of tasks robots can undertake but also contribute to a more seamless integration with human activities. From manufacturing to daily household chores, the impact of enhanced cognitive capabilities will be pervasive.

Collaboration Redefined: Humans and Robots in Synergy

A pivotal shift in the future will be the redefinition of collaboration between humans and robots. Beyond autonomous task performance, robots will seamlessly integrate into human workflows, forming collaborative partnerships. Sectors such as healthcare will see robots working hand-in-hand with medical professionals, providing assistance in surgeries, rehabilitation, and patient care. This collaborative approach will optimize efficiency, with robots handling repetitive tasks, allowing humans to focus on intricate and creative aspects of their work.

Swarms and Collective Intelligence: Revolutionizing Problem-Solving

The rise of swarm robotics will usher in a new era of collective intelligence. Inspired by social insects, swarms of robots will collaborate intelligently to solve complex problems and adapt to changing environments. This collective behavior holds promise for applications ranging from environmental monitoring to disaster response. The ability of swarms to efficiently address challenges that individual robots find daunting will redefine the possibilities of autonomous systems working in unison.

Ethics, Regulation, and Responsible Innovation: Navigating Challenges Ahead

The future of AI robotics will inevitably lead to heightened ethical considerations and the formulation of comprehensive regulatory frameworks. As robots become more integrated into society, discussions on privacy, data security, and the ethical use of AI will take center stage. Policymakers and industry leaders will collaborate to strike a delicate balance between innovation and ethical considerations, ensuring that the deployment of robotic systems aligns with societal values and principles. The responsible development and use of AI in robotics will shape a future where technological advancement goes hand in hand with ethical standards.

Evolving Dynamics: Humans and AI-Powered Robots

The relationship between humans and AI-powered robots is undergoing a profound evolution, marked by a shift from mere automation to true collaboration. In the early stages, robots were often confined to performing repetitive and labor-intensive tasks, reducing human involvement in mundane operations. However, the trajectory of this relationship is moving beyond mere task execution. Modern AI-powered robots, equipped with advanced learning algorithms, are becoming capable collaborators, working alongside humans in shared spaces. This evolution is evident in industries such as manufacturing, where collaborative robots, or cobots, are designed to operate alongside human workers, optimizing efficiency and productivity. This collaborative paradigm is fostering a symbiotic relationship, where the strengths of humans and robots complement each other.

Collaborative Synergy: Redefining Workspaces and Workflows

The evolving relationship between humans and AI-powered robots is redefining traditional workspaces and workflows. Unlike earlier industrial robots, which operated in isolated environments, new generations of robots are designed to interact directly with humans. This interaction is characterized by robots understanding human gestures, responding to voice commands, and adapting to dynamic changes in the environment. In fields like healthcare, robots are emerging as valuable assistants, aiding medical professionals in surgeries, diagnostics, and patient care. The collaborative synergy extends beyond physical tasks to include decision-making processes, with robots providing data-driven insights that augment human expertise. This collaborative evolution is reshaping job roles and fostering an environment where humans and robots collaborate harmoniously, each leveraging their unique capabilities.

Challenges and Opportunities: Navigating the Human-Robot Partnership

As the relationship between humans and AI-powered robots deepens, it brings forth both challenges and opportunities. Ethical considerations become paramount, necessitating careful navigation of issues such as job displacement, data privacy, and the responsible use of AI. Striking a balance between increased automation and preserving human employment becomes a critical challenge for policymakers and industry leaders. However, the evolving partnership also presents opportunities for upskilling and reskilling the workforce to adapt to the changing nature of work. Collaborative efforts between humans and robots can unlock new levels of efficiency, innovation, and problem-solving. The coexistence of humans and robots holds the potential to create a workplace that maximizes the strengths of both, fostering a future where AI-powered robots augment human capabilities and contribute to a more productive and dynamic global workforce.

The evolving relationship between humans and AI-powered robots represents a paradigm shift from automation to collaboration. This shift is evident in various industries, where robots are not only executing tasks but actively engaging with and assisting humans. Navigating the challenges and opportunities of this evolving partnership requires a thoughtful approach that prioritizes ethical considerations, workforce development, and the responsible deployment of AI technology. The future holds a landscape where humans and robots collaborate seamlessly, contributing to a more efficient, innovative, and interconnected world.

CONCLUSION

This chapter illuminates the transformative evolution in the relationship between humans and AI-powered robots, marking a paradigm shift from traditional automation to a collaborative era. The narrative reveals the emergence of robots not merely as tools for task execution but as active collaborators, working in tandem with humans across diverse industries. This symbiotic synergy is characterized by robots that understand human gestures, respond to voice commands, and contribute meaningfully to decision-making processes. As industries and job roles undergo reshaping, ethical considerations rise to prominence, necessitating a thoughtful balance between increased automation and the preservation of human employment. However, amidst the challenges, opportunities arise for upskilling and reskilling the workforce, fostering a future where humans and robots collaborate seamlessly, maximizing their respective strengths. Envisioning a collaborative future, this chapter emphasizes the imperative of responsible innovation, ethical deployment of technology, and workforce development to ensure a harmonious coexistence between humans and the intelligent machines shaping our shared future.

REFERENCES

Andras, E., Mazzone, E., van Leeuwen, F. W. B., De Naeyer, G., van Oosterom, M. N., Beato, S., Buckle, T., O'Sullivan, S., van Leeuwen, P. J., Beulens, A., Crisan, N., D'Hondt, F., Schatteman, P., van Der Poel, H., Dell'Oglio, P., & Mottrie, A. (2020). Artificial intelligence and robotics: A combination that is changing the operating room. *World Journal of Urology, 38*(10), 2359–2366. doi:10.1007/s00345-019-03037-6 PMID:31776737

Arinez, J. F., Chang, Q., Gao, R. X., Xu, C., & Zhang, J. (2020). Artificial intelligence in advanced manufacturing: Current status and future outlook. *Journal of Manufacturing Science and Engineering, 142*(11), 142. doi:10.1115/1.4047855

Bai, Q., Li, S., Yang, J., Song, Q., Li, Z., & Zhang, X. (2020). Object detection recognition and robot grasping based on machine learning: A survey. *IEEE Access : Practical Innovations, Open Solutions, 8*, 181855–181879. doi:10.1109/ACCESS.2020.3028740

Belk, R. (2021). Ethical issues in service robotics and artificial intelligence. *Service Industries Journal, 41*(13-14), 860–876. doi:10.1080/02642069.2020.1727892

Cooke, P. (2020). Gigafactory logistics in space and time: Tesla's fourth gigafactory and its rivals. *Sustainability (Basel), 12*(5), 2044. doi:10.3390/su12052044

Elallid, B. B., Benamar, N., Hafid, A. S., Rachidi, T., & Mrani, N. (2022). A comprehensive survey on the application of deep and reinforcement learning approaches in autonomous driving. *Journal of King Saud University. Computer and Information Sciences, 34*(9), 7366–7390. doi:10.1016/j.jksuci.2022.03.013

He, S., Leanse, L. G., & Feng, Y. (2021). Artificial intelligence and machine learning assisted drug delivery for effective treatment of infectious diseases. *Advanced Drug Delivery Reviews, 178*, 113922. doi:10.1016/j.addr.2021.113922 PMID:34461198

Howard, J. (2019). Artificial intelligence: Implications for the future of work. *American Journal of Industrial Medicine, 62*(11), 917–926. doi:10.1002/ajim.23037 PMID:31436850

Jiang, X., Satapathy, S. C., Yang, L., Wang, S.-H., & Zhang, Y.-D. (2020). S.-.H. Wang, Y.-.D. Zhang, "A survey on artificial intelligence in Chinese sign language recognition,". *Arabian Journal for Science and Engineering*, *45*(12), 9859–9894. doi:10.1007/s13369-020-04758-2

Lammie, C., & Azghadi, M. R. (2020). Memtorch: a simulation framework for deep memristive cross-bar architectures. 2020 IEEE international symposium on circuits and systems (ISCAS). IEEE.

Le Nguyen, T., & Do, T. T. H. (2019). Artificial intelligence in healthcare: a new technology benefit for both patients and doctors. *2019 Portland International Conference on Management of Engineering and Technology (PICMET)*, IEEE. 10.23919/PICMET.2019.8893884

Lee, W. J., Wu, H., Yun, H., Kim, H., Jun, M. B., & Sutherland, J. W. (2019). Predictive maintenance of machine tool systems using artificial intelligence techniques applied to machine condition data. *Procedia CIRP*, *80*, 506–511. doi:10.1016/j.procir.2018.12.019

Linaza, M. T., Posada, J., Bund, J., Eisert, P., Quartulli, M., Döllner, J., Pagani, A., Olaizola, I. G., Barriguinha, A., & Moysiadis, T. (2021). Data-driven artificial intelligence applications for sustainable precision agriculture. *Agronomy (Basel)*, *11*, 1227. doi:10.3390/agronomy11061227

Ning, H., Yin, R., Ullah, A., & Shi, F. (2021). A survey on hybrid human-artificial intelligence for autonomous driving. *IEEE Transactions on Intelligent Transportation Systems*, *23*(7), 6011–6026. doi:10.1109/TITS.2021.3074695

Panesar, S., Cagle, Y., Chander, D., Morey, J., Fernandez-Miranda, J., & Kliot, M. (2019). Artificial intelligence and the future of surgical robotics. *Annals of Surgery*, *270*(2), 223–226. doi:10.1097/SLA.0000000000003262 PMID:30907754

Sarker, S., Jamal, L., Ahmed, S. F., & Irtisam, N. (2021). Robotics and artificial intelligence in healthcare during COVID-19 pandemic: A systematic review. *Robotics and Autonomous Systems*, *146*, 103902. doi:10.1016/j.robot.2021.103902 PMID:34629751

Soori, M., & Arezoo, B. (2020). Recent development in friction stir welding process: A review. *SAE International Journal of Materials and Manufacturing*, 18.

Soori, M., & Arezoo, B. (2023). Dimensional, geometrical, thermal and tool deflection errors compensation in 5-Axis CNC milling operations. *Australian Journal of Mechanical Engineering*, 1–15. doi:10.1080/14484846.2023.2195149

Soori, M. & Arezoo, B. (2013). Machine learning and artificial intelligence in CNC machine tools, a review. *Sustain. Manuf. Service Econ.*

Soori, M., Arezoo, B., & Habibi, M. (2013). Dimensional and geometrical errors of three-axis CNC milling machines in a virtual machining system. *Computer Aided Design*, *45*(11), 1306–1313. doi:10.1016/j.cad.2013.06.002

Soori, M., Arezoo, B., & Habibi, M. (2014). Virtual machining considering dimensional, geometrical and tool deflection errors in three-axis CNC milling machines. *Journal of Manufacturing Systems*, *33*(4), 498–507. doi:10.1016/j.jmsy.2014.04.007

Soori, M., Arezoo, B., & Habibi, M. (2016). Tool deflection error of three-axis computer numerical control milling machines, monitoring and minimizing by a virtual machining system. *Journal of Manufacturing Science and Engineering*, *138*(8), 138. doi:10.1115/1.4032393

Soori, M., Arezoo, B., & Habibi, M. (2017). Accuracy analysis of tool deflection error modelling in prediction of milled surfaces by a virtual machining system. *International Journal of Computer Applications in Technology*, *55*(4), 308–321. doi:10.1504/IJCAT.2017.086015

Soori, M., Asmael, M., & Solyalı, D. (2022). Radio frequency identification (RFID) based wireless manufacturing systems, a review. *Independent Journal of Management & Production*, *13*(1), 258–290. doi:10.14807/ijmp.v13i1.1497

Chapter 2
AI–Powered Tools

P. Selvakumar
(iD) https://orcid.org/0000-0002-3650-4548
Nehru Institute of Technology, India

Sivaraja Muthusamy
Nehru Institute of Technology, India

D. Satishkumar
Nehru Institute of Technology, India

P. Vigneshkumar
KGiSL Institute of Technology, India

C. Selvamurugan
(iD) https://orcid.org/0000-0003-3447-1970
Dhaanish Ahmed Institute of Technology, India

P. Satheesh Kumar
Dr. N.G.P. Institute of Technology, India

ABSTRACT

The field of artificial intelligence has a long history and is always developing. It focuses on intelligent agents, which are devices composed of sensors that sense their environment and act in a way that maximizes the chance of achieving a goal. In this essay, the authors discuss the foundations of modern AI along with several examples of its use. Artificial intelligence approaches, including machine learning, deep learning, and predictive analysis, have been the subject of recent research with the goal of enhancing human learning, planning, reasoning, and initiative. Artificial intelligence driven literature review applications that can help save time and effort, enhance your comprehension and writing abilities, and generate high caliber research. Give them a try and discover how they might greatly enhance your

DOI: 10.4018/979-8-3693-2615-2.ch002

literature review procedure. On the other hand, they can be an invaluable resource for assisting in locating pertinent research publications, recognizing key ideas, and monitoring the evolution of research over time.

INTRODUCTION

Let's start by providing a brief overview of artificial intelligence technologies before discussing the different kinds that are already on the market. It is obvious that artificial intelligence will play a significant role in all future key advancements and fundamentally alter how things are done now. It is probably a vital supporting role in every major industry. Businesses currently need to be aware of the potential advantages AI may have on their operations. (Mathieu, M. et.al, 2015) These tools are helpful since humans are still in charge of making decisions; they can handle predetermined jobs. These tools not only assist in developing processes, but they also significantly improve networks and workflows. These days, artificial intelligence (AI) is a "hot topic" because of apps like ChatGPT that demonstrate the technology's increasing power and capabilities.

It has been evident in recent months that "generative" AI is no longer limited to the domain of academic research or Silicon Valley tech companies. This is due to the introduction of a new generation of these tools. And far from being just the newest "viral sensation," artificial intelligence (AI) has evolved into a technology that everyone, regardless of industry, can use to transform their daily operations and work processes. What are the resources that everyone needs to become familiar with in order to fully comprehend the capabilities of AI as it exists today? I've highlighted some of the most significant ones so may use them right now. If you're like most people and have only recently been interested in AI after seeing one of the newest viral apps, may be asking how they relate to what is often known as "artificial intelligence." The majority of people would have first encountered the phrase in science fiction, where for more than a century we have been captivated by tales of sentient robots, sometimes amiable, sometimes not. Actually, the technology that comes to mind when we refer to artificial intelligence nowadays is called "machine learning." This is merely a reference to software algorithms that can learn, getting increasingly better at doing a single task as they are exposed to more data, as opposed to robotics that mimics every facet of natural intelligence as seen in motion pictures.

Artificial Intelligence and Its Impact on Everyday Life

In recent years, artificial intelligence (AI) has become more and more ingrained in our everyday lives in ways that we may not even be aware of. Many people are still unaware of its impact and how reliant on it we are because it has spread so extensively. From morning to night, a lot of our daily activities are powered by AI technology. Many of us start our days as soon as we wake up by grabbing our phone or laptop (Chen, G. et.al, 2015). This is fundamental to our decision-making, planning, and information-gathering processes and has grown to be second nature to us. Upon turning on our gadgets, we can instantly access AI functions such as: Online banking, email, applications, social networking, Google search, facial ID and picture recognition, digital voice assistants like Apple's Siri and Amazon's Alexa, driving aids including route planning, traffic updates, and weather forecasts, and shopping centers. Every aspect of our contemporary internet lives is impacted, both personally and professionally. (M. Mathieu, et.al, 2015) In the realm of business, global communication and interconnectedness have always been crucial (Chen, G. et.al, 2015). Making the most of data science and artificial intelligence is crucial, and there is no end to its potential expansion.

Performance of Artificial Intelligence

The foundation of artificial intelligence is data collection. Then, by manipulating this data, insights, patterns, and knowledge can be found. Building on each of these building components and using the finished products in novel and uncharted situations is the goal. High-level programming, datasets, databases, computer architecture, and sophisticated machine learning techniques are necessary for this kind of technology. Computational thinking, software engineering, and an emphasis on problem solving are among the factors that determine the success of particular projects. There are numerous uses for artificial intelligence, from straightforward chatbots in customer support apps to sophisticated machine learning systems for large corporations. This large field includes technology like:

Machine Learning (ML). Machine learning (ML) describes the use of statistical models and algorithms to create computer systems that can learn and adapt without explicit guidance. supervised, unsupervised, and reinforcement learning are the three primary categories of inferences and analysis that machine learning (ML) distinguishes from data patterns.

Narrow AI. This is essential to contemporary computer systems; it refers to those that have been trained, or learned, to do particular functions without having that functionality specifically built in. Mobile virtual assistants, like those on the Apple iPhone and Android personal assistants like Google Assistant, and recommendation engines that offer recommendations based on past searches or purchases are two examples of limited artificial intelligence.

Artificial General Intelligence (AGI). The lines separating reality from science fiction can seem to blur at times. The potential for artificial general intelligence

Figure 1. Key components of AI

KEY COMPONENTS OF AI

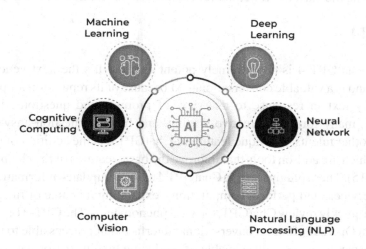

(AGI) has been demonstrated by the robots in television shows like Westworld, The Matrix, and Star Trek. These machines are capable of comprehending and learning any task or procedure that is typically performed by a person.

Strong AI. This phrase and AGI are frequently used interchangeably. It should, in the opinion of some academics and researchers studying artificial intelligence, only be applicable when robots are capable of sensibility or consciousness.

Natural Language Processing (NLP). This is a difficult topic in computer science AI because it needs a lot of data. To educate intelligent machines to comprehend human writing and speech, expert systems and data interpretation are needed. Applications of NLP are being employed more and more in call center and hospital environments, for instance.

Deepmind. Large IT companies are creating cloud services to reach markets like leisure and recreation as they want to take the lead in the machine learning space. For instance, IBM's Watson is a supercomputer that is well-known for participating in a televised Watson and Jeopardy! Challenge, while Google's Deepmind has developed a computer programme called AlphaGo to play the board game Go. Watson raised public awareness about the possible future of AI by responding to inquiries with recognizable speech recognition and response using natural language processing (NLP).

ARTIFICIAL INTELLIGENCE TOOLS

Let's examine the various AI automation solutions that are now in use in the sector.

ChatGPT4

OpenAI's ChatGPT 4 is an extremely potent chatbot. It's the next generation of ChatGPT and is a valuable conversational AI because of its reputation for producing high-quality text in response to a variety of prompts and questions. Language translation, many forms of creative writing, and helpful question-answering are among its other talents. A unique feature of ChatGPT 4 is the addition of ChatGPT plugins, which are add-on tools that improve the basic features of the chatbot (Chen, G. et.al, 2015). These plugins enable ChatGPT 4 to access updated information, utilize external services, and perform computations, expanding its range of functions. To begin, let's quickly review ChatGPT, the viral phenomenon. The GPT-3 big language model from OpenAI offers a conversational interface that is accessible to the public as a free research preview. It is capable of producing text in any format—including computer code—in response to text prompts like queries or directions. Tool used in:

- Customizes educational materials and teaching approaches to suit the distinct requirements and learning speed of individual students, enhancing the overall learning process.
- Possesses the power to analyze intricate data, providing valuable insights, summaries, and recommendations, which aid in making well-informed decisions in various fields.
- Capable of adjusting its communication style to align with the user's personality, interests, and knowledge level, resulting in more natural and engaging conversations.

Google Bard

The principal rival of ChatGPT is Google Bard, an AI chatbot created by Google that is frequently referred to as conversational generative. Google Bard can engage in dynamic dialogues and speak with users. It makes use of cutting-edge AI technologies such as Imagen, MusicLM, LaMDA, and PaLM). New forms of interaction with text, images, videos, and audio are created with the use of these technologies. The tool's capabilities are improved by this. Tool used in: Enables real-time translation of conversations, ensuring smooth communication and understanding between individuals conversing in different languages and promoting global connections. Adjusts its responses and suggestions to match the context of your conversations and tasks, offering timely and valuable information that enriches your user experience. GoogleBard, powered by LaMDA, an AI language tool, helps in generating a range of content, spanning from poems and stories to essays, serving as an innovative tool for writing and content creation.

Chatsonic

Created by Writesonic, ChatSonic is an intelligent artificial intelligence tool. It leverages AI to assist businesses in creating more meaningful customer conversations. It has pre-written dialogue to sound human and makes use of sophisticated machine learning. (Wulff J. and Black M. J., 2015) In addition to creating text and images, ChatSonic also searches Google for relevant information to provide more insightful responses. Tool used in: Generate eye-catching and shareable social media posts that grabs attention, boosts interactions, and elevates your brand's visibility on digital platforms. Uses advanced technology to understand emotions in text messages, allowing it to respond in a way that's more empathetic and personalized for better interaction. Enables AI-driven chatbots to deliver swift, precise, and customized customer assistance, ultimately enhancing customer satisfaction and cutting down on support expenses.

DALL-E 3

DALL-E 3 is the latest iteration of OpenAI's groundbreaking AI model, specializing in image generation based on textual descriptions. It expands the AI's creative abilities by producing a wide range of images based on detailed descriptions, covering everything from basic objects to intricate scenes and abstract ideas. DALL-E 3 leads the way in AI-powered visual creativity. It understands text better and generates images to produce a variety of contextually relevant visual content. Tool used in: Transform your cinematic concepts into reality by crafting impactful movie posters customized to different genres, using DALL-E 3's ability to visualize scenes and evoke emotions through imagery. It generates visual solutions for complex challenges, assisting in architectural design, conceptualizing fashion ideas, and prototyping products. The AI model helps in language comprehension by visually representing idioms, expressions, and vocabulary, enhancing overall understanding and learning.

Midjourney

The most well-known AI art generating program is called Midjourney; it creates art with artificial intelligence. It's similar to commissioning artwork from a virtual artist. To generate art, it makes use of complex AI algorithms, especially deep learning neural networks. Patterns, aesthetic aspects, and styles found in existing artworks can be analyzed and comprehended by these algorithms. Tool used in: Design and visualize realistic product prototypes, from fashion accessories to furniture pieces, using Midjourney's ability to generate 3D models from text descriptions. Utilize Midjourney's feature to craft lifelike portraits, vividly illustrating your characters' individuality, expressions, and unique personalities, breathing life into your stories. Reimagine historical events and create realistic scenes from the past, using Midjourney's ability to generate images based on historical research and descriptions.

Alli AI

Because it simplifies SEO tasks for businesses, Alli AI distinguishes out as a superior AI solution for SEO. Finding technologies that make their job easier is essential for busy marketers. (Ruiz-Sarmiento, J.-R. et.al, 2015). With this tool, can also test several keywords on your pages and, once the testing is complete, automatically retain the term that performs the best. Making educated decisions and precisely assessing the ROI of your SEO efforts is made simple with this tool's tracking and reporting features. Capable of helping in the creation of impactful PPC advertisements, refining landing pages, and overseeing bids to enhance your campaign's effectiveness while maximizing your return on investment (ROI). Tool used in: Conducts thorough data analysis, offering valuable insights to effortlessly enhance and optimize marketing strategies. Customizes content for each person, ensuring that the messages they receive are very personalized.

SlidesAI

SlidesAI, also known as the best AI PowerPoint generation tool, is a platform or piece of software that helps users create and improve presentations by utilizing artificial intelligence (AI) features. SlidesAI creates PowerPoint presentation content automatically using AI algorithms. This material may consist of text, pictures, graphs, and even design ideas. The AI creates slides based on user-provided key points or presentation topics. Tool used in: Translates content into multiple languages, allowing global accessibility and localized presentations, ensuring a wider reach and understanding across different regions. Provides design suggestions and layout improvements, by using AI to refine aesthetics and optimize visual impact for more engaging presentations. Extract key points and summarize lengthy documents into concise and informative presentation slides, saving time and effort.

HubSpot Free AI Content Writer Tool

HubSpot's AI content writer tool is a powerful tool that can help businesses generate content for many programs and channels. It can be used to create AI-written, human-reviewed content, which can save businesses time and money. The tool works by using a variety of data sources, including your website content, social media posts, and blog posts, to learn about your business and your target audience. Once it has learned about your business, it can generate content that is tailored to your specific needs. Tool used in: Generates content in over 20 languages. Creates content for a variety of industries, including healthcare, technology, and education. Capable of generating content for different purposes like marketing, sales, and customer support.

Paradox

An HRM tool called Paradox.ai assists businesses in selecting the top applicants for available positions. It accomplishes this through the employment of Olivia, a helpful AI helper. Employers can recruit more workers, identify the greatest job candidates, and get answers to frequently asked issues with the help of Paradox.ai. Tool used in: Utilizes predictive analytics to forecast employee performance, helping with smarter recruitment decisions. Integration with diverse HR technologies to simplify functions such as interview scheduling and candidate data management. Olivia, the chatbot from Paradox, interacts with candidates, making communication easier and enhancing their overall experience.

Synthesia

Synthesia stands out as a top artificial intelligence (AI) video generating tool for creating videos from written content. This program provides a quick and easy way to create professional-looking videos. It serves companies looking for marketing or training materials, educational institutions making educational movies, and people making videos for their own or their clients' usage. Tool used in: Creates personalized videos for each of your customers, with different avatars, voices, and messages that can increase engagement and customer satisfaction. Utilizes AI to transform the text into videos, eliminating the need for self-recording or hiring actors. Being proficient in over 120 languages enables smooth video translation, expanding your content's global reach and engaging a wider spectrum of viewers.

aiXcoder

Since it can immediately produce code at the method level from natural language input, aiXcoder stands out as the best AI coding tool. Moreover, it offers smart code completion for several lines of code or complete sections. This utility helps users write code and get instantaneous, context-driven feedback. It also helps users incorporate suggested code from aiXcoder effortlessly. Tool used in: It can be added to commonly used software like IntelliJ IDEA, CLion, GoLand, PyCharm, WebStorm, and Visual Studio Code. Works with different programming languages like Java, Python, C++, C#, JavaScript, TypeScript, Go, and several others. Functions without needing the internet, allowing to use it anytime, even when you're not connected online.

TabNine

Over a million developers worldwide, representing several industries, rely on TabNine as their AI code assistant. It provides customized code recommendations that produce excellent, industry-standard code while also significantly increasing efficiency and streamlining tedious coding jobs. StackOverflow Q&A, your complete codebase (Enterprise feature), and reliable open-source code with permissive licenses are all used to train TabNine's Large Language Model (LLM). It produces code that is more secure, more relevant, and of higher quality than ChatGPT 4 or other available tools. Tool used in: Using TabNine for code refactoring enhances code clarity, making it more understandable and maintainable, which is particularly beneficial when collaborating on or modifying code composed by others.Capable of generating code documentation which proves exceptionally efficient, particularly for developers dealing with extensive codebases, saving substantial time and effort. By using it, one can address standard coding errors, ensuring more reliable code and a decreased chance of bugs.

DeepBrain AI

Broadcasting, education, and the service sector are just a few of the industries that DeepBrain AI's cutting-edge conversational artificial intelligence technology has shown to be highly applicable in. Using the hyper-realistic AI avatars that DeepBrain offers to give your training videos a more lifelike appearance, the primary objective of DeepBrain AI is to establish a relationship with your audience. Tool used in: Enables anomaly detection in manufacturing processes, optimizing quality control, and reducing defects through AI-powered analysis. AI technology in DeepBrain helps in financial forecasting by analyzing market trends and supporting more informed investment decisions. Performs anomaly detection in manufacturing processes, optimizing quality control, and reducing defects through AI-powered analysis.

SecondBrain

Previously known as MagicChat, SecondBrain assists in creating ChatGPT-like bots that are deeply knowledgeable about your company or product, meaning that your customer support and sales efforts will be strengthened. The option to train your bot to respond to inquiries about your services from website visitors by using a variety of material sources, including files, documents, and webpages. Tool used in: It lets the chatbot learn on its own, getting smarter as it interacts with real customers, and constantly growing its knowledge base. It allows to customize chatbot conversations using information about customers, such as purchase history or browsing behavior.

Textio

Textio is an AI-driven talent acquisition platform that helps create job descriptions and offers suggestions for conducting performance evaluations. Its main objective is to do away with prejudice in the recruiting process and promote the development of a more varied and inclusive workforce. Textio helps hiring teams use inclusive, brand-consistent language to quickly enhance job postings and candidate outreach. Tool used in: Uses augmented writing to predict how well job listings, emails, or other written content will perform by analyzing language patterns and past data. Applies machine learning and data analysis to offer insights into how certain phrases or language choices might impact the target audience, helping users craft more impactful communication. Textio easily links up with popular writing tools like Slack, Gmail, and Microsoft Word, simplifying its integration into your established workflow.

Wordtune

Wordtune is an artificial intelligence (AI) reading and writing assistant that can correct grammar, comprehend context and meaning, recommend paraphrases or different writing styles, and create written content based on context. Wordtune easily combines with popular work tools like Google Chrome, iOS, and Microsoft Word, improving writing and editing for a more productive work environment. Tool used in: Wordtune gets better as use it, learning from how write and giving personalized suggestions, making it a smart tool that keeps improving to help write better. Supports multiple languages, expanding its utility beyond English-speaking users, making it valuable for a diverse global audience looking to refine their writing in different languages.

Figstack

A complete suite of artificial intelligence capabilities is offered by Figstack to help developers better understand and document code. Its wide range of features, which include a natural language interpreter that can comprehend code in almost any programming language, are intended to make the development process simpler. Tool used in: Figstack is not just a diagramming tool, but a versatile visual collaboration platform that allows users to create diagrams, flowcharts, mind maps, and wireframes, making it ideal for various industries and purposes. With its cloud-based nature, Figstack is accessible from any device with an internet connection, allowing users to work on their diagrams and collaborate from anywhere. Allows many people to work together on a diagram or project at the same time. This helps teams collaborate better, making it easier for people working from different locations to work together smoothly.

Descript

With Descript, can quickly create high-quality videos with an all-in-one video editing tool driven by AI. It allows to add various images, headings, descriptions, and even moving layers. Easily incorporate voiceovers into your videos with Descript, which offers a variety of stock voices and the ability to record your own. Additionally, consumers can obtain a limited-time free plan. Tool used in: Through its integration with hosting platforms, Descript enables users to seamlessly release and distribute their podcast episodes directly from the platform. Descript's Overdub feature creates artificial voices for editing audio. This means can easily edit and replace parts of the audio by generating new speech that sounds like, based on your existing recordings. Descript's AI-powered transcription engine provides high accuracy, making it easy to convert audio to text, enabling quick editing and collaboration on written content.

INK

Three essential elements are integrated into this AI-powered application: an SEO optimization tool, a content planner, and an AI writer. These components can be used singly or in combination as the INK Suite, a complete product. This program aims to simplify the writing process by combining several tools that writers usually use to provide a smooth user experience that includes all the different parts of the writing process (Turan, M. et.al, 2017). The "Semantic Editor" in this tool assists users by analyzing their content to help them grasp how relevant and understandable it is, suggesting improvements for better clarity and tone. It includes a "Headline Performance Score" feature to evaluate the effectiveness of headlines and offers recommendations for enhancing their impact. The word count goals functionality here assists writers in maintaining target word counts and ensures the content aligns with specified requirements.

LyricStudio

This adaptable AI program assists musicians and songwriters in crafting unique lyrics appropriate for any kind of music. With its "Smart Suggestions" feature, LyricStudio provides a selection of lyric options when select your theme. It also helps in identifying rhymes for specific words. By creating an account, may begin by requesting a free trial. Tool used in: Analyzes lyrics and suggest matching phrases, rhymes, and melodies that suit the style and emotions of your song. Through LyricStudio, pinpoint your strengths and areas that need improvement; like songwriting, receiving insightful guidance and feedback to shape and refine your individual musical expression. This tool can sync with music composition software, helping combine lyrics with melodies to ensure they fit together perfectly.

ClickUp

ClickUp stands as a versatile productivity platform, empowering both individuals and teams to seamlessly organize, manage, and collaborate on projects. Offering an extensive array of features such as task lists, calendars, time tracking, and adaptable dashboards, ClickUp serves as a centralized hub, streamlining workflows and elevating productivity. (Gottfredson, L. S. 1998). Whether you're a solo entrepreneur or part of a large organization, ClickUp tailors its capabilities to meet your distinct needs, facilitating efficient and effective goal achievement. Tool used in: Customizes your work areas to manage projects in various ways using tools like Kanban boards, Gantt charts, and Mind Maps. Helps automate tasks and works well with other apps, making your workflow smoother and more organized. Allows real-time teamwork by letting everyone edit, comment, and chat simultaneously, even if they're in different locations.

Canva: AI Slide Creator

Canva's AI Slide Creator offers a user-friendly and efficient solution for creating presentations by utilizing artificial intelligence. It simplifies the process by suggesting design elements, layouts, and content placement, allowing users to craft visually appealing slides with ease. Through smart suggestions and automated design assistance, Canva's AI Slide Creator simplifies the presentation-making experience, empowering users to produce professional and engaging slides effortlessly. Tool used in: Employs machine learning to analyze user preferences, offering customized design suggestions for each slide, and streamlining the customization process. This AI-powered tool not only recommends layout and content but also auto-adjusts design elements, enabling real-time modifications and ensuring coherence in presentations. By integrating user-provided data and preferences, Canva's AI Slide Creator customizes slide templates, making it an ideal solution for professionals seeking personalized and efficient presentation design.

GitHub Copilot

GitHub Copilot is an AI-powered coding assistant developed by OpenAI in collaboration with GitHub. It functions as a code completion tool within integrated development environments (IDEs) and provides real-time suggestions, generating code snippets based on the context of the developer's work. By using machine learning, Copilot assists programmers by offering code suggestions, improving coding efficiency, and helping in the development process by predicting, generating, and completing code lines based on the task at hand. Tool used in: It is an adaptable tool supporting multiple coding languages, offering context-specific code suggestions and snippets to expedite development tasks.

Designs.ai

Designs.ai is an all-in-one online platform that provides easy-to-use tools powered by artificial intelligence, allowing users to create stunning graphics, edit videos, and generate mockups. With its user-friendly interface and diverse templates, Designs.ai empowers users to effortlessly craft professional-quality designs for various purposes, from social media graphics to presentations, and more, without the need for extensive design expertise. Tool used in: Designs.ai's AI capabilities can automatically resize designs for different platforms, update content with ease, and even remove backgrounds from images, saving precious time and effort. Users can maintain a unified look across different platforms by using customizable templates and design tools, ensuring their brand has a clear and consistent visual identity. Offers collaborative tools, enabling to easily share your designs with your team. It allows real-time feedback and edits, ensuring everyone stays updated and works together seamlessly on the same designs. GitHub Copilot rapidly produces recurring code structures like loops, function definitions, and data arrangements, ensuring time-saving efficiency and uniformity in your codebase. The AI-powered assistant not only supports coding but also functions as an educational resource, exposing developers to diverse coding patterns and best practices.

Copyleaks

Copyleaks is a smart tool that helps find out if any text has been copied from other sources on the internet. Using advanced technology and artificial intelligence, it checks and compares texts thoroughly, giving accurate results. It's useful for teachers, writers, and businesses who want to ensure their work is original and not copied from elsewhere. Tool used in: Detects plagiarism in multiple languages, making it a valuable tool for global users and diverse content across various linguistic backgrounds. Examines diverse file formats like Word documents, PDFs, and web pages enhances its adaptability and ease of use for various content varieties. Offers API integration, allowing smooth incorporation into various platforms, such as Learning Management Systems (LMS) and content management systems, streamlining the plagiarism-checking process for educational institutions and businesses.

SEO.ai

SEO.ai is an innovative platform that utilizes artificial intelligence to optimize online content and improve search engine rankings (Turan, M. et.al, 2017). Using advanced algorithms and machine learning, SEO.ai simplifies keyword analysis, content improvement, and SEO strategies, offering a comprehensive solution for businesses aiming to maximize their online visibility and organic traffic. Tool used in: SEO.ai incorporates semantic analysis to understand user intent and context, enabling customized content creation that aligns more closely with search queries, boosting relevance. Offers an extensive range of SEO analysis tools, encompassing keyword research, competitor evaluation, and backlink monitoring, enabling to make informed, data-driven choices for improving your SEO tactics. Compatibility with more than 50 languages, SEO.ai enables the creation and optimization of content to reach and engage a diverse, global audience.

DeepDream

DeepDream, a creation of Google, is an AI technique using neural networks to transform images, making them appear dreamy and surreal. It takes ordinary pictures and enhances them, bringing out patterns and shapes and producing abstract and vibrant visuals that seem almost hallucinatory. The process involves repeatedly modifying the image to reveal the network's interpretation, resulting in unique and sometimes psychedelic-like imagery (Zhang, Y et.al, 2016). Tool used in: Enables exploration through the diverse layers within a neural network, each producing distinct visual interpretations of the input image, resulting in a range of artistic styles. Beyond enhancing images, DeepDream's flexibility allows for creative applications like generating music, composing poetry, and editing videos, showcasing its diverse range of uses in artistic endeavors. DeepDream's role in image processing and creative experimentation offers valuable perspectives to the evolution of AI and machine learning.

Writer.ai

With Writer.ai, may write in a wide range of creative text formats, including code, poetry, screenplays, songs, emails, and letters. Writer.ai is an AI-powered writing tool. Its capabilities extend beyond simple writing support; it can also be used to generate ideas, produce a variety of artistic content types, and provide educational feedback. Writer.ai is becoming more adept at handling a wide range of jobs as it develops. Tool used in: Automates repetitive writing tasks, such as summarizing lengthy documents, transcribing audio recordings, and generating code from natural language descriptions. Offers AI-driven content optimization, suggesting improvements in structure, readability, and tone, thereby enhancing the overall quality and impact of the written material. The AI algorithms within Writer.ai help in overcoming writer's block, generating new ideas, and creating engaging content across diverse formats.

Pikazo

Pikazo, an art creation platform driven by AI, allows users to create their own unique and original artwork using a variety of tools and features. Through Pikazo, crafting images, paintings, and diverse art forms becomes effortless with simple clicks. The platform is user-friendly, making it accessible even for individuals without prior art expertise. Tool used in: Offers a wide variety of artistic styles like realistic, surreal, abstract art, and more. Try out different methods and materials to find your own unique way of expressing yourself through art. Bring your artistic creations to life by converting them into prints, canvases, or digital assets. Pikazo simplifies the process of sharing and displaying your artwork to a wider audience. Pikazo utilizes advanced deep neural networks and a distinctive style merging method, allowing users to combine the visual style from one image with the content of another, leading to the creation of surreal and personalized artistic expressions.

Divi AI

Divi AI is an integrated AI tool within Divi Builder that empowers WordPress users to create spectacular images, write fantastic content, and improve their websites on the fly using artificial intelligence. It provides unique features such as understanding the context of your website and the pages you're working on, allowing it to generate content that is relevant and engaging (Parkes D. C. and Wellman M. P., 2015). Tool used in: The tool uses AI to grasp the website's context, enabling it to produce engaging, SEO-friendly content customized for particular pages, ultimately supporting efficient and effective content creation. Provides real-time feedback and suggestions, guiding users on improving website elements, ensuring content relevance, and maintaining audience engagement. Utilizes AI algorithms to auto-generate high-quality, relevant images, assisting users in creating visually striking content without the need for external graphic design tools.

Scikit Learn

Scikit-learn is one of the most well-known ML libraries. It supports a wide range of computations related to supervised and unsupervised learning. Bunching, k-implies, decision trees, direct and calculated relapses, etc. are a few examples. Instrument used in: It expands upon two core Python libraries, NumPy and SciPy. It involves a number of calculations for common tasks in artificial intelligence and data mining, like bunching, relapsing, and order. As a matter of fact, a few lines of code can handle activities like ensemble methods, information switching, and feature extraction. Scikit-learn is a more-than-adequate tool for someone just starting out in machine learning, until start implementing increasingly sophisticated calculations..

Tensorflow

In the unlikely event that work in the field of artificial intelligence, have probably heard of, tried, or used some form of deep learning algorithm. Tensorflow is interesting because can write a Python program that can be organized and executed on your CPU or GPU. Thus, to continue operating on GPUs, do not need to compose at the C++ or CUDA level. Instrument utilized in: With the help of a network of multi-layered hubs, can quickly assemble, train, and transmit fake neural networks using enormous datasets. This is what allows Google to understand words spoken aloud in its voice-acknowledgment app or identify inquiries from images.

Theano

Theano runs almost parallel to the Keras library, which is an abnormal state neural systems library, and is brilliantly folded over it. The main advantage of Keras is that it is a modest Python deep learning package that may be used instead of TensorFlow or Theano. Tool used in: It was designed to make it as easy and quick as possible for creative labor to actualize significant learning models. It continues to operate on Python 2.7 or 3.5 and can reliably run on CPUs and GPUs. Theano is unique in that it makes use of the GPU on the PC. Because of this, it can process information escalated counts up to several times faster than it could if it relied solely on the CPU. Because of its speed, Theano is very beneficial for deep learning and other computationally demanding tasks.

MxNet

Through the use of "forgetful backprop," it is possible to trade computation time for memory, which is particularly helpful for recurrent nets on very long sequences. Constructed with scalability (multi-GPU and multi-machine training) very easily accessible. Tool used in: Many cool features, such as the ease with which custom layers in high-level languages can be written. It is not directly controlled by a large corporation, in contrast to practically all other major frameworks, which is advantageous for an open-source, community-developed framework. TVM support, which enables operating on a wide range of new device types and will further enhance deployment support.

Auto ML

Among all the frameworks and tools mentioned above, Auto ML is perhaps one of the most powerful and is a relatively new addition to a machine learning engineer's toolkit. In machine learning tasks, optimizations are crucial, as the introduction explains. Even though they have many financial advantages, finding the ideal hyperparameters is a difficult undertaking. This is particularly true for black-box systems like neural networks, since the deeper a network gets, the harder it is to figure out what matters (Turan M. et.al, 2017). We have now moved into a new field of meta where software helps to build software. The AutoML package is widely used by machine learning engineers for model optimization. Tool used in: Apart from the evident time savings, this can be quite beneficial for someone with little to no experience in machine learning who doesn't have the intuition or know-how to independently modify particular hyperparameters.

AI-POWERED TOOLS FOR LITERATURE REVIEW

Research Rabbit

Researchers can identify, arrange, and evaluate research papers more easily with the aid of the AI-powered app Research Rabbit. Anyone can utilize it for free, without needing any prior research knowledge. AI is used by Research Rabbit to search the web for scientific publications that are relevant. Furthermore, can save and arrange items in your own personal library using the AI-powered tool. First register for an account in order to utilize Research Rabbit. Begin looking for research articles as soon as register. It is possible to search by author, topic, or keyword. Include the papers you find interesting in your collections when you've located them. Arrange your research papers into collections. As many collections as require can be made.

Rayyan

Researchers can undertake systematic literature reviews with the use of Rayyan, an AI-powered tool. Finding, assessing, and integrating all of the research that is currently accessible on a given topic is the thorough research process known as a systematic literature review. Even your largest reviews can be swiftly read thanks to it. Work with your team, create reports, and de-duplicate, filter, and arrange references with its help. Using labels and ratings, applying inclusion and exclusion criteria, importing references from different sources, and exporting your data for additional analysis are all possible with Rayyan.

Scholarcy

Your academic reading may be aided by the AI-powered app Scholarcy. Together with creating flashcards and bibliographies, it may automatically summarize articles. Find relevant research and highlight important details in articles with the aid of Scholarcy. Must register for an account before using Scholarcy. It is now possible for to add articles to your library after creating an account. Include scholarly articles from a range of sources, including your personal collection, PubMed, and Google Scholar. can use Scholarcy's features after adding academic publications to your library. Click the "Summarize" button to quickly summarize an article. After that, Scholarcy will provide an article summary in a matter of seconds. All things considered, Scholarcy is a useful resource that can assist with your academic reading. Scholarcy is an excellent choice if want to boost productivity, enhance comprehension, and save time.

Lateral

An AI-powered app called Lateral can support in your academic study. It can assist in organizing your findings, locating pertinent information, and producing superior articles. Lateral analyzes academic papers using artificial intelligence to pinpoint important ideas, connections, and patterns. can then utilize this information to identify pertinent research, arrange your findings, and produce superior articles. It facilitates the searching, organizing, and saving of data from article collections. To locate specific material in your collections, can perform natural language queries, import articles from a variety of sources, and make comments and tags. Lateral allows to access your material from anywhere at any time, control the workflow for your literature review, and keep track of your sources and citations.

Scite

A free, open-source AI tool called Scite aids researchers in finding and comprehending academic publications. Scite does this by offering Smart Citations, which indicate if the article offers opposing or supporting evidence and show the context of a citation. Just visit the Scite website and create an account to begin using Scite. can begin looking for research publications as soon as create an account. can click on an item to read its Smart Citations after Scite displays a list of articles that meet your search criteria. Scite can be used to evaluate the caliber and dependability of the literature, stay away from referencing dubious sources, and locate data supporting or refuting any idea.

Consensus

Consider experimenting with Consensus AI, a new search engine that leverages artificial intelligence to extract and summarize results straight from peer-reviewed research, if you're searching for a quick and simple approach to obtain information from scientific research. Consensus AI is a potent tool that may assist in quickly and easily locating evidence-based solutions from scientific research. To focus your search and discover more possibilities, may also make use of a number of tools. Consensus AI can assist with locating trustworthy information from scientific studies more quickly and easily.

Semantic Scholar

It's a free academic search engine driven by artificial intelligence that makes it simpler and faster than ever to locate pertinent academic publications. It understands your research needs and provides with smart filters, citation analysis, and important insights from publications using machine learning and natural language processing. Semantic Scholar allows to compare various approaches and outcomes, find the most important and latest scientific literature in your subject, and monitor the influence of your own papers.

Iris AI

Finding pertinent articles, summarizing articles, and coming up with research ideas are just a few of the duties our AI-powered research assistant can assist with. A broad range of intelligent filters, reading list analysis, auto-generated summaries, autonomous data extraction, and systematization are all included in this all-inclusive platform for processing your research.

AI Tools to Use for Content Creation

Over the past few years, there has been a lot of discussion on the development of artificial intelligence. The development of content is one area where AI has truly excelled. They are an excellent resource for content producers and digital marketers that want to operate efficiently by creating ideas and optimizing workflows. For businesses, they can prove incredibly cost-effective, offering scalable content solutions for multi-channel marketing. More than 75% of marketers admit to using AI tools to some degree. What's more, around 19% of businesses use AI tools to generate content. Even if you're yet to be convinced about the effectiveness of AI-generated content, you'll need to think seriously about adopting them if want to remain competitive.

Canva: Best for AI Image Generation

Principal Elements can design high-quality graphics using Canva without having any advanced coding experience. Furthermore, with only a few simple commands, may create stunning pictures thanks to the strong image generating capabilities. To ensure that your created imagery is prepared for its intended use, you can experiment with aspect ratios and add after effects.

The Benefits of AI Content Creation Tools

Below are just some of the reasons why should be considering making AI tools a staple of your content planning and creation.

Time-Saving

According to one survey, a typical 500-word blog post takes around 4 hours to complete. This doesn't even take into account compiling a briefing, proofreading the first draft, and waiting on any revision requests to be completed. With a content creation tool, can streamline the process considerably.

Increased Efficiency and Productivity

Whether outsource writing assignments or handle content creation in-house, an AI content generator can boost productivity. These powerful tools make quick work of blog ideation, while social media posts and compelling CTAs can be produced in seconds.

Cost Savings

Looking to slash costs? Outsourcing content to freelancers is relatively inexpensive. However, expect to pay upwards of $175 for a 1,500-word article. can use AI writing assistant tools to control your outgoings. These tools can help refine the ideation and briefing process. Alternatively, can leverage the potential of artificial intelligence to make content creation alleviate pressures on internal teams.

Scalability

If want to create and distribute content at scale, you'll need to invest a significant amount of time and money. Alternatively, use AI tools to automate processes for a more streamlined approach to content creation.

Improved SEO Performance

Long gone are the days when had to rely on external support for SEO services. With the right AI tools at your disposal, can carry out exhaustive keyword research and access optimization insights in seconds. Look for keyword trends, generate fresh and competitive content ideas, and play around with on-page elements to ensure your content is as optimized as it can be.

Enhanced Content Quality

Although current AI content creation tools require the human touch to produce high-quality output, they're getting much more efficient at their job. One of the main ways AI tools help with quality is by delivering consistent results. While tone of voice and context often needs a little finessing, the time saved on generating content from scratch can be put to better use during the editing stage.

CONCLUSION

Businesses that have embraced artificial intelligence from the start have seen higher success. In a short amount of time, AI not only powers the company but also increases its efficiency. The introduction of digital processes to replace paper-based procedures can be likened to the current trend of artificial intelligence in business. Just as this revolution resulted in a significant shift, artificial intelligence (AI) will have a profound impact on all industries. Enroll in the top Artificial Intelligence training in Bangalore offered by Intellipaat to learn how to use the AI technologies. Real-world projects and case studies are combined with structured learning offered by Intellipaat. The AI-powered solutions that assist companies in finding more effective ways to focus on efficiency and increase revenue are at the core of this shift. But one enormous effort that businesses must do is identifying the right artificial intelligence solutions for their requirements. Utilizing the appropriate AI tool can help companies achieve greater strides in cost reduction and net profit growth. Machine learning is a strong technique that creates a plethora of opportunities and is employed in practically every field. Individuals can launch their careers in machine learning with a certification in the discipline. Artificial intelligence (AI)-driven literature review applications that can help save time and effort, enhance your comprehension and writing abilities, and generate high-caliber research. Give them a try and discover how they might greatly enhance your literature review procedure. Remember that AI-powered apps cannot take the role of human judgment. On the other hand, they can be an invaluable resource for assisting in locating pertinent research publications, recognizing key ideas, and monitoring the evolution of research over time.

REFERENCES

Chen, G., Clarke, D., Giuliani, M., Gaschler, A., & Knoll, A. (2015). Combining unsupervised learning and discrimination for 3d action recognition. *Signal Processing*, *110*, 67–81. doi:10.1016/j.sigpro.2014.08.024

Mathieu, M., Couprie, C., & LeCun, Y. (2015). *Deep multi-scale video prediction beyond mean square error*. arXiv preprint arXiv:1511.05440.

Parkes, D. C., & Wellman, M. P. (2015). Economic reasoning and artificial intelligence. *Science*, *349*(6245), 267–272. doi:10.1126/science.aaa8403 PMID:26185245

Ruiz-Sarmiento, J.-R., Galindo, C., & Gonzalez-Jimenez, J. (2015). Scene object recognition for mobile robots through semantic knowledge and probabilistic graphical models. *Expert Systems with Applications, 42*(22), 8805–8816. doi:10.1016/j.eswa.2015.07.033

Turan, M., Almalioglu, Y., Araujo, H., Konukoglu, E., & Sitti, M. (2017). A non-rigid map fusion-based direct slam method for endoscopic capsule robots. *International Journal of Intelligent Robotics and Applications, 1*(4), 399–409. doi:10.1007/s41315-017-0036-4 PMID:29250588

Turan, M., Almalioglu, Y., Gilbert, H., Sari, A. E., Soylu, U., & Sitti, M. (2017). Endo-vmfusenet: deep visual-magnetic sensor fusion approach for uncalibrated, unsynchronized and asymmetric endoscopic capsule robot localization data. *arXiv preprint arXiv:1709.06041*

Turan, M., Almalioglu, Y., Konukoglu, E., & Sitti, M. (2017). *A deep learning based 6 degree-of-freedom localization method for endoscopic capsule robots.* arXiv preprint arXiv:1705.05435.

Turan, M., Shabbir, J., Araujo, H., Konukoglu, E., & Sitti, M. (2017). A deep learning based fusion of rgb camera information and magnetic localization information for endoscopic capsule robots. *International Journal of Intelligent Robotics and Applications, 1*(4), 442–450. doi:10.1007/s41315-017-0039-1 PMID:29250590

Wong, T. Y., & Bressler, N. M. (2016). Artificial intelligence with deep learning technology looks into diabetic retinopathy screening. *Journal of the American Medical Association, 316*(22), 2366–2367. doi:10.1001/jama.2016.17563 PMID:27898977

Wulff, J., & Black, M. J. (2015). Efficient sparse-to-dense optical flow estimation using a learned basis and layers. *Proceedings of the IEEE Conference on Computer Vision and Pattern Recognition,* (pp. 120–130). IEEE. 10.1109/CVPR.2015.7298607

Zhang, Y., Robinson, D. K., Porter, A. L., Zhu, D., Zhang, G., & Lu, J. (2016). Technology roadmapping for competitive technical intelligence. *Technological Forecasting and Social Change, 110,* 175–186. doi:10.1016/j.techfore.2015.11.029

Chapter 3
Forging Trust:
A Futuristic Exploration of AI and ML in Trust Manufacturing

Dhinakaran Damodaran

(iD) https://orcid.org/0000-0002-3183-576X

Vel Tech Rangarajan Dr. Sagunthala R&D Institute of Science and Technology, India

Selvaraj Damodaran

Panimalar Engineering College, India

M. Thiyagarajan

Vel Tech Rangarajan Dr. Sagunthala R&D Institute of Science and Technology, India

L. Srinivasan

Dr. N.G.P. Institute of Technology, India

ABSTRACT

This chapter delves into the dynamic integration of artificial intelligence (AI) and machine learning (ML) in redefining trust in the digital age. Traditional trust-building methods encounter transformation through the capabilities of AI and ML, presenting both challenges and opportunities. The exploration encompasses ethical considerations in trustworthy algorithms, transparency in automated decision systems, and the crucial role of data in risk prediction. It delves into the reshaping of security and privacy paradigms, addressing bias, and fostering collaboration between humans and machines in trust manufacturing. The chapter also investigates the role of emotional intelligence in AI-enabled trust. In presenting a comprehensive overview, it contributes to the discourse on potential future scenarios where AI and ML revolutionize the establishment and scaling of trust in an interconnected digital landscape.

DOI: 10.4018/979-8-3693-2615-2.ch003

INTRODUCTION

The combination of ML and AI has created new opportunities in the field of trust manufacturing in the quickly changing technological landscape. This chapter delves into the potential future scenarios where AI and ML play pivotal roles in shaping and enhancing trust. From algorithmic decision-making to predictive analytics, the integration of these technologies holds the promise of revolutionizing how trust is established, maintained, and scaled. In the dynamic and interconnected landscape of the digital age, the concept of trust has undergone a profound transformation, taking on multifaceted dimensions that extend far beyond traditional, face-to-face interactions (Hadian et al., 2020). Trust in the digital age is a nuanced and complex amalgamation of reliability, security, transparency, and user experience. In essence, it is a subjective yet critical element that underpins interactions, transactions, and relationships in the virtual realm. One key aspect of defining trust in the digital age involves acknowledging the diverse contexts in which it operates. Unlike the tangible and observable cues present in physical interactions, digital trust is often established through virtual interfaces, algorithms, and online platforms. It encompasses the confidence users place in the reliability of digital services, the security of their data, and the transparency of the systems with which they interact.

Transparency emerges as a foundational pillar in this definition of digital trust. Users now seek clear, comprehensible information about how their data is collected, processed, and utilized (Sankar et al., 2023). The digital trust equation incorporates a user's ability to understand and navigate the policies, terms of service, and privacy settings associated with online platforms. Consequently, the more transparent and comprehensible these digital environments are, the more likely users are to trust and engage with them. Furthermore, trust in the digital age is profoundly influenced by the psychological and emotional dimensions of user experience. Users need to feel confident that their interactions online align with their expectations and values. This involves a delicate balance between the convenience and personalization offered by digital services and the assurance that their privacy and security are prioritized. As we define trust in the digital age, it becomes evident that traditional markers of trust, such as face-to-face interactions and physical presence, are no longer the sole determinants. Instead, trust in the digital age is built upon a foundation of intangible yet powerful elements—algorithmic reliability, secure data practices, and transparent communication. Recognizing and understanding these nuances is essential for businesses, organizations, and digital platforms aiming to foster and maintain trust in an increasingly interconnected and virtual world.

BACKGROUND

Challenges in Traditional Methods

Traditional methods of trust-building face several challenges in the contemporary digital landscape. One significant limitation lies in their scalability, as traditional approaches often rely on personal interactions, making it difficult to extend trust-building processes to larger, digital ecosystems (Srinivasan et al., 2023). Moreover, subjectivity and bias are inherent issues, with trust assessments being influenced by personal experiences and judgments, potentially leading to inconsistencies and unfair evaluations. Additionally, traditional methods tend to be slow to adapt to change, struggling to keep pace with the rapid evolution of technology and markets. Another challenge is the limited accessibility of traditional trust networks. Reliance on physical presence or established circles may create barriers for individuals or businesses attempting to enter these networks, hindering collaboration and growth opportunities. The vulnerability to deception is also a concern, as trust built on face-to-face interactions may not always accurately reflect true intentions, leaving room for manipulative practices.

Resistance to change is a common hurdle, as traditional environments may be reluctant to adopt new technologies or methods, whether due to comfort with familiar practices, lack of awareness, or concerns about security. Inefficient information processing is another drawback, with traditional methods struggling to handle information overload and manual data processing. This is in stark contrast to modern technologies like AI and machine learning, which excel in efficiently processing large datasets to derive meaningful insights. Traditional trust mechanisms face difficulty in adapting to remote transactions, where physical presence is lacking. Establishing trust in digital interactions requires a different set of considerations, and traditional methods may not be well-equipped to navigate the complexities of online transactions.

Traditional Methods vs. AI-Driven Approaches

In comparing traditional methods of trust-building with the emerging landscape of AI-driven approaches, the pivotal role of data in trust-building becomes a focal point of distinction and innovation. Traditional methods often relied on interpersonal relationships, reputation, and historical experiences as the primary drivers of trust. These mechanisms, though effective in certain contexts, lacked the scalability and precision demanded by the fast-paced digital era. In contrast, AI-driven approaches bring forth a paradigm shift by leveraging vast amounts of data to inform and enhance the trust-building process. The crucial role of data in trust building within AI-driven frameworks is two-fold: it serves as the fuel that powers intelligent algorithms, and it provides a dynamic, real-time source of insights into user behaviors and preferences. AI-driven trust-building methods capitalize on diverse data sources, ranging from user interactions and feedback to broader trends in the digital landscape. This multifaceted data allows algorithms to discern patterns, predict user expectations, and dynamically adjust trust parameters. As a result, the trust established in AI-driven systems is not solely reliant on historical interactions but is continuously refined based on the most recent and relevant data available. The precision and adaptability afforded by data-driven approaches enable a more nuanced understanding of user trustworthiness. Algorithms can evaluate a myriad of factors, including user behavior, transaction history, and even sentiment analysis, to generate a more accurate and personalized trust profile. This personalized approach, grounded in real-time data, enhances the reliability and effectiveness of trust-building mechanisms in AI-driven systems.

However, the crucial role of data in AI-driven trust building also raises ethical considerations and privacy concerns (Dhinakaran et al., 2022). Striking the right balance between harnessing the power of data for trust enhancement and safeguarding user privacy becomes imperative. Establishing transparent data practices, obtaining informed consent, and incorporating ethical frameworks are essential elements to ensure that the use of data aligns with user expectations and regulatory standards. In essence, the evolution from traditional to AI-driven trust-building signifies a transformative reliance on data as the cornerstone of trust in the digital age. The ability of AI systems to harness, analyze, and learn from data in real-time not only amplifies the efficiency of trust mechanisms but also necessitates a responsible and ethical approach to data utilization to maintain the delicate balance between innovation and user protection (Joe Prathap et al., 2022).

The Figure 1 visually narrates the evolution of trust-building methods, tracing the journey from traditional approaches to the contemporary landscape shaped by artificial intelligence (AI). This detailed explanation will provide insights into each component of the diagram, elaborating on the key concepts and transitions depicted. Figure 1 commences with the foundation of traditional trust-building methods, characterized by interpersonal relationships, reputation, and historical experiences. These time-honored mechanisms have been the cornerstones of trust in various contexts, relying on personal connections, established credibility, and past interactions to foster confidence. Acknowledging the strengths of traditional methods, the diagram recognizes their inherent challenges. While effective in certain contexts, traditional trust-building lacks scalability and precision demanded by the rapidly evolving digital era. As the pace of interactions accelerates in the digital space, the limitations of these methods become apparent, necessitating a shift toward more dynamic and adaptable approaches. Figure 1 then introduces the pivotal role of data in trust-building. This marks a crucial turning point in the narrative, as it highlights

Figure 1. Transformation of traditional to AI-driven approach

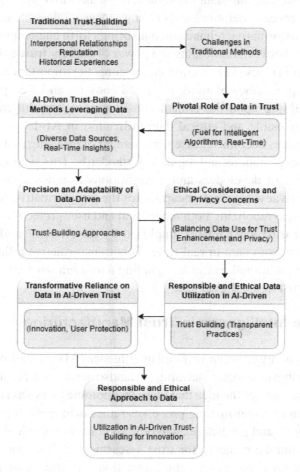

the transformative impact of data-driven insights on trust. Data is depicted as the fuel that powers intelligent algorithms and serves as a dynamic, real-time source of insights into user behaviors and preferences.

The central theme of the diagram unfolds with the introduction of AI-driven trust-building methods. This marks a paradigm shift, leveraging vast amounts of data to inform and enhance the trust-building process. Unlike traditional methods, AI-driven approaches capitalize on diverse data sources, including user interactions, feedback, and broader digital trends. Figure 1 emphasizes the precision and adaptability afforded by data-driven approaches. Algorithms, fueled by real-time data, can discern patterns, predict user expectations, and dynamically adjust trust parameters (Dhinakaran et al., 2023). This ensures that trust in AI-driven systems is not solely reliant on historical interactions but is continuously refined based on the most recent and relevant data available. As the narrative unfolds, the diagram addresses the ethical considerations and privacy concerns inherent in AI-driven trust-building. The transformative reliance on data brings forth challenges, necessitating a delicate balance between harnessing the power of data for trust enhancement and safeguarding user privacy. This section underscores the importance of establishing transparent data practices, obtaining informed consent, and incorporating ethical frameworks. Building upon the ethical considerations, the diagram emphasizes the need for responsible and ethical data utilization in AI-driven trust-building. Transparent practices become an essential element in ensuring that the use of data aligns with user expectations and regulatory standards. This section highlights the dual responsibility of developers and organizations to innovate while protecting user rights and privacy. Figure 1 encapsulates the transformative reliance on data in AI-driven trust, symbolizing both innovation and user protection. It underscores the efficiency of trust mechanisms amplified by AI systems' ability to harness, analyze, and learn from data in real-time. This transformative reliance represents a departure from traditional methods, signaling a new era where data serves as the cornerstone of trust in the digital age.

Collaborative Intelligence in Trust Manufacturing

In the dynamic synergy between artificial intelligence (AI) and the human element, a nuanced relationship emerges, redefining the landscape of trust. Augmenting human trust with AI encapsulates the idea that these technologies serve as supportive tools, enhancing human decision-making processes rather than replacing them. Through data-driven insights and predictive analytics, AI contributes valuable information, augmenting the human capacity for trust assessment and decision-making. This collaborative approach fosters a symbiotic relationship where human intuition and ethical considerations align with AI's analytical prowess, creating a more robust and informed foundation for trust. Collaborative intelligence in trust manufacturing signifies a shift towards collective problem-solving, where humans and AI work in tandem to build and sustain trust. AI systems bring scalability and efficiency, processing vast amounts of data and identifying patterns, while human intuition and contextual understanding contribute the qualitative insights necessary for nuanced trust judgments. This collaboration extends beyond mere coexistence, emphasizing a shared responsibility for cultivating and maintaining trust in a digitally-driven ecosystem.

The role of emotional intelligence in AI-enabled trust introduces a human-centric dimension to technological interactions. Recognizing that trust is not solely a rational construct, emotional intelligence becomes integral in designing AI systems that can comprehend and respond to human emotions. By incorporating emotional cues and context into algorithms, AI systems can better understand user sentiments, adapt to individual preferences, and tailor trust-related interactions in a more empathetic and human-like manner. This infusion of emotional intelligence into AI not only enhances user experiences but also contributes to the establishment of deeper and more meaningful digital trust.

Smart manufacturing processes, leveraging ML models, hold the potential to significantly reduce the duration required for pre-production validation of new processes. While the development of accurate and dependable ML models is pivotal, a substantial challenge emerges in instilling trust among users, including plant operators, engineers, and technicians, in the outputs generated by these models. Goldman et al. (2023) aims to establish AI-based trustworthy manufacturing systems. This means giving these platforms the ability to explain their thought processes and results—like predictions—automatically. They utilize explainable AI techniques to tackle two important production problems: predicting the quality of ultrasonic welds and minimizing spatial diversity in body-in-white (BIW) procedures. Class activation maps are produced to provide insight into the accuracy of assumptions of a transducer weld yield, classifying it as either positive or negative. The influence of signal inputs along with their structure on the neural network's projections is well explained by these maps. Furthermore, saliency maps built around contrastive gradients are used to assess the resilience of the classifier. The study also clarifies a connectionist network that is intended to predict, from variances in underbody stages, the spatial standard of body-in-white framework points. The justifications offered are intended to improve engineers' comprehension of the impact that particular underbody points have on framer point variations. The research highlights the critical importance of using explainable AI techniques in the modern manufacturing sector through several different uses.

The examination of trust dynamics between users and robots is a focal point of this investigation, with a particular focus on the context of robots assuming decision-making roles in football games as referees. Kaustav et al. (2021) focused on online study encompassing 104 participants, aiming to establish a connection between the variables of "Trust" and "Preference" concerning humanoid, contrasting with "AI" and "mechanical" linesmen. The results of the study indicate a optimistic association amongst "Trust" and "Preference" for humanoid, while such a correlation was not observed for "AI" and "mechanical" linesmen. Despite the absence of significant trust differences across various types of linesmen, participants displayed a distinct preference for human linesmen over their mechanical and humanoid counterparts. The qualitative aspect of the study supplements these quantitative findings, delving into the reasons underpinning people's preferences. Notably, when the appearance of a linesman deviates from human likeness, individuals tend to place less emphasis on trust issues and instead prioritize factors as key determinants of their linesman preference.

In the study of Campagna et al. (2023), a scenario within the chemical industry was simulated, involving a robot collaborating with a human in the process of mixing chemicals. The primary objective was to examine the inspiration of proximity in addition to the apparent risk on the trust levels of participants engaged in this collaborative setting. Through a conducted experiment, the researchers sought to analyze how the physical closeness of the robot to the human and the perceived level of risk inherent in the task affected the participants' trust. The results unveiled a noteworthy trend, with a higher trust score observed in scenarios characterized by low proximity and low risk. This indicates that participants generally exhibited greater trust when the robot was situated farther away from them and when the task involved lower perceived risk. Furthermore, a thorough statistical analysis of the data showed that risk significantly affected trust levels more than proximity did. The findings not only contribute valuable insights to the understanding of trust dynamics in human-robot collaborations but also suggest avenues for further exploration in this domain. Specifically, the study implies the potential for leveraging tools such as artificial intelligence to modulate the behavior of the robot in accordance with the established level of trust amongst the hominoid as well as the machine. This concept opens up opportunities for future research and development, aiming to refine the interaction between humans and robots in scenarios characterized by varying levels of proximity and risk within the chemical industry and beyond.

The Becker et al. (2021), introduces the concepts of Cyber-Physical Systems of Tools (COT) and Artificial Intelligence (AI)-based process analysis in the context of their future application in a microelectronics production environment. The focus is on two reference processes: Surface Mount Device utilizing automated manufacturing equipment employing a high-mix/low volume approach. The paper aims to outline a comprehensive concept for digitizing manufacturing processes. This involves creating a digital representation of the processes, which can subsequently be utilized to establish a secure and efficient distributed manufacturing flow. The overarching goal is to enhance both product and process quality through the integration of these technological advancements. The Gocev et al. (2020), employs semantic technologies to articulate the capabilities of machines and align them with production requests. This approach is taken to tackle the issue of flexibility in manufacturing processes. Additionally, AI planning techniques are harnessed to formulate the sequence of productions. The application of OWL rationales, a well-known explanation method in OWL ontologies, clarifies how machine capabilities match product needs along with the reasoning underlying the manufacturing sequences that are formed. The author presents empirical results derived from the application of these methods within an experimental production facility.

Throughout the chip creation procedure, machine learning and intense learning have proven indispensable to constructing tools for designing electronics automation. As contrasted to conventional approaches, DL in EDA offers noteworthy advantages, such as greater generalization, faster turnaround times, and equivalent quality. However, using DL techniques in EDA creates built-in weaknesses that are susceptible to backdooring as well as adversarial perturbation assaults. Although deep learning is widely used in computer-aided design processes and has shown promising results, there is a risk associated with flawed DL-based EDA products. These hacked instruments could unintentionally spread undiagnosed design flaws, cause problems with fabrication, and eventually reduce integrated circuit (IC) yields downwards. Liu et al. (2021), clarifies existing issues and recent developments in protecting and strengthening lithography hotspot detectors utilizing lithographic hotspot identification as an instance study. Additionally, the analysis offers guidance on future paths to improve the robustness as well as security of DL utilized for the field of electronic mechanization.

The Anantha Lakshmi et al. (2023), proposes an innovative approach that synergizes Blockchain technology with Machine Learning (ML) models to assess faults in SMS (smart manufacturing systems). By harnessing the immutability and transparency inherent in blockchain and the predictive capabilities of ML, this approach not only enhances fault detection but also facilitates traceability, contributing to the overall resilience of smart manufacturing. Given the industrial sector's escalating data generation, monitoring systems have become pivotal for effective management and decision-making. The integration of the Internet of Things (IoT) into this proposed methodology further strengthens its capabilities. IoT, being sensor-based and at the forefront of technology, provides a robust framework for monitoring the manufacturing process. In this research, temperature, humidity, gyroscope, and accelerometer IoT sensors are employed to collect environmental data. The sensor data, being unstructured, extensive, and real-time, undergoes processing using various big data approaches. Their technique helps identify flaws in the production process by using the Random Forest classifier strategy to identify outliers in data from sensors. The effectiveness of this approach is demonstrated through an examination conducted in the context of South Korean vehicle production. Notably, the system employs a strategy to fortify data trust, mitigating the risk of genuine data alterations with fabricated data and system interactions.

Jacob et al. (2020), conducts an investigation into the influence of robotic agent transparency on the particular conviction levels of human operators. Multiple experiments were designed and implemented, utilizing a attractive portable robot beneath trafficked control. Throughout the experiments, data, including trust levels, was systematically collected. The findings of their study suggest that trust is more readily eroded than it is built. Moreover, the results highlight that the impact of

51

agent transparency on operator trust becomes more pronounced as tasks increase in complexity. In essence, the transparency of the robotic agent has a more substantial effect on the trust levels of human operators when engaged in tasks that demand greater cognitive and operational intricacy. These outcomes shed light on the nuanced relationship between agent transparency and human trust in the context of human-robot interactions. Pramod and R. Raman (2020), examine important facets of students' technological readiness, namely how knowledge of artificial intelligence (AI) offerings in finance affects students' propensity to use AI to make investment decisions. The study explicitly looks at a number of important factors, such as an inventive mindset, a positive outlook on technological advances, tech apprehension, and the influence of technological trust on the intention to utilize AI technologies for financial investments. Postgraduate learners with existing financial technology employment participate in the research.

The study's findings demonstrate a favorable correlation between the intention to employ financial robots or artificial intelligence (AI) technologies for investment decision-making and a good attitude toward innovation and a greater understanding. On the other hand, anxiety with technology has the opposite effect, meaning that the willingness to use AI tools for making economic choices is hampered. A lack of faith in technology is a significant barrier that significantly affects the desire to use AI technologies. Furthermore, students' decisions regarding the usage of AI tools for investing decisions are significantly influenced by perceived usefulness and simplicity of use of these tools. This comprehensive investigation clarifies the intricate relationship among students' intents to use AI technologies for money, including their views, consciousness, and faith in innovation. In response to the growing amount of data being produced in the industrial sector, Shahbazi et al. (2021), have launched a study project. They understand the need of having a robust system for tracking for decision-making and administration. Their suggested method incorporates state-of-the-art technologies, particularly the IoT, which uses sensors to keep an eye on the production process. The South Korean automobile manufacturing industry is the main subject of this study. The suggested system combines ML, IoT, as well as tracking technologies. IoT sensors are used to gather environmental data. These enormous, unstructured data kinds are produced in real-time. Shahbazi et al. analyze this abundance of data by utilizing a variety of methods based on big data. The suggested approach stands out due to its use of a hybrid forecasting model that makes use of the Random Forest categorization method. By locating and removing anomalies from the sensor data, this model helps the manufacturing system as a whole discover faults.

In Shah et al.'s (2023) study, machine learning and blockchain platforms are combined to expedite the timely delivery of medications and look into drug safety. Their approach is establishing a supply chain management system on a permissioned distributed ledger with role-based authentication using Blockchain technology. This design guards against unregistered individuals interfering with the network's data by ensuring its security along with integrity. The decentralized blockchain-based system seeks to improve accountability, accountability, and uniformity among all pharmaceutical item units in the supply chain. This openness will likely promote cooperation and confidence among interested parties. ML techniques will be used, utilizing information gathered from the distributed ledger, in order to forecast any delivery delays. This predictive capability enables Supply Chain Managers to take pre-emptive actions, ultimately reducing supply chain expenses. Their study proposes a holistic approach that combines Blockchain's secure and transparent supply chain management with the predictive power of ML algorithms. By ensuring the integrity of pharmaceutical supply chain data and predicting potential delays, this integrated system strives to optimize drug safety and facilitate the efficient distribution of medicines.

With the viewpoint of institutions in mind, Zhou et al. (2021), investigate faith in AI in the manufacturing sector. The study delineates three principal aspects of institutions that emanate from the theory of institutions. These dimensions are conceptualized as follows: trust in the AI promoter, centralized control, and managerial involvement. The main hypothesis states that trust in AI is positively impacted by each of the three institutional factors. The paper also presents theories on the controlling effect that AI self-efficacy plays on these three characteristics of institutions. Following the introduction of an AI-based assessment platform for identifying problems and seclusion in process machinery assistance, a survey was conducted within a large petroleum-based enterprise in eastern China. The results show that trust in AI correlates positively with the dedication of management, centralized management, and credibility in the AI advertiser. Interestingly, the study shows that when users have high AI self-assurance, the influence of leadership loyalty and confidence in the AI promoter is amplified.

Mardiani et al. (2023), delves into the intricate interplay between technological advancements in AI decision provision schemes and the dynamics of trust, accountability, and technology within Indonesian manufacturing organizations. Employing a cross-sectional quantitative research approach, the study gathers responses from a representative sample of professionals encompassing various organizational levels, age groups, and functions. The findings of the study reveal a substantial level of trust in AI systems, with this trust being significantly influenced by the dependability and transparency of the technology. The establishment of robust perceived accountability frameworks is identified as a key factor fostering judicious decision-making among professionals. The impact of technological developments, particularly in the realms of Explainable AI and bias prevention, is highlighted as a significant contributor to trust and responsibility in the context of AI decision support systems. The study's demographic analysis adds a nuanced layer to the interpretation of the results, offering practitioners and policymakers practical insights. This information is valuable for supporting the ethical integration of AI in Indonesia's industrial sector. In essence, the study provides a comprehensive exploration of the complex dynamics involving trust, accountability, and technology in the context of AI decision support systems within the manufacturing landscape in Indonesia.

In order to reduce waste in the production process, Arunmozhi et al. (2022), presents a novel concept called "Trust Threshold Limit (TTL)" as a crucial component intended to control the excessive consumption of integrated machinery, tools, power, and other expensive functions. In the context of the business's trading practices, this study focuses on the use of AI in autonomous blockchain technology with smart contracts. The study clarifies the benefits of this strategy for handling market risk evaluations in an efficient manner, particularly in times of socioeconomic crisis. The created model incorporates cost calculations, transportation time, as well as energy assessments and includes real-time instances. The findings demonstrate the effectiveness of artificial intelligence (AI) in improving choice reliability in the created intelligent contract-based Automotive Manufacturing System. The aforementioned threshold level's subjective restriction includes energy, cost, and various other control operations in the production and acquisition process of assembly. The research lists difficulties including cloud synchronization with visual user interfaces and personalization.

In essence, the research contributes to the understanding of how AI, Blockchain, and smart contracts can be leveraged to optimize decision-making and control functions in manufacturing processes. The introduced concept of Trust Threshold Limit aims to enhance efficiency and reduce wastage by setting limits on resource usage, showcasing potential applications in the automotive assembly domain. The authors aim to develop a multifaceted approach based on the results of qualitative data collection about the adoption of AI-powered technologies that are user-friendly by employees. To make it easier to identify useful suggestions for businesses looking to increase acceptability and trust in the method of implementation, this structure incorporates important components such as the workability, AI-autonomy level, as well as implementation level. Making use of the framework's findings, the researchers have also started working on the first draft of a socio-technical AI help tool. This tool is intended to provide practical guidance and assistance in navigating the complexities of implementing AI-based systems in a manner that prioritizes employee satisfaction and fosters a positive organizational environment.

AI AND ML IN DECISION-MAKING

In the realm of decision-making, the integration of AI and ML introduces a transformative landscape with its own set of critical considerations. At the forefront is the imperative of cultivating trustworthy algorithms and practicing ethical AI. Trustworthy algorithms are fundamental to fostering user confidence, demanding a commitment to fairness, accountability, and transparency. This involves not only the technical aspects of algorithmic design but also an ethical underpinning that ensures the decision-making process aligns with societal values and norms. Explainability and transparency represent additional keystones in the integration of AI and ML into decision systems. The complexity of many modern algorithms often results in them being perceived as "black boxes," making it challenging for end-users to comprehend the rationale behind the decisions they make. Achieving explainability is essential in demystifying these algorithms, enabling users to understand, validate, and trust the outcomes of automated decisions. Transparency not only fosters user understanding but also facilitates scrutiny, promoting accountability in AI-driven decision processes.

Balancing precision with fairness stands as a delicate yet crucial challenge in AI and ML decision-making. Precision, often measured by the accuracy of predictions, is a paramount goal, but not at the expense of fairness. The potential for bias in algorithms can lead to discriminatory outcomes, reinforcing existing inequalities. Striking a balance involves meticulous considerations in algorithmic design to minimize biases and ensure that decision systems are equitable across diverse demographics. It demands an intentional effort to avoid perpetuating or exacerbating societal disparities through automated decision-making processes. In essence, the successful integration of AI and ML in decision-making hinges on the development of algorithms that are not only technically robust but also ethically sound. Trustworthy algorithms, characterized by ethical considerations, transparency, and explainability, form the foundation for user acceptance and confidence. Simultaneously, the pursuit of precision must be tempered by a commitment to fairness, emphasizing the ethical responsibility of technologists, data scientists, and policymakers to navigate the intricate landscape of AI-driven decision systems with a keen awareness of societal implications.

Figure 2 shows the relationship between artificial intelligence and machine learning in manufacturing systems. IoT devices are physical objects that are embedded with sensors and software that allow them to collect and transmit data (Keerthana et al., 2023). In a manufacturing setting, IoT devices can be used to collect data on everything from the temperature and humidity of the environment to the performance of machines and production lines. Environmental sensors collect data on the physical environment, such as temperature, humidity, and air quality.

Figure 2. AI and ML in manufacturing system

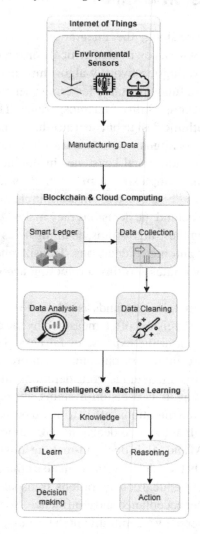

This data can be used to monitor the manufacturing environment and ensure that it is within optimal conditions. For example, environmental sensors can be used to detect if the temperature is too high or the humidity is too low, which could impact the quality of the products being manufactured. Manufacturing data is any data that is related to the manufacturing process, such as production data, machine data, and quality control data. This data can be used to track the progress of production, identify bottlenecks, and improve the efficiency and quality of the manufacturing process. Blockchain is a distributed ledger technology that can be used to create a secure and tamper-proof record of transactions. Cloud computing is the delivery of computing services over the internet. In a manufacturing setting, blockchain and cloud computing can be used to store and share manufacturing data in a secure and efficient way. A smart ledger is a type of blockchain that is designed to store and manage data related to business transactions. Smart ledgers can be used to track the movement of goods through a supply chain, manage inventory, and automate payments. Data collection is the process of gathering data from various sources, such as IoT devices, manufacturing equipment, and human operators. The data collected can be used for a variety of purposes, such as monitoring the manufacturing process, identifying trends, and making predictions (Selvaraj et al., 2023). Data cleaning is the process of identifying and correcting errors and inconsistencies in data. Data cleaning is important because it ensures that the data is accurate and reliable.

Artificial intelligence (AI) and machine learning (ML) are two technologies that can be used to analyze manufacturing data and extract insights. AI and ML can be used to identify patterns, trends, and anomalies in the data. This information can then be used to improve the efficiency and quality of the manufacturing process. Knowledge is the information and insights that are extracted from manufacturing data using AI and ML. This knowledge can be used to make better decisions about the manufacturing process, such as how to optimize production lines, how to improve product quality, and how to reduce costs. The learning process involves using AI and ML to analyze manufacturing data and extract knowledge. This knowledge can then be used to improve the efficiency and quality of the manufacturing process. The reasoning process involves using knowledge to make decisions about the manufacturing process. For example, reasoning can be used to decide how to optimize production lines, how to improve product quality, and how to reduce costs. Decision making is the process of selecting the best course of action based on the knowledge and reasoning that has been generated. Decision making is important because it allows manufacturers to respond to changes in the market and to improve their operations. Action is the process of implementing the decisions that have been made. Action is important because it allows manufacturers to make changes to their operations and to improve their products and services. Overall, the Figure 2 shows how AI and ML can be used to improve the efficiency and quality of manufacturing systems. By collecting and analyzing manufacturing data, AI and ML can be used to extract knowledge that can then be used to make better decisions about the manufacturing process.

Here are some specific examples of how AI and ML are being used in manufacturing systems today:

Predictive maintenance: AI and ML can be used to predict when machines and equipment are likely to fail. This allows manufacturers to schedule maintenance in advance, which can help to avoid unplanned downtime and disruptions to production.

Quality control: AI and ML can be used to inspect products for defects. This can help manufacturers to identify and remove defective products before they reach customers.

Process optimization: AI and ML can be used to analyze manufacturing data and identify areas where the process can be improved. This can help manufacturers to reduce costs and improve efficiency.

Demand forecasting: AI and ML can be used to forecast demand for products. This information can be used by manufacturers to plan production and inventory levels.

AI and ML are still relatively new technologies, but they have the potential to revolutionize the manufacturing industry. By improving the efficiency and quality of manufacturing systems, AI and ML can help manufacturers to reduce costs, improve their products and services, and gain a competitive advantage.

Predictive Trust Analytics

In the era of AI and machine learning, predictive trust analytics emerges as a critical component, revolutionizing how trust is assessed and managed. At its core is the utilization of big data to extract nuanced insights, providing a proactive approach to trust-building and risk mitigation. Harnessing big data for trust insights represents a fundamental shift from reactive to proactive trust management. Organizations can have a thorough grasp of consumer behaviors, trends, and personal preferences by integrating and evaluating enormous volumes of data from many sources. This allows for the identification of trust indicators and the development of predictive models that can anticipate trust-related dynamics before they manifest. The application of advanced analytics to big data enables the extraction of meaningful patterns, empowering decision-makers with the foresight needed to cultivate and maintain trust in a dynamic digital landscape.

Anticipating and mitigating risks is a key focus of predictive trust analytics. By leveraging machine learning algorithms, organizations can identify potential risks and vulnerabilities in real-time. This proactive risk assessment enables the implementation of preemptive measures to address emerging threats to trust, whether they stem from cybersecurity concerns, changing user behaviors, or external factors impacting the digital ecosystem (Harini et al., 2023). This anticipatory approach not only safeguards against potential trust breaches but also enhances overall resilience in the face of evolving challenges. Real-time trust monitoring is a cornerstone of predictive trust analytics, ensuring continuous vigilance over trust dynamics. Through real-time data processing and analysis, organizations can monitor trust-related metrics and respond promptly to deviations or anomalies. This dynamic monitoring capability allows for agile decision-making, enabling organizations to adapt strategies in response to evolving trust landscapes. Real-time trust monitoring is particularly crucial in the

Figure 3. A pathway to building acceptance and trust in AI systems

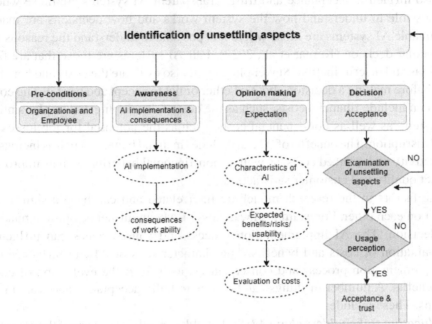

context of rapidly changing digital environments, where trust can be influenced by real-time events, user sentiments, and emerging trends. The research model for the acceptance and trust building process when implementing AI systems, as shown in the Figure 3, is a complex and multifaceted one. It involves a number of factors, including the pre-conditions, the AI implementation process and consequences, the characteristics of AI, and the evaluation of costs and benefits. The pre-conditions for acceptance and trust building in AI systems include awareness, opinion making, and decision. Awareness of AI and its potential benefits and risks is essential for stakeholders to be able to make informed decisions about its adoption. Opinion making is the process of forming a judgment about AI, based on one's own knowledge and, as well as the information that is available. Decision is the process of choosing whether or not to accept and trust AI.

The AI implementation process and consequences can also have a significant impact on acceptance and trust. A well-managed implementation process that is transparent and inclusive can help to build trust in AI. However, a poorly managed implementation process can lead to negative experiences and erode trust. The consequences of AI implementation can also affect acceptance and trust. For example, if AI leads to job losses or other negative outcomes, this can reduce acceptance and trust. The characteristics of AI, such as its transparency, explainability, and fairness, can also influence acceptance and trust. Transparent AI systems are those where it is possible to understand how the system works and how decisions are made. Explainable AI systems are those where it is possible to understand the reasons for a particular decision (Kumar et al., 2023). Fair AI systems are those that are free from bias and discrimination. Stakeholders will also evaluate the costs and benefits of AI before making a decision about whether or not to accept and trust it. The costs of AI can include financial costs, such as the cost of developing and implementing the system, as well as non-financial costs, such as the potential for job losses or social disruption. The benefits of AI can include financial benefits, such as increased productivity and reduced costs, as well as non-financial benefits, such as improved product quality and customer satisfaction.

The factors in the research model are interrelated and can have a significant impact on each other. For example, awareness of AI can influence opinion making and decision. The AI implementation process and consequences can influence the evaluation of costs and benefits. The characteristics of AI can influence the AI implementation process and consequences, as well as the evaluation of costs and benefits. A number of things can be done to build acceptance and trust in AI systems. These include:

Educating stakeholders about AI: Stakeholders need to be aware of the potential benefits and risks of AI in order to make informed decisions about its adoption.

Ensuring that AI is implemented in a transparent and inclusive way: Stakeholders should be involved in the AI implementation process and should have a clear understanding of how the system works and how decisions are made.

Designing AI systems that are transparent, explainable, and fair: AI systems should be transparent so that people can understand how they work. AI systems should be explainable so that people can understand the reasons for particular decisions. AI systems should be fair so that they are free from bias and discrimination.

Communicating effectively about AI: Organizations need to communicate effectively with stakeholders about AI, including its potential benefits and risks, as well as the steps that are being taken to ensure that AI is used responsibly.

The research model for the acceptance and trust building process when implementing AI systems is a complex one that involves a number of factors. By understanding these factors and taking steps to address them, organizations can build acceptance and trust in AI systems, which can lead to a number of benefits, such as increased productivity, reduced costs, and improved product quality and customer satisfaction.

AI-Enhanced Security and Privacy

AI-enhanced security and privacy form a formidable alliance in the digital age, ushering in advancements that redefine the parameters of trust and protection. Central to this paradigm is the imperative of securing trust in the digital ecosystem. Securing trust in the digital ecosystem involves the integration of AI technologies to fortify digital landscapes against evolving threats. AI-driven security systems can analyze vast datasets, detect patterns indicative of potential breaches, and respond swiftly to emerging risks. This not only enhances the resilience of digital infrastructures but also instills confidence in users, assuring them that their interactions within the digital realm are safeguarded by intelligent and adaptive security measures. Privacy-preserving AI models represent a groundbreaking development in the intersection of artificial intelligence and personal data protection (Aswin et al., 2020). As concerns over data privacy escalate, AI models are being designed with privacy-centric features, allowing for the extraction of valuable insights without compromising individual privacy (Dhinakaran et al., 2020). Techniques such as federated learning, homomorphic encryption, and differential privacy enable the training of models on decentralized, encrypted data sources, ensuring that sensitive information remains confidential while still contributing to the improvement of AI algorithms.

The Figure 4 visually narrates the conceptual flow of AI-enhanced security and privacy in the digital ecosystem. Addressing bias and discrimination in security algorithms is an essential aspect of fostering trust and inclusivity in AI-enhanced security systems. The inherent biases present in datasets used to train security algorithms can result in discriminatory outcomes, disproportionately impacting certain demographics. It takes a concentrated effort to identify and address these biases in the creation and application of AI-driven security solutions. This includes diversifying datasets, putting fairness-aware algorithms into place, and continuously evaluating and reducing biases. This commitment to fairness not only aligns with ethical considerations but also ensures that security systems treat all users equitably. In essence, the integration of AI into security and privacy domains represents a transformative leap toward a more robust and user-centric digital ecosystem (Udhaya Sankar et al., 2020). By securing trust, preserving privacy, and addressing biases, AI-enhanced security and privacy measures contribute to the creation of a digital landscape where individuals can engage with confidence, knowing that their data is protected, and security measures are ethically designed and free from discrimination.

Figure 4. AI-enhanced security and privacy in the digital ecosystem: A conceptual flow

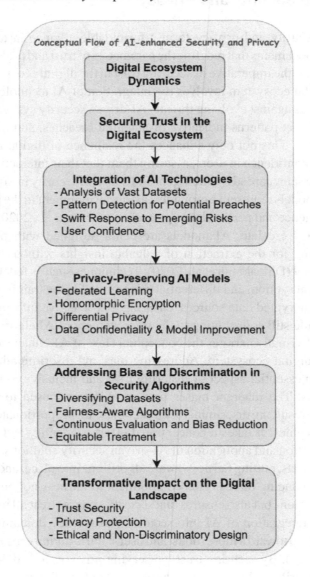

OPPORTUNITIES FOR INNOVATION AND NEW BUSINESS MODELS

In the dynamic landscape of AI-driven trust manufacturing, several opportunities for innovation and the development of new business models emerge. As organizations leverage the transformative capabilities of artificial intelligence (AI) and machine learning (ML) to reshape trust-building processes, they simultaneously unlock avenues for creative solutions, enhanced user experiences, and the establishment of novel business models.

AI-Powered Personalization:

One significant opportunity lies in the realm of AI-powered personalization. As AI algorithms analyze vast datasets to discern user preferences and behaviors in real-time, businesses can tailor their offerings with unprecedented precision. This personalized approach not only fosters a deeper sense of trust by aligning services with individual expectations but also opens new avenues for businesses to differentiate themselves in a crowded digital marketplace.

Predictive Analytics for Proactive Trust Management:

The predictive analytics capabilities embedded in AI-driven trust systems provide organizations with a unique opportunity to adopt a proactive stance in trust management. By anticipating potential trust-related challenges and risks, businesses can implement preemptive strategies, mitigating issues before they escalate. This not only enhances the resilience of trust mechanisms but also positions organizations as leaders in building and maintaining user confidence.

Data-Driven Insights for Strategic Decision-Making:

The abundance of data generated by AI-driven trust systems offers organizations valuable insights into user behavior, preferences, and market trends. This wealth of information becomes a strategic asset, guiding decision-makers in making informed choices about product development, marketing strategies, and overall business direction. The integration of data-driven insights into strategic decision-making processes can lead to more agile, responsive, and competitive business models.

Ethical AI Consulting and Assurance Services:

The ethical considerations surrounding AI-driven trust-building present a unique opportunity for businesses to offer consulting and assurance services. Organizations can specialize in guiding others through the ethical challenges of implementing AI in trust systems, ensuring transparent data practices, and adhering to ethical frameworks. By becoming leaders in ethical AI, businesses can establish themselves as trustworthy partners in the broader AI ecosystem.

Blockchain Integration for Enhanced Security and Transparency:

The marriage of AI-driven trust systems with blockchain technology presents an innovative opportunity. Blockchain's inherent security features and transparency mechanisms complement the trust-building efforts powered by AI. Businesses can explore new business models centered around blockchain integration, offering enhanced security and transparency in digital transactions, thereby further fortifying trust in digital interactions.

Trust-as-a-Service (TaaS) Models:

A paradigm shift towards offering trust as a service (TaaS) represents a futuristic business model. Organizations can position themselves as providers of comprehensive trust solutions, integrating AI-driven algorithms, data analytics, and ethical frameworks into a unified service. TaaS models can cater to businesses across industries, offering scalable and customizable trust solutions tailored to specific needs.

Cross-Industry Collaborations for Trust Ecosystems:

Building trust ecosystems that transcend industry boundaries presents an exciting opportunity for innovation. Organizations can explore collaborations with partners in different sectors to create comprehensive trust networks. This cross-industry approach allows for the sharing of best practices, data insights, and the establishment of universal standards for AI-driven trust, fostering a collective effort to enhance trust across digital platforms.

Trust-Based Certification and Accreditation Services:

With the increasing importance of trust in digital interactions, businesses can establish themselves as certifiers and accreditors of trustworthy AI systems. Offering certification services that verify the ethical practices, transparency, and reliability of AI-driven trust mechanisms can become a niche market. Businesses can position themselves as authorities in ensuring the highest standards of trustworthiness.

FUTURE RESEARCH DIRECTIONS

As the field of AI-driven trust manufacturing continues to evolve, several avenues for future research directions emerge, offering opportunities for scholars, practitioners, and policymakers to deepen their understanding and contribute to the advancement of this dynamic domain. The following research directions highlight key areas that warrant exploration and investigation:

Ethical Implications of AI-Driven Trust:

Delve into the ethical considerations surrounding the deployment of AI in trust-building mechanisms. Explore the ethical implications of algorithmic decision-making, the potential biases embedded in AI systems, and the impact on user trust. Research could focus on developing frameworks for responsible AI use and assessing the ethical dimensions of AI-driven trust across diverse cultural and societal contexts.

Human-AI Collaboration in Trust Assessment:

Investigate how humans and AI systems can collaboratively assess and build trust. Explore the optimal balance between human intuition and AI-driven analytics in decision-making processes. Research could aim to develop models that enhance collaborative intelligence, ensuring that AI complements human judgment while respecting user values and preferences.

Explainability and Transparency in AI Systems:

Further research on the explainability and transparency of AI algorithms in the context of trust. Explore novel methods for making AI-driven decision systems more interpretable for end-users, ensuring that the inner workings of algorithms are accessible and understandable. Investigate the impact of explainability on user trust and acceptance of AI-driven trust mechanisms.

User-Centric Trust Metrics and Evaluation:

Develop comprehensive metrics and evaluation frameworks for assessing user-centric trust in AI systems. Explore how users perceive and evaluate trustworthiness in different digital interactions and contexts. Research could focus on the development of standardized metrics that consider user preferences, cultural variations, and evolving expectations in the dynamic digital landscape.

Long-Term Trust Dynamics:

Investigate the long-term dynamics of trust in AI-driven systems. Examine how trust evolves over extended periods, considering factors such as user experience, system reliability, and adaptability. Research could explore the development of trust models that account for the temporal aspect of digital interactions and the changing nature of user expectations.

Cross-Domain Trust Transferability:

Explore the transferability of trust across different domains and industries. Investigate whether trust established in one context can be effectively transferred to another, and how AI systems can adapt to varying trust dynamics. Research could provide insights into developing adaptable AI-driven trust models that maintain effectiveness across diverse applications.

Trust and Security in Blockchain Integration:

Research the intersection of trust and security in the integration of AI-driven trust systems with blockchain technology. Explore how blockchain's security features can enhance the trustworthiness of AI systems and contribute to transparent and accountable digital interactions. Investigate potential challenges, benefits, and optimal configurations for this integration.

User Empowerment and Control in Trust Relationships:

Examine ways to empower users and provide them with greater control over their trust relationships with AI systems. Research could focus on developing user-centric tools and interfaces that allow individuals to customize and adjust trust parameters, providing a more transparent and user-driven approach to AI-driven trust-building.

Cross-Cultural Variations in Trust Perceptions:

Investigate how cultural factors influence trust perceptions in AI-driven systems. Explore cross-cultural variations in the understanding of trust and the impact of cultural nuances on user acceptance. Research could contribute to the development of culturally sensitive AI-driven trust mechanisms that adapt to diverse cultural contexts.

Legal and Regulatory Frameworks for AI-Driven Trust:

Explore the development of legal and regulatory frameworks specific to AI-driven trust mechanisms. Investigate how existing regulations may need to be adapted or new regulations created to ensure ethical and responsible deployment of AI in trust-building. Research could contribute to the establishment of guidelines that balance innovation with user protection.

By addressing these future research directions, the academic and industry communities can advance the understanding of AI-driven trust manufacturing, fostering innovation, and contributing to the development of ethical, transparent, and user-centric digital ecosystems.

Human Impact and Societal Implications

Exploring the human impact and societal implications of the chapter reveals a multifaceted landscape shaped by the integration of AI and ML in trust systems. As these technologies redefine trust dynamics, they wield a profound influence on user trust and confidence, ushering in a paradigm where the relationship between individuals and digital systems undergoes transformation. Simultaneously, considerations of accessibility and inclusion come to the forefront, prompting an examination of how these advanced systems cater to diverse user demographics and contribute to, or potentially alleviate, digital divides. Beyond user interactions, the societal fabric is woven into this narrative, with employment and workforce dynamics undergoing shifts that necessitate a careful exploration of the evolving job market. Ethical considerations form a crucial thread throughout, demanding an assessment of the responsible implementation of AI in trust systems and its far-reaching implications for individuals and society at large.

Practical Guidelines for Implementing Ethical AI in Trust Systems

Implementing ethical AI in trust systems necessitates clear guidelines. Establish a robust ethical framework, prioritizing fairness, accountability, and privacy. Actively identify and mitigate biases in algorithms to ensure equitable treatment. Transparency is key—communicate how AI influences trust systems for user understanding. Implement privacy-preserving measures to safeguard sensitive information. Continuous monitoring and collaboration with diverse stakeholders, including regular training, are crucial for responsible AI deployment. These guidelines aim to create a trustworthy AI environment, respecting user rights and upholding ethical standards, essential for building and maintaining trust in AI-driven systems.

HIGHLIGHTING CHALLENGES AND TRENDS

The chapter adeptly navigates the intricate landscape of challenges and trends, offering a comprehensive exploration that contributes valuable insights to the overarching theme. Through a discerning lens, it highlights the nuanced challenges posed by the integration of AI and ML in trust systems, dissecting their implications with precision. Simultaneously, the chapter serves as a compass for emerging trends, providing a forward-looking perspective that illuminates the path toward the future of trust manufacturing. This dual focus not only enriches the reader's understanding of the current hurdles but also equips them with foresight into the evolving dynamics of trust in the digital era. The seamless integration of these analyses positions the chapter as a significant source of knowledge, shedding light on the complexities inherent in the subject matter while contributing a valuable mosaic of insights to the broader theme.

CONCLUSION

The chapter navigates the transformative intersection of AI, ML, and trust in the digital age. It begins by defining trust in contemporary digital contexts, acknowledging its subjective nature and exploring the crucial role of transparency, security, and user experience. Traditional trust-building methods are then compared to AI-driven approaches, emphasizing the limitations and opportunities inherent in each. The subsequent sections delve into the multifaceted dimensions of AI as well as ML in decision-making, highlighting the importance of trustworthy algorithms grounded in ethical considerations. The exploration extends to the critical need for explainability and transparency in automated decision systems, addressing the challenges of algorithmic opacity and fostering user comprehension.

The chapter then shifts its focus to predictive trust analytics, where big data becomes a cornerstone for insights, risk anticipation, and real-time monitoring. It examines how AI algorithms leverage vast datasets to forecast trust-related dynamics, providing a proactive approach to trust-building and risk mitigation. AI-enhanced security and privacy are explored as a dynamic duo, emphasizing the significance of securing trust in the digital ecosystem. The integration of privacy-preserving AI models and efforts to address bias and discrimination in security algorithms underlines the chapter's commitment to ethical considerations and user protection. The human element is woven into the narrative, emphasizing the collaborative relationship between humans and AI in trust manufacturing. Augmenting human trust with AI, collaborative intelligence, and the incorporation of emotional intelligence in AI systems are discussed, showcasing a balanced and symbiotic approach to trust-building.

REFERENCES

Arunmozhi Manimuthu, V. G. (2022). Design and development of automobile assembly model using federated artificial intelligence with smart contract. *International Journal of Production Research, 60*(1), 111–135. doi:10.1080/0020 7543.2021.1988750

Becker, K.-F., Voges, S., Fruehauf, P., Heimann, M., Nerreter, S., Blank, R., & Erdmann, M. (2021). Implementation of Trusted Manufacturing & AI-Based Process Optimization into Microelectronic Manufacturing Research Environments. *IMAPSource Proceedings 2021 (IMAPS Symposium).* iMaps. 10.4071/1085-8024-2021.1

Campagna, G., & Rehm, M. (2023). *Analysis of Proximity and Risk for Trust Evaluation in Human-Robot Collaboration.* 2023 32nd IEEE International Conference on Robot and Human Interactive Communication (RO-MAN), Busan, Korea. 10.1109/RO-MAN57019.2023.10309470

Das, K., Wang, Y., & Green, K. E. (2021). Are robots perceived as good decision makers? A study investigating trust and preference of robotic and human linesman-referees in football. *Paladyn : Journal of Behavioral Robotics, 12*(1), 287–296. doi:10.1515/pjbr-2021-0020

Dhinakaran, D., & Joe Prathap, P. M. (2022). Protection of data privacy from vulnerability using two-fish technique with Apriori algorithm in data mining. *The Journal of Supercomputing, 78*(16), 17559–17593. doi:10.1007/s11227-022-04517-0

Dhinakaran, D., Selvaraj, D., Dharini, N., Raja, N. S. E., & Priya, C. S. L. (2023). Towards a Novel Privacy-Preserving Distributed Multiparty Data Outsourcing Scheme for Cloud Computing with Quantum Key Distribution. *International Journal of Intelligent Systems and Applications in Engineering, 12*(2), 286–300.

Dhinakaran, D., Udhaya Sankar, S. M., Edwin Raja, S., & Jeno Jasmine, J. (2023). Optimizing Mobile Ad Hoc Network Routing using Biomimicry Buzz and a Hybrid Forest Boost Regression - ANNs [IJACSA]. *International Journal of Advanced Computer Science and Applications, 14*(12). doi:10.14569/IJACSA.2023.0141209

Gocev, I., Grimm, S., & Runkler, T. (2020). Supporting Skill-based Flexible Manufacturing with Symbolic AI Methods. *IECON 2020 The 46th Annual Conference of the IEEE Industrial Electronics Society, Singapore,* (pp. 769-774). IEEE. 10.1109/IECON43393.2020.9254797

Goldman, C. V., Baltaxe, M., Chakraborty, D., Arinez, J., & Diaz, C. E. (2023). Interpreting learning models in manufacturing processes: Towards explainable AI methods to improve trust in classifier predictions. *Journal of Industrial Information Integration*, *33*, 100439. doi:10.1016/j.jii.2023.100439

Hadian, H., Chahardoli, S., Golmohammadi, A. M., & Mostafaeipour, A. (2020). A Practical Framework for Supplier Selection Decisions with an Application to the Automotive Sector. *International Journal of Production Research*, *58*(10), 2997–3014. doi:10.1080/00207543.2019.1624854

Harini, M., Prabhu, D., Udhaya Sankar, S. M., Pooja, V., & Kokila Sruthi, P. (2023). Levarging Blockchain for Transparency in Agriculture Supply Chain Management Using IoT and Machine Learning. 2023 World Conference on Communication & Computing (WCONF), RAIPUR, India. pp. 1-6. 10.1109/WCONF58270.2023.10235156

Jacob, T. Cassady, Chris Robinson, and Dan O. Popa. (2020). Increasing user trust in a fetching robot using explainable AI in a traded control paradigm. In *Proceedings of the 13th ACM International Conference on PErvasive Technologies Related to Assistive Environments (PETRA '20)*. Association for Computing Machinery, New York, NY, USA. 10.1145/3389189.3393740

Joe Prathap, P. M. (2022). Preserving data confidentiality in association rule mining using data share allocator algorithm. *Intelligent Automation & Soft Computing*, *33*(3), 1877–1892. doi:10.32604/iasc.2022.024509

Keerthana, M., Ananthi, M., Harish, R., Udhaya Sankar, S. M., & Sree, M. S. (2023). IoT Based Automated Irrigation System for Agricultural Activities. *2023 12th International Conference on Advanced Computing (ICoAC)*, Chennai, India. 10.1109/ICoAC59537.2023.10249426

Kumar, K. Y., Kumar, N. J., Udhaya Sankar, S. M., Kumar, U. J., & Yuvaraj, V. (2023). *Optimized Retrieval of Data from Cloud using Hybridization of Bellstra Algorithm*. 2023 World Conference on Communication & Computing (WCONF), RAIPUR, India. 10.1109/WCONF58270.2023.10234974

Lakshmi, G. A., Gummadi, A., & Changala, R. (2023). Block Chain and Machine Learning Models to Evaluate Faults in the Smart Manufacturing System. *International Journal of Scientific Research in Science and Technology*, *10*(5), 247–255. doi:10.32628/IJSRST2321438

Li, J., Zhou, Y., Yao, J., & Liu, X. (2021). An empirical investigation of trust in AI in a Chinese petrochemical enterprise based on institutional theory. *Scientific Reports*, *11*(1), 13564. doi:10.1038/s41598-021-92904-7 PMID:34193907

Liu, K., Zhang, J. J., Tan, B., & Feng, D. (2021). *Can We Trust Machine Learning for Electronic Design Automation?* 2021 IEEE 34th International System-on-Chip Conference (SOCC), Las Vegas, NV, USA. 10.1109/SOCC52499.2021.9739485

Mardiani, E., Judijanto, L., & Rukmana, A. Y. (2023). Improving Trust and Accountability in AI Systems through Technological Era Advancement for Decision Support in Indonesian Manufacturing Companies. *West Science Interdisciplinary Studies.*, *1*(10), 1019–1027. doi:10.58812/wsis.v1i10.301

Sai Aswin, B. G., Vishnubala, S., Dhinakaran, D., Kumar, N. J., Udhaya Sankar, S. M., & Mohamed Al Faisal, A. M. (2023). *A Research on Metaverse and its Application.* 2023 World Conference on Communication & Computing (WCONF), Raipur, India. 10.1109/WCONF58270.2023.10235216

Selvaraj, D., Udhaya Sankar, S. M., Pavithra, S., & Boomika, R. (2023). Assistive System for the Blind with Voice Output Based on Optical Character Recognition. In: Gupta, D., Khanna, A., Hassanien, A.E., Anand, S., Jaiswal, A. (eds) *International Conference on Innovative Computing and Communications. Lecture Notes in Networks and Systems.* Springer, Singapore. 10.1007/978-981-19-3679-1_1

Shah, Y., Verma, Y., Sharma, U., Sampat, A., & Kulkarni, V. (2023). *Supply Chain for Safe & Timely Distribution of Medicines using Blockchain & Machine Learning. 5th International Conference on Smart Systems and Inventive Technology (ICSSIT)*, Tirunelveli, India. 10.1109/ICSSIT55814.2023.10061049

Shahbazi, Z., & Byun, Y.-C. (2021). Smart Manufacturing Real-Time Analysis Based on Blockchain and Machine Learning Approaches. *Applied Sciences (Basel, Switzerland)*, *11*(8), 3535. doi:10.3390/app11083535

Srinivasan, L., Selvaraj, D., & Udhaya Sankar, S. M. (2023). Leveraging Semi-Supervised Graph Learning for Enhanced Diabetic Retinopathy Detection. *SSRG International Journal of Electronics and Communication Engineering.*, *10*(8), 9–21. doi:10.14445/23488549/IJECE-V10I8P102

Udhaya Sankar, S. M., Kumar, N. J., Dhinakaran, D., Kamalesh, S. S., & Abenesh, R. (2023). *Machine Learning System for Indolence Perception. 2023 International Conference on Innovative Data Communication Technologies and Application (ICIDCA)*, Uttarakhand, India. 10.1109/ICIDCA56705.2023.10099959

V. B. S., Pramod, D., & Raman, R. (2022). *Intention to use Artificial Intelligence services in Financial Investment Decisions.* 2022 International Conference on Decision Aid Sciences and Applications (DASA), Chiangrai, Thailand. 10.1109/DASA54658.2022.9765183

Werens, S., & von Garrel, J. (2023). Implementation of artificial intelligence at the workplace, considering the work ability of employees. *Tatup, 32*(2), 43-9. https://www.tatup.de/index.php/tatup/article/view/7064

Chapter 4
Applications of Artificial Intelligence in Thrust Manufacturing:
Enhancing Precision and Efficiency

P. Ramkumar
Sri Sairam College of Engineering, Bangalore, India

P. Hosanna Princye
ⓘD https://orcid.org/0000-0001-5056-980X
Sri Sairam College of Engineering, India

D. Satishkumar
Nehru institute of Technology Coimbatore, India

A. Ahila
Sri Sairam College of Engineering, India

Cynthia Anbuselvi Thangaraj
S.E.A. College of Engineering and Technology, India

ABSTRACT

Dive deep into the many ways AI is revolutionizing thrust manufacturing in the aerospace sector with this in-depth study piece. Many parts of production are being rethought by artificial intelligence technology, from design optimization to real-time monitoring. Precision, efficiency, and creativity in the production of thrust systems are showcased in this study, which investigates the revolutionary influence of AI on predictive maintenance, quality control, supply chain management, and human-machine collaboration.

DOI: 10.4018/979-8-3693-2615-2.ch004

INTRODUCTION

The aircraft industry is always on the lookout for new innovations to satisfy the needs for efficiency and precision. In this portion, the essay is introduced by emphasizing the significant impact that AI is having on the aerospace industry's thrust manufacturing and its future development. Thrust manufacturing's incorporation of Artificial Intelligence (AI) is causing a revolutionary wave in the aerospace sector, which is known for its dogged quest of efficiency and innovation. Recent advances in (AI) have opened up new possibilities for improving thrust system production in terms of accuracy, process optimization, and overall efficiency. This article explores the wide range of AI applications that are changing the thrust manufacturing industry in response to the rising performance and reliability standards for aerospace components. Artificial intelligence (AI) is revolutionizing various aspects of aircraft technology, including generative design, predictive maintenance, quality control, and real-time monitoring (Li, 2018).

At the intersection of artificial intelligence and thrust manufacturing, there is great potential to revolutionize the design, production, and maintenance of components while also automating mundane jobs. Artificial intelligence (AI) is causing a sea change in manufacturing by facilitating the exploitation of data, machine learning, and sophisticated algorithms (Angelopoulos et al., 2020).

To demonstrate how these technologies are improving several parts of the aerospace production process, this article will examine important AI uses in thrust manufacturing. When it comes to creating innovative designs or keeping an eye on production lines in real-time, AI is quickly becoming an essential tool for excellence. In what follows, we'll examine some concrete uses, such as generative design for better parts, predictive maintenance for more dependability, AI-driven anomaly detection for quality control, predictive analytics for supply chain optimization, real-time monitoring for decision support at a pinch, and human-robot collaboration in manufacture. For the purpose of thrust bearing fault diagnostics, information regarding the specific problems is necessary. This led to the construction of a test apparatus and the subsequent execution of an experiment to replicate the thrust bearing's failure (Heo, 2016).

We are entering a new era of possibilities as we investigate AI's potential uses in thrust manufacturing; this is because AI is being wed to aerospace engineering. These innovations not only address the demands of the aerospace industry right now, but they also set the stage for a future where efficiency and accuracy soar to new heights, influencing the course of aerospace technology well into the next century.

GENERATIVE DESIGN FOR THRUST COMPONENTS

Generative design has changed the game for aerospace engineers when it comes to designing critical components. It's a state-of-the-art AI solution for thrust manufacturing. Using state-of-the-art algorithms and machine learning, engineers can now uncover optimal solutions that may outperform traditional design methods by exploring an extended variety of design alternatives.

In generative design, artificial intelligence algorithms are used to iteratively explore and develop many design options within the limitations and parameters that are predefined. This method enables engineers to optimize the form, mass, and distribution of materials for components in thrust production, with the goal of improving performance, efficiency, and dependence. By 2040, three-quarters of Europe's vehicles—including ships, cars, and planes—will run on hydrogen instead of fossil fuels (Joung et al., 2020).

Conceptually and analytically evaluating an infinite variety of design changes is beyond the capabilities of the human intellect. When it comes to analyzing large datasets and quickly evaluating multiple design variations, generative design truly shines. With this skill, engineers may push the limits of traditional design and find solutions that are both unorthodox and incredibly efficient.

Generative design is great at making components that are customized for certain operational needs, which is highly useful in thrust production where performance and accuracy are paramount. Optimizing configurations for thrust components can be achieved by engineers through simulation and analysis of performance data from several design possibilities.

The generative design process is strong because it is iterative. Incorporating input and data from actual use into designs allows engineers to make iterative improvements. Components that undergo this iterative process eventually fulfill or surpass expectations because of the continuous improvement cycle it induces.

Blending generative design with additive manufacturing opens up new options for complicated geometries and structures that could be difficult, if not impossible, to create using more conventional methods. Combining generative design with additive printing is a natural progression for aerospace manufacturing technology.

PREDICTIVE MAINTENANCE FOR THRUST SYSTEMS

Aerospace engineering is a high-stakes, ever-changing field, and one area where AI is having a revolutionary impact is predictive maintenance. Aerospace engineers can now anticipate and resolve possible problems before they worsen thanks to predictive maintenance, which is especially designed for propulsion systems and uses sophisticated machine learning algorithms to examine both current and past data. Thrust system dependability, downtime, and operating efficiency can all be improved using this preventative method. Numerous coastal regions across the globe will have access to solar, wind, and wave energy resources, allowing them to diversify their energy sources and so decrease the prices and variability of renewable power and their integration into existing power systems (Stoutenburg et al., 2010).

Using artificial intelligence algorithms, predictive maintenance may identify patterns and outliers in operational data to foretell when equipment may break. If propulsion systems are to operate reliably and continuously, this methodology is especially important in thrust manufacture.

The use of analytics on data in real-time is fundamental to predictive maintenance. In thrust systems, sensors and monitoring equipment record vital signs including vibration, temperature, and pressure in real time. The data is analyzed by AI systems, which look for patterns that could mean problems are on the horizon. Discovering anomalies early on is when predictive maintenance algorithms really shine (McCormick, 2007). A quick way to spot outliers is to set up baseline operational routines. Engineers are able to take preventative measures by responding to these early warning signs of trouble before they affect the thrust system's performance or safety. Uptime and reliability maximization are the key objectives of predictive maintenance for thrust systems. In order to prevent expensive and unanticipated interruptions to aircraft operations, engineers can arrange maintenance operations during planned downtime by addressing possible issues early on.

Scheduling maintenance at regular intervals or responding only to obvious symptoms of wear and tear are commonplace in more conventional approaches. But with predictive maintenance, you may take a data-driven, optimized approach. Not only can predictive maintenance improve reliability, but it also helps save a lot of money because it is proactive. Significant repair and operational disruption expenses might result from unanticipated maintenance incidents. By enhancing the allocation of maintenance resources and reducing the occurrence of unexpected failures, predictive maintenance helps to lower these costs.

QUALITY CONTROL AND ANOMALY DETECTION

Anomaly detection and quality control are crucial components in many domains, such as data analysis, software development, and manufacturing. They find out whether things aren't up to par and flag any discrepancies that could mean problems with processes and products. An item or service is said to have passed muster if it has undergone quality control (Bao et al., 2019). Quality control is the practice of keeping an eye on and adjusting different parts of a process so that it runs smoothly and without errors. The goal of anomaly detection is to find patterns or occurrences that don't fit the mould. It finds extensive application in systems for monitoring, data analysis, and cyber security (Corallo et al., 2020).

SUPPLY CHAIN OPTIMIZATION WITH PREDICTIVE ANALYTICS

Forecasting future occurrences and optimizing different parts of the supply chain are the goals of supply chain optimization with predictive analytics. This approach makes use of data analysis, statistical algorithms, and machine learning approaches (You et al., 2015). Organizations can utilize this method to make better decisions, increase efficiency, decrease costs, and boost supply chain performance overall. With the help of AI, predictive analytics may be used to optimize the supply chain for aerospace. The section demonstrates how AI improves supply chain responsiveness and efficiency for thrust manufacturing by enhancing inventory management, predicting demand, and evaluating supplier reliability.

REAL-TIME MONITORING AND DECISION SUPPORT

Crucial components in many sectors include real-time monitoring and decision support systems, which enable organizations to respond to events in real-time, make educated decisions, and improve operational efficiency overall. Manufacturing processes may be instantly analyzed with the help of AI when combined with real-time monitoring systems. In order to create a more responsive and adaptable industrial environment, this section explains how AI helps with decision-making by giving operators access to data in real-time.

HUMAN-MACHINE COLLABORATION WITH COLLABORATIVE ROBOTS

Collaborative robots, often known as cobots, are creating waves in the automation and technology industry as game-changing instruments that usher in a new age of human-machine cooperation. The use of different robotic architectures than robotic manipulators, along with machine learning techniques, is suggested for future works in human-robot collaboration applications (Semeraro et al., 2023). The goal of creating cobots was to eliminate the need for humans to work in tandem with machines, as opposed to the standalone nature of conventional industrial robots. The importance of collaborative robots in human-machine interaction is discussed in this essay, along with the pros, cons, and possible future advancements of this exciting area. When it comes to tedious and repetitive jobs, collaborative robots are masters. Overall efficiency and production are improved when cobots take over mundane tasks, allowing human workers to concentrate on more creative and sophisticated parts of their professions.

When compared to conventional robots, cobots are safer because they can work in close quarters with people. Thanks to their high-tech sensors and visual systems, they can identify when people are around and react accordingly, making accidents much less likely. Through the automation of labor-intensive processes, this team effort improves workplace safety and allays ergonomic fears.

When it comes to shifting jobs and production needs, collaborative robots are naturally adaptable. Their adaptability to changing production settings is a result of their speed and ease of programming and reconfiguration.

CYBERSECURITY IN AI-INTEGRATED MANUFACTURING

Though it has ushered in many improvements, the introduction of artificial intelligence (AI) into production processes has also brought about new worries, particularly about data security. It is critical to protect these systems from cyber threats as AI becomes more common in production settings. Process optimization, product design, and overall efficiency are three areas where AI is commonly used in manufacturing. The key to staying ahead of the competition is safeguarding the intellectual property that is built into AI models and algorithms. Cybersecurity procedures prevent unauthorized individuals from gaining access to sensitive information and stealing it. The identified studies were analyzed to extract relevant information on the impact of AI-based attacks and to provide insights for structuring defense measures (De Azambuja et al., 2023).

Information and systems that are interdependent are crucial to the manufacturing process. There is a risk of production delays, downtime, and financial losses when cyberattacks target manufacturing systems that incorporate AI. Cybersecurity solutions are crucial for protecting vital manufacturing processes from unauthorized access, manipulation, or disruption. Artificial intelligence systems used in manufacturing need massive volumes of data for both training and making decisions. Cyber risks, such data modification or malicious code insertion, can damage data integrity, which in turn can cause AI-driven choices to be flawed. For AI applications to rely on accurate and trustworthy data, stringent cybersecurity protections are required. Organisations in the manufacturing sector frequently handle confidential information pertaining to product blueprints, manufacturing procedures, and company tactics.

In order to prevent competitors or other bad actors from acquiring vital information through espionage or cyber-espionage, cyber security is of the utmost importance.

IoT devices are among the many interconnected nodes in the network that is common in AI-integrated manufacturing. Intruders in cyberspace may be able to access these gadgets. To protect these devices from being exploited or accessed without authorization, cybersecurity measures are implemented.

The increasing connectivity of manufacturing systems raises concerns about cybersecurity. This section addresses the importance of AI in monitoring and securing sensitive data, safeguarding against potential cyber threats and ensuring the integrity of manufacturing processes. Design files, customer records, and production schedules are all examples of sensitive data that could be at danger of unauthorized access (Wu et al., 2018). In order to keep sensitive information safe and stop data breaches, cybersecurity precautions are essential. The goal of cybercriminals is to influence AI models in their favor by making them do things they shouldn't or by generating false findings. Worst case scenario: this causes problems with the product's design, quality, or safety. A quite intricate supply chain is frequently involved in manufacturing processes. Because of the potential impact on suppliers and partners, the entire ecosystem is at risk from cybersecurity threats. The whole robustness of the supply chain depends on ensuring cybersecurity across it.

CONCLUSION

In conclusion, the applications of Artificial Intelligence in thrust manufacturing are diverse and impactful. From design optimization to real-time monitoring and cybersecurity, AI technologies are reshaping the aerospace industry's approach to precision and efficiency. As manufacturers embrace these applications, it becomes evident that AI is a driving force in the evolution of thrust manufacturing, promising a future characterized by heightened performance, reliability, and innovation in aerospace systems.

REFERENCES

Angelopoulos, A., Michailidis, E. T., Nomikos, N., Trakadas, P., Hatziefremidis, A., Voliotis, S., & Zahariadis, T. (2020). Tackling Faults in the Industry 4.0 Era-A Survey of Machine-Learning Solutions and Key Aspects. *Sensors (Basel)*, 20(1), 109. doi:10.3390/s20010109 PMID:31878065

Bao, Y., Tang, Z., Li, H., & Zhang, Y. (2019). Computer vision and deep learning–based data anomaly detection method for structural health monitoring. *Structural Health Monitoring*, *18*(2), 401–4215. doi:10.1177/1475921718757405

Corallo, A., Lazoi, M., & Lezzi, M. (2020). Cybersecurity in the context of industry 4.0: A structured classification of critical assets and business impacts. *Computers in Industry*, *114*, 103165. doi:10.1016/j.compind.2019.103165

De Azambuja, A. J. G., Plesker, C., Schützer, K., Anderl, R., Schleich, B., & Almeida, V. R. (2023). Artificial Intelligence-Based Cyber Security in the context of Industry 4.0—A survey. *Electronics (Basel)*, *12*(8), 1920. doi:10.3390/electronics12081920

Heo, S. (2016). Climate Change and Concerted Actions by Mankind. *J. Korean Soc. Trends Perspectibes*, *96*, 214–220.

Joung, T. H., Kang, S. G., Lee, J. K., & Ahn, J. (2020). The IMO initial strategy for reducing Greenhouse Gas (GHG) emissions, and its follow-up actions towards 2050. *J. Int. Marit. Saf. Environ. Aff. Shipp.*, *4*(1), 1–7. doi:10.1080/25725084.2019.1707938

Li, J. (2018). hua: Cyber security meets artificial intelligence: A survey. *Front. Inf. Technol. Electron. Eng.*, *19*(12), 1462–1474. doi:10.1631/FITEE.1800573

McCormick, M. E. (2007). *Ocean Wave Energy Conversion*. Dover Publications Inc.

Semeraro, F., Griffiths, A., & Cangelosi, A. (2023). Human–robot collaboration and machine learning: A systematic review of recent research. *Robotics and Computer-integrated Manufacturing*, *79*, 102432. doi:10.1016/j.rcim.2022.102432

Stoutenburg, E. D., Jenkins, N., & Jacobson, M. Z. (2010). Power output variations of co-located offshore wind turbines and wave energy converters in California. *Renewable Energy*, *35*(12), 2781–2791. doi:10.1016/j.renene.2010.04.033

Wu, D., Ren, A., Zhang, W., Fan, F., Liu, P., Fu, X., & Terpenny, J. (2018). Cybersecurity for digital anufacturing. *Journal of Manufacturing Systems*, *48*, 3–12. doi:10.1016/j.jmsy.2018.03.006

You, Z., Si, Y.-W., Zhang, D., Zeng, X., Leung, S. C. H., & Li, T. (2015). A decision-making framework for precision marketing. *Expert Systems with Applications*, *42*(7), 3357–3367. doi:10.1016/j.eswa.2014.12.022

Chapter 5
Revolutionizing Thrust Manufacturing:
The Synergy of Real-Time Data and AI Advancements

K. R. Senthilkumar

(iD) https://orcid.org/0000-0001-7426-5376
Sri Krishna Arts and Science College, India

ABSTRACT

This chapter explores the transformative impact of integrating real-time data and artificial intelligence (AI) in the field of thrust manufacturing, particularly within the aerospace and automotive industries. As manufacturing processes evolve, the synergy between real-time data and AI advancements emerges as a catalyst for unparalleled efficiency, precision, and innovation. The chapter examines the foundational role of real-time data in providing a granular view of operations, complemented by the sophisticated capabilities of AI—from automation to adaptive intelligence. Through case studies, the document showcases successful applications of this synergy in optimizing production, predictive maintenance, and quality control. Despite the promise, challenges such as data security and workforce upskilling are acknowledged. The chapter concludes by envisioning a future where the convergence of real-time data and AI defines the landscape of intelligent thrust manufacturing, presenting opportunities for smart factories and adaptive supply chains.

DOI: 10.4018/979-8-3693-2615-2.ch005

INTRODUCTION

The manufacturing sector has always been at the forefront of technological advancements, and the synergy between real-time data and AI is ushering in a new era of transformation[REMOVED REF FIELD]. This paper delves into the ways in which thrust manufacturing, a critical component in aerospace and automotive industries, is being revolutionized by harnessing the power of real-time data and AI (Ge et al. 2023).

Real-Time Data: The Foundation of Precision

The ability to collect, process, and analyze real-time data has become a cornerstone in modern manufacturing. We examine how real-time data acquisition from sensors, IoT devices, and other sources provides manufacturers with a granular view of operations, enabling precise control and decision-making (Sayedahmed, Fahmy, and Hefny 2021).

AI Advancements: From Automation to Intelligence

Artificial intelligence has transcended traditional automation, evolving into a sophisticated tool that learns, adapts, and augments human capabilities. This section explores the role of AI in thrust manufacturing, including predictive maintenance, quality control, and adaptive manufacturing processes (Dondos and Papanagopoulos 1996).

THE SYNERGY UNLEASHED

The true transformative power lies in the convergence of real-time data and AI. By combining the insights derived from real-time data with AI algorithms, manufacturers gain a holistic understanding of their processes. This synergy leads to improved efficiency, reduced downtime, and enhanced product quality (Prhashanna and Dormidontova 2020).

Case Studies: Success Stories in Thrust Manufacturing

Highlighting real-world applications, this section presents case studies showcasing companies that have successfully implemented the synergy of real-time data and AI in thrust manufacturing. From optimized production lines to predictive maintenance, these examples demonstrate tangible benefits and return on investment (Saikia, Gaurav, and Rakshit 2020).

Challenges and Considerations

While the potential is immense, the implementation of real-time data and AI in thrust manufacturing is not without challenges. We discuss considerations such as data security, integration complexities, and workforce upskilling, providing insights into overcoming these hurdles (Ennaji, Vergütz, and El Allali 2023).

Future Prospects: Towards Intelligent Manufacturing

Looking ahead, we explore the future prospects of thrust manufacturing in an era dominated by real-time data and AI. From smart factories to adaptive supply chains, we discuss how continued advancements will shape the landscape of manufacturing. The revolutionizing force of real-time data and AI in thrust manufacturing is undeniable. The synergy of these technologies not only enhances current processes but also opens the door to unprecedented possibilities. As industries strive for greater efficiency and innovation, the integration of real-time data and AI advancements stands as a beacon of progress in the realm of manufacturing.

1. Review Methodology:

The review methodology employed in this paper combines a comprehensive analysis of existing literature, case studies, and real-world applications to explore the impact of integrating real-time data and artificial intelligence (AI) in thrust manufacturing. The following key steps were undertaken in the methodology (Brenes, Johanssen, and Chukhrova 2022; Mahbub and Shubair 2023; Xu et al. 2022):

1. **Literature Review:**
    ◦   Conducted an extensive review of academic articles, research papers, and industry publications related to real-time data and AI in manufacturing.
    ◦   Examined the foundational concepts, theories, and best practices in the integration of these technologies.
2. **Case Studies:**
    ◦   Selected relevant case studies from the aerospace and automotive industries that highlight successful applications of real-time data and AI in thrust manufacturing.
    ◦   Analyzed these case studies to extract insights, challenges, and outcomes.

3.   Real- Time- Definition, Principle, Prerequisites

Real-Time:

Real-time refers to the processing of data or events as they occur, without any noticeable delay. In a real-time system, the response time is constrained within a specific time frame, ensuring that the system provides timely and instantaneous results. Real-time processing is crucial in applications where immediate and accurate responses are essential, such as in control systems, communication systems, and certain types of data analysis (Alqahtani and Kumar 2024).

The fundamental principle of real-time systems is to guarantee that the system can respond to events or input within a predefined and often deterministic time frame. This involves efficient processing, minimal latency, and predictable performance. Real-time systems are designed to meet deadlines consistently, ensuring that tasks are completed within their specified time constraints. The principle is often associated with ensuring reliability, predictability, and accuracy in the system's responses to external stimuli (Etengu et al. 2023).

Prerequisites: Achieving real-time processing involves several prerequisites and considerations:

1.   **Deterministic Timing:**
     ◦   Real-time systems must have deterministic timing characteristics, meaning that the time it takes to execute a task or respond to an event is known and consistent.
2.   **Fast Response Time:**
     ◦   The system should be capable of responding quickly to input or events. This requires optimizing algorithms, minimizing processing delays, and employing efficient hardware.
3.   **Predictable Performance:**
     ◦   Predictability is crucial in real-time systems. Variability in system performance can lead to missed deadlines, impacting the reliability of the system.
4.   **Reliable Hardware and Software:**
     ◦   The hardware and software components of the system must be reliable to ensure consistent and error-free operation. Redundancy and fault tolerance are often employed to enhance reliability.
5.   **Task Scheduling:**
     ◦   Efficient task scheduling mechanisms are essential to prioritize and manage tasks within the system. This includes techniques such as priority-based scheduling and real-time operating systems.

Figure 1. Real time operating system

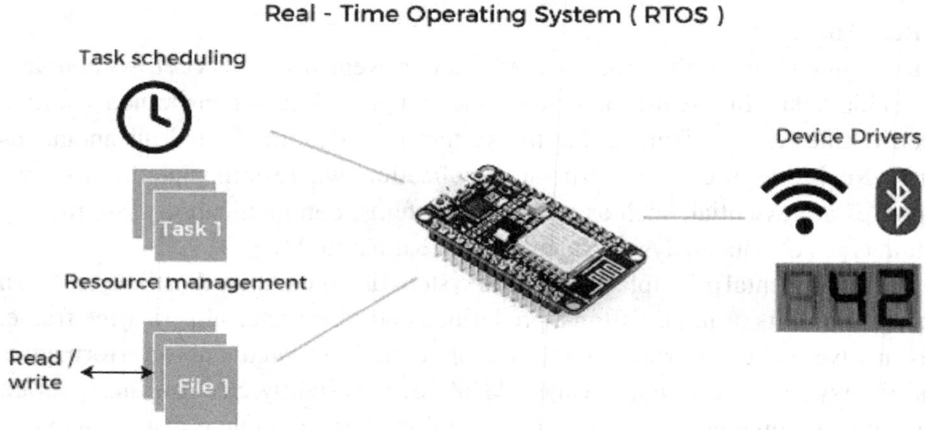

6. **High Throughput:**
   ○ Real-time systems often require high throughput to handle a large number of events or tasks concurrently. This involves optimizing the system's processing capabilities.

7. **Real-Time Operating System (RTOS):**
   ○ The use of a real-time operating system tailored for time-sensitive applications is common. An RTOS provides features like task scheduling, prioritization, and synchronization to meet real-time requirements.

8. **Sensors and Input Devices:**
   ○ In applications like control systems, the use of responsive sensors and input devices is crucial to providing real-time feedback to the system.

9. **Communication Protocols:**
   ○ Reliable and low-latency communication protocols are necessary, especially in distributed real-time systems, to ensure timely exchange of information between components.

REAL-TIME MANUFACTURING TECHNOLOGY: MONITORING, FEEDBACK, CONTROLLING, AND MACHINE LEARNING

Real-time manufacturing technology plays a pivotal role in enhancing efficiency, productivity, and quality in modern industrial processes (Lim et al. 2023). This technology integrates various components, including monitoring systems, feedback mechanisms, control systems, and machine learning, to create a dynamic and responsive manufacturing environment. Here's an overview of each component (Etengu et al. 2023; Pughazendi, Rajaraman, and Mohammed 2023):

Monitoring:

Definition: Monitoring in real-time manufacturing involves the continuous and instantaneous observation of various parameters, processes, and equipment on the shop floor.

Principle: Sensors, IoT devices, and other data acquisition tools are deployed to collect real-time data. This data encompasses factors such as temperature, pressure, speed, and other relevant metrics.

Prerequisites: Reliable sensors, robust data communication infrastructure, and a centralized monitoring system are essential. The monitoring system should be capable of collecting, processing, and visualizing data in real time.

Feedback:

Definition: Feedback mechanisms in real-time manufacturing involve the analysis of monitored data to provide timely insights and corrective actions.

Principle: Algorithms analyze real-time data and compare it against predefined parameters or quality standards. Any deviations trigger immediate feedback to adjust processes and maintain optimal conditions.

Prerequisites: Intelligent algorithms, a feedback loop, and a responsive control system are crucial. Fast data processing capabilities are essential to provide timely feedback to the manufacturing processes.

Controlling:

Definition: Real-time control in manufacturing refers to the ability to adjust and optimize processes instantaneously based on feedback and monitoring data.

Principle: Control systems use feedback information to make real-time adjustments to equipment, parameters, or other variables to ensure that the manufacturing process operates within specified tolerances.

Prerequisites: Robust control algorithms, actuators, and a responsive control infrastructure are necessary. Integration with monitoring systems facilitates seamless data exchange for effective control.

Machine Learning:

Definition: Machine learning (ML) in real-time manufacturing involves the use of algorithms that can learn and adapt based on data patterns, optimizing processes and decision-making over time.

Principle: ML algorithms analyze historical and real-time data to identify patterns, predict outcomes, and optimize manufacturing processes. This can lead to improved efficiency, reduced downtime, and enhanced product quality.

Prerequisites: Access to quality training data, the integration of machine learning models into the manufacturing environment, and the ability to continuously update and refine these models.

SUBTRACTIVE MANUFACTURING/MACHINING

Definition: Subtractive manufacturing, also known as machining, is a process in which a part or product is produced by removing material from a workpiece (de Oliveira, J.A. Bastos-Filho, and Oliveira 2022). This is achieved through various machining techniques, such as cutting, milling, drilling, turning, or grinding. The goal is to shape the raw material into the desired form by removing excess material, often starting with a larger workpiece and progressively removing material until the final shape is achieved (Sharifi and Shokouhyar 2021; Solairaj, Sugitha, and Kavitha 2023).

Principles:

1.   **Material Removal:** The fundamental principle of subtractive manufacturing is the controlled removal of material from a workpiece. This can be done through processes like cutting, milling, or drilling, depending on the desired outcome.
2.   **Precision Machining:** Subtractive manufacturing emphasizes precision in shaping the workpiece. Computer Numerical Control (CNC) machines are commonly used to achieve high levels of accuracy and repeatability in the machining process.

Figure 2. Subtractive manufacturing

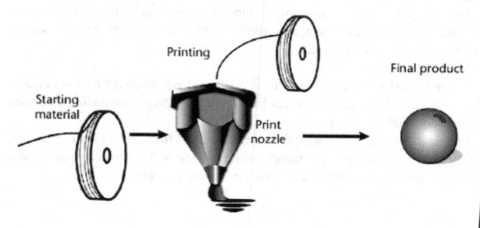

3. **Tooling:** Different machining processes require specific tools. These tools, such as cutting tools, end mills, drills, and lathe tools, are selected based on the material being machined and the desired outcome.

4. **Computer-Aided Design (CAD) and Computer-Aided Manufacturing (CAM):** Designs are typically created using CAD software, and CAM software is used to generate toolpaths and control CNC machines. This integration ensures accurate translation of digital designs into physical products.

5. **Workholding:** Proper clamping and securing of the workpiece are critical to prevent movement during machining. Workholding devices like vises, chucks, and fixtures are used to ensure stability.

6. **Material Considerations:** The choice of materials affects the machinability and tool selection. Metals, plastics, composites, and other materials can be machined, each requiring specific tools and cutting parameters.

7. **Tolerances and Surface Finish:** Machining processes can achieve tight tolerances and specific surface finishes, meeting the required specifications for the final product.

Common Subtractive Manufacturing Processes:

1. **Turning:** Involves rotating a workpiece while a cutting tool removes material. Common in the production of cylindrical parts.

2. **Milling:** Utilizes rotating cutting tools to remove material from a stationary workpiece. Can produce complex shapes and features.

3. **Drilling:** Involves creating holes in a workpiece using a rotating drill bit.

4. **Grinding:** Uses abrasive wheels to remove material and achieve precise dimensions and surface finishes.

5. **Laser Cutting:** Involves using a laser to cut through materials, often used for intricate designs and thin materials.

6. **Waterjet Cutting:** Utilizes a high-pressure stream of water or water mixed with abrasive particles to cut through various materials.

Applications: Subtractive manufacturing is widely used in industries such as aerospace, automotive, medical, and consumer goods. It is suitable for producing prototypes, custom parts, and high-precision components where accuracy and material properties are critical (Azhari et al. 2018; Ma and Zhang 2022; Pughazendi, Rajaraman, and Mohammed 2023).

FUTURE PROSPECT

The future prospects of subtractive manufacturing (machining) are influenced by ongoing technological advancements, industry trends, and the evolving needs of manufacturing sectors. Several key areas highlight the potential future developments in subtractive manufacturing:

1. **Advanced Materials:**
   - As new and advanced materials are developed, subtractive manufacturing processes will need to adapt. Machining technologies will evolve to handle materials with enhanced properties, such as lightweight composites, advanced alloys, and high-performance plastics.

2. **Industry 4.0 Integration:**
   - Subtractive manufacturing is likely to become more integrated into Industry 4.0, with the widespread use of smart manufacturing technologies. This includes the incorporation of sensors, real-time monitoring, data analytics, and connectivity to enhance efficiency, reduce downtime, and enable predictive maintenance.

3. **Digitalization and Simulation:**
   - Increased integration of digital technologies, such as digital twins and simulation tools, will enable manufacturers to optimize machining processes virtually before physical production. This can lead to improved accuracy, reduced waste, and faster product development cycles.

4. **Additive-Subtractive Hybrid Manufacturing:**
   - The combination of subtractive and additive manufacturing processes (hybrid manufacturing) is gaining traction. This approach allows for the advantages of both techniques, with additive processes creating complex structures, followed by subtractive processes to achieve precise details and surface finishes. This hybrid approach is likely to be increasingly adopted for complex and customized components.

5. **Enhanced Automation and Robotics:**
   - Automation and robotics will play a more significant role in subtractive manufacturing. Advanced robotic systems will be employed for tasks such as tool changes, material handling, and even autonomous machining. This can lead to increased efficiency, cost savings, and improved safety in manufacturing environments.

6. **Precision and Miniaturization:**
   - With the growing demand for miniaturized and highly precise components, subtractive manufacturing processes will need to push the limits of precision. Advances in machining technologies, tooling,

and control systems will contribute to achieving tighter tolerances and improved surface finishes.

7. **Environmental Sustainability:**
   ○ There will likely be an increased focus on sustainability in manufacturing. Efforts to reduce waste, optimize energy consumption, and implement eco-friendly machining practices will become more prominent. This may involve the development of more sustainable cutting fluids, recycling of machining by-products, and energy-efficient machining technologies (Senapati and Rawal 2023).

8. **Educational and Workforce Development:**
   ○ With the evolving nature of subtractive manufacturing technologies, there will be a growing need for skilled professionals who can operate and program advanced machining equipment. Educational programs and workforce development initiatives will play a crucial role in preparing individuals for the changing landscape of subtractive manufacturing (Jiang et al. 2021).

In conclusion, the future of subtractive manufacturing holds exciting prospects driven by technological innovations, digitalization, and a focus on sustainability. As manufacturing continues to evolve, the adaptability and integration of subtractive manufacturing processes will contribute to the advancement of various industries.

CONCLUSION

In conclusion, the landscape of subtractive manufacturing, or machining, is poised for significant advancements and transformative changes. Several key factors contribute to the evolving nature and promising future of subtractive manufacturing:

1. **Technological Advancements:**
   ○ Ongoing innovations in machining technologies, including precision machining tools, cutting-edge materials, and advanced control systems, are driving the industry toward higher levels of efficiency, accuracy, and productivity.

2. **Integration with Industry 4.0:**
   ○ The integration of subtractive manufacturing into the broader framework of Industry 4.0 is enhancing connectivity, automation, and data-driven decision-making. Real-time monitoring, data analytics, and smart manufacturing principles are becoming integral to optimizing machining processes.

3. **Hybrid Manufacturing Approaches:**
   ◦ The emergence of hybrid manufacturing, combining both subtractive and additive processes, reflects a holistic approach to production. This allows manufacturers to leverage the benefits of both techniques, offering greater flexibility and the ability to produce complex, customized components efficiently.

4. **Precision and Miniaturization Demands:**
   ◦ Increasing demands for miniaturized and highly precise components are pushing the boundaries of subtractive manufacturing. Advances in machining technologies are enabling manufacturers to achieve tighter tolerances and superior surface finishes, meeting the evolving requirements of various industries.

5. **Automation and Robotics:**
   ◦ Automation and robotics are playing an increasingly pivotal role in subtractive manufacturing. From robotic tool changes to autonomous machining processes, automation is enhancing efficiency, reducing human intervention, and contributing to safer and more streamlined manufacturing environments.

6. **Sustainability Focus:**
   ◦ The future of subtractive manufacturing is expected to place a greater emphasis on sustainability. Efforts to minimize waste, optimize energy consumption, and implement environmentally friendly practices in machining processes are becoming integral to industry practices.

7. **Workforce Development:**
   ◦ The evolving nature of subtractive manufacturing technologies underscores the importance of workforce development and education. Training programs that equip individuals with the skills to operate, program, and adapt to advanced machining equipment will be crucial for sustaining industry growth.

8. **Global Competitiveness:**
   ◦ The continual evolution of subtractive manufacturing technologies contributes to the global competitiveness of industries. The ability to adopt and implement cutting-edge machining practices positions manufacturers to stay ahead in a rapidly changing and competitive marketplace.

In summary, the future of subtractive manufacturing holds great promise as it embraces technological advancements, integrates with broader industry trends, and addresses the growing demands for precision, customization, and sustainability. The adaptability of subtractive manufacturing processes positions them as a key player in shaping the future of modern manufacturing across diverse sectors.

REFERENCES

Alqahtani, H., & Kumar, G. (2024). Machine Learning for Enhancing Transportation Security: A Comprehensive Analysis of Electric and Flying Vehicle Systems. *Engineering Applications of Artificial Intelligence, 129*, 107667. https://www.sciencedirect.com/science/article/pii/S0952197623018511. doi:10.1016/j.engappai.2023.107667

Azhari, M. E., Toumanari, A., Latif, R., & El Moussaid, N. (2018). Round Estimation Period for Cluster-Based Routing in Mobile Wireless Sensor Networks. *International Journal of Advanced Intelligence Paradigms, 10*(4), 374–390. https://www.scopus.com/inward/record.uri?eid=2-s2.0-85048034340&doi=10.1504%2FIJAIP.2018.092034&partnerID=40&md5=c9e1d8c8ea8aed7e82c2f82de9960806. doi:10.1504/IJAIP.2018.092034

Brenes, R. F., Johannssen, A., & Chukhrova, N. (2022). An Intelligent Bankruptcy Prediction Model Using a Multilayer Perceptron. *Intelligent Systems with Applications, 16*. https://www.sciencedirect.com/science/article/pii/S2667305322000734

de Oliveira, J., Bastos-Filho, C., & Oliveira, S. (2022). Non-Invasive Embedded System Hardware/Firmware Anomaly Detection Based on the Electric Current Signature. *Advanced Engineering Informatics, 51*. https://www.sciencedirect.com/science/article/pii/S1474034621002676

Dondos, A., & Papanagopoulos, D. (1996). Three Models for Chain Conformation of Block Copolymers in Solution and in Solid State. *Journal of Polymer Science. Part B, Polymer Physics, 34*(7), 1281–1288. doi:10.1002/(SICI)1099-0488(199605)34:7<1281::AID-POLB10>3.0.CO;2-6

Ennaji, O., Vergütz, L., & El Allali, A. (2023). Machine Learning in Nutrient Management: A Review. *Artificial Intelligence in Agriculture, 9*, 1–11. https://www.sciencedirect.com/science/article/pii/S258972172300017X

Etengu, R., Tan, S. C., Chuah, T. C., Lee, Y. L., & Galán-Jiménez, J. (2023). AI-Assisted Traffic Matrix Prediction Using GA-Enabled Deep Ensemble Learning for Hybrid SDN. *Computer Communications, 203*, 298–311. https://www.sciencedirect.com/science/article/pii/S0140366423000920. doi:10.1016/j.comcom.2023.03.014

Ge, W., Lueck, C., Suominen, H., & Apthorp, D. (2023). Has Machine Learning Over-Promised in Healthcare?: A Critical Analysis and a Proposal for Improved Evaluation, with Evidence from Parkinson's Disease. *Artificial Intelligence in Medicine, 139*, 102524. https://www.sciencedirect.com/science/article/pii/S0933365723000386. doi:10.1016/j.artmed.2023.102524 PMID:37100503

Jiang, X., Lin, G.-H., Huang, J.-C., Hu, I.-H., & Chiu, Y.-C. (2021). Performance of Sustainable Development and Technological Innovation Based on Green Manufacturing Technology of Artificial Intelligence and Block Chain. *Mathematical Problems in Engineering, 2021*, 1–11. https://www.scopus.com/inward/record.uri?eid=2-s2.0-85104505059&doi=10.1155%2F2021%2F5527489&partnerID=40&md5=545252da8e6b7413a04ff7dd925c3c8d. doi:10.1155/2021/5527489

Lim, J. Y., Lim, J. Y., Baskaran, V. M., & Wang, X. (2023). A Deep Context Learning Based PCB Defect Detection Model with Anomalous Trend Alarming System. *Results in Engineering, 17*, 100968. https://www.sciencedirect.com/science/article/pii/S2590123023000956. doi:10.1016/j.rineng.2023.100968

Ma, X., & Zhang, Y. (2022). Digital Innovation Risk Management Model of Discrete Manufacturing Enterprise Based on Big Data Analysis. *Journal of Global Information Management, 30*(7), 1–14. https://www.scopus.com/inward/record.uri?eid=2-s2.0-85114305136&doi=10.4018%2FJGIM.286761&partnerID=40&md5=bed6df90a735598eedf350ce3a3b8268. doi:10.4018/JGIM.286761

Mahbub, M., & Shubair, R. M. (2023). Contemporary Advances in Multi-Access Edge Computing: A Survey of Fundamentals, Architecture, Technologies, Deployment Cases, Security, Challenges, and Directions. *Journal of Network and Computer Applications, 219*, 103726. https://www.sciencedirect.com/science/article/pii/S1084804523001455. doi:10.1016/j.jnca.2023.103726

Prhashanna, A., & Dormidontova, E. E. (2020). Micelle Self-Assembly and Chain Exchange Kinetics of Tadpole Block Copolymers with a Cyclic Corona Block. *Macromolecules, 53*(3), 982–991. https://www.scopus.com/inward/record.uri?eid=2-s2.0-85079191834&doi=10.1021%2Facs.macromol.9b02398&partnerID=40&md5=363787355c3e21af9e1a1e2187b3e234. doi:10.1021/acs.macromol.9b02398

Pughazendi, N., Rajaraman, P., & Mohammed, M. (2023). Graph Sample and Aggregate Attention Network Optimized with Barnacles Mating Algorithm Based Sentiment Analysis for Online Product Recommendation. *Applied Soft Computing, 145*. https://www.sciencedirect.com/science/article/pii/S1568494623005501

Saikia, P. G. & Rakshit, D. (2020). Designing a Clean and Efficient Air Conditioner with AI Intervention to Optimize Energy-Exergy Interplay. *Energy and AI, 2*. https://www.sciencedirect.com/science/article/pii/S266654682030029X

Sayedahmed, H. A. M., Fahmy, I. M. A., & Hefny, H. A. 2021. "Impact of Fuzzy Stability Model on Ad Hoc Reactive Routing Protocols to Improve Routing Decisions." In *Advances in Intelligent Systems and Computing*, Springer, Cham, 441–54. https://link.springer.com/chapter/10.1007/978-3-030-58669-0_40 (October 14, 2020).

Senapati, B., & Rawal, B. S. (2023). Quantum Communication with RLP Quantum Resistant Cryptography in Industrial Manufacturing. *Cyber Security and Applications*. Science Direct. https://www.sciencedirect.com/science/article/pii/S2772918423000073

Sharifi, Z., & Shokouhyar, S. (2021). Promoting Consumer's Attitude toward Refurbished Mobile Phones: A Social Media Analytics Approach. *Resources, Conservation and Recycling*, *167*, 105398. https://www.sciencedirect.com/science/article/pii/S0921344921000057. doi:10.1016/j.resconrec.2021.105398

Solairaj, A., Sugitha, G., & Kavitha, G. (2023). Enhanced Elman Spike Neural Network Based Sentiment Analysis of Online Product Recommendation. *Applied Soft Computing*, *132*, 109789. https://www.sciencedirect.com/science/article/pii/S1568494622008389. doi:10.1016/j.asoc.2022.109789

Xu, H., Chai, L., Luo, Z., & Li, S. (2022). Stock Movement Prediction via Gated Recurrent Unit Network Based on Reinforcement Learning with Incorporated Attention Mechanisms. *Neurocomputing*, *467*, 214–228. https://www.sciencedirect.com/science/article/pii/S0925231221014508. doi:10.1016/j.neucom.2021.09.072

Chapter 6
The Usage of Artificial Intelligence in Manufacturing Industries:
A Real-Time Application

Renugadevi Ramalingam
R.M.K. Engineering College, India

Malathi Murugesan
E.G.S. Pillay Engineering College, India

S. Vaishnavi
Manipal Institute of Technology, India

N. Priyanka
https://orcid.org/0009-0004-2007-2973
Vellore Institute of Technology, India

S. Nalini
SRM Institute of Science and Technology, India

T. Chandrasekar
https://orcid.org/0000-0003-3591-2205
Kalasalingam Academy of Research and Education, India

ABSTRACT

Artificial intelligence (AI) has the capacity to revolutionize the manufacturing sector. Positive effects include things like more output, lower costs, better quality, and less downtime. Large factories are just one group of people who can take advantage of this technology. It is important for many smaller firms to understand how simple it

DOI: 10.4018/979-8-3693-2615-2.ch006

is to obtain high-quality, affordable AI solutions. AI has a wide range of potential applications in manufacturing. It enhances defect identification by automatically classifying faults in a variety of industrial products using sophisticated image processing techniques. Artificial intelligence has various potential applications in manufacturing since industrial IoT and smart factories generate enormous amounts of data every day. To better analyze data and make choices, manufacturers are increasingly using artificial intelligence solutions like deep learning neural networks and machine learning (ML). One common use of artificial intelligence in manufacturing is predictive maintenance.

INTRODUCTION

In order to stay abreast of technological advancements, manufacturers must delve into a pivotal factor propelling factories towards the future: machine learning. Now, let's discuss the primary applications and groundbreaking innovations that machine learning technology is offering in the year 2024. In essence, machine learning algorithms leverage training data to fuel an algorithm, enabling the software to address a specific problem. This data source can range from real-time data captured by Internet of Things (IoT) sensors on a factory floor to alternative collection methods. Machine learning encompasses diverse techniques, including neural networks and deep learning. Neural networks emulate biological neurons, discerning patterns in datasets to tackle problems. Deep learning involves multiple layers of neural networks, with the initial layer handling raw data input and transmitting processed information sequentially through subsequent layers.

In addition to the imperative role of advanced robotics in facilitating automated assembly, machine learning plays a crucial role in optimizing various tasks, including quality assurance, non-destructive testing (NDT) analysis, pinpointing the origins of defects, and more. An illustrative concept, the "factory in a box," serves as a means of simplifying a larger manufacturing facility, and in some instances, it's a literal interpretation. Nokia, for instance, has adopted portable manufacturing units in the form of retrofitted shipping containers equipped with sophisticated automated assembly machinery. These portable containers offer flexibility, enabling manufacturers to conduct on-site product assembly, eliminating the need for extensive product transportation.

A PROTOTYPE OF AN AI DEVELOPMENT GUIDE FOR PRODUCT OWNERS

The development of AI projects can be conceptualized as a three-layer pyramid, with each layer representing a different level of focus and complexity. At the top layer are well-established solutions that have undergone extensive study and application. Moving down the pyramid, the middle layer consists of projects involving advanced applications and implementations, while the bottom layer encompasses scientific research projects that contribute to the cutting-edge understanding of AI. The detailed picture of this prototype is shown in Figure 1.

The intricacy of each AI project is contingent upon the level of detail and customization desired. This complexity may be addressed either in the initial stages of project development or as an integral component of AI consulting services. In other words, the level of customization and the incorporation of intricate details can significantly impact the overall complexity and depth of an AI initiative, with the choice of when to address these aspects depending on the project's specific requirements and objectives.

The next stage involves identifying and defining the Machine Learning (ML) problem that needs to be addressed and resolved. This process should consider the technological capabilities of various Artificial Intelligence subfields, including

Figure 1. AI project complexity prototype

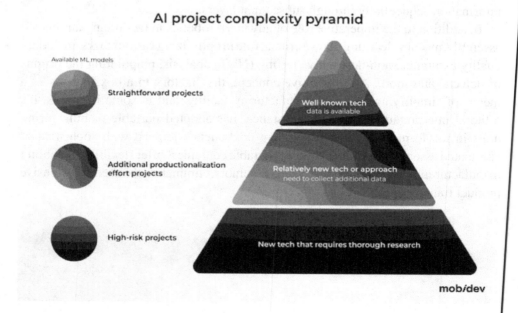

AI project complexity pyramid

Available ML models

Straightforward projects

Well known tech
data is available

Additional productionalization
effort projects

Relatively new tech or approach
need to collect additional data

High-risk projects

New tech that requires thorough research

mob*i*dev

Computer Vision, Natural Language Processing, Speech Recognition, Forecasting, Generative AI, and more. Different approaches can be employed during this stage, and, in general, there are three main components to consider when discussing the actual machine learning solution.

Problem Definition:

Clearly articulate and define the problem that the machine learning solution aims to solve. This involves specifying the desired outcome or prediction and understanding the business or operational context in which the solution will be applied. For example, if working on a computer vision project, the problem definition might involve image classification, object detection, or segmentation.

Feature Engineering:

Identify and select the relevant features or input variables that will be used to train the machine learning model. Feature engineering is a critical step in preparing the data for model training, and it involves choosing the right data attributes that contribute meaningful information to the model. In Natural Language Processing, for instance, feature engineering could involve selecting relevant linguistic features for text analysis.

Model Selection and Evaluation Metrics:

Choose the appropriate machine learning model that aligns with the problem at hand. Consider the strengths and weaknesses of various algorithms within the selected subfield. Additionally, define the evaluation metrics that will be used to assess the performance of the model. For instance, in speech recognition, the choice of a suitable model might involve selecting between recurrent neural networks (RNNs) or convolutional neural networks (CNNs), and evaluation metrics could include accuracy or word error rate. By carefully addressing these three components, the machine learning problem can be thoroughly understood, and a well-defined and effective solution can be crafted based on the capabilities of relevant AI subfields.

Pretrained Models:

Pretrained models are algorithms that have been previously trained to address specific business challenges using a predefined dataset. These models can be adapted to custom data and fine-tuned according to specific output requirements. This fine-tuning process is typically handled by a data science team. MobiDev boasts extensive experience in deploying various algorithms, such as those related to speech processing, computer vision, demand forecasting, recommendation engines, and more. These pretrained models serve as effective tools for solving targeted business problems.

Foundation Models:

Foundation models have emerged more recently and are trained on vast amounts of data. Examples include language models like CHAT GPT, which can be adapted for various downstream tasks. Similar generative models exist for audio, video, still images, and other domains. Foundation models are often accessible through

Figure 2. Advantages and disadvantages of different AI and ML models

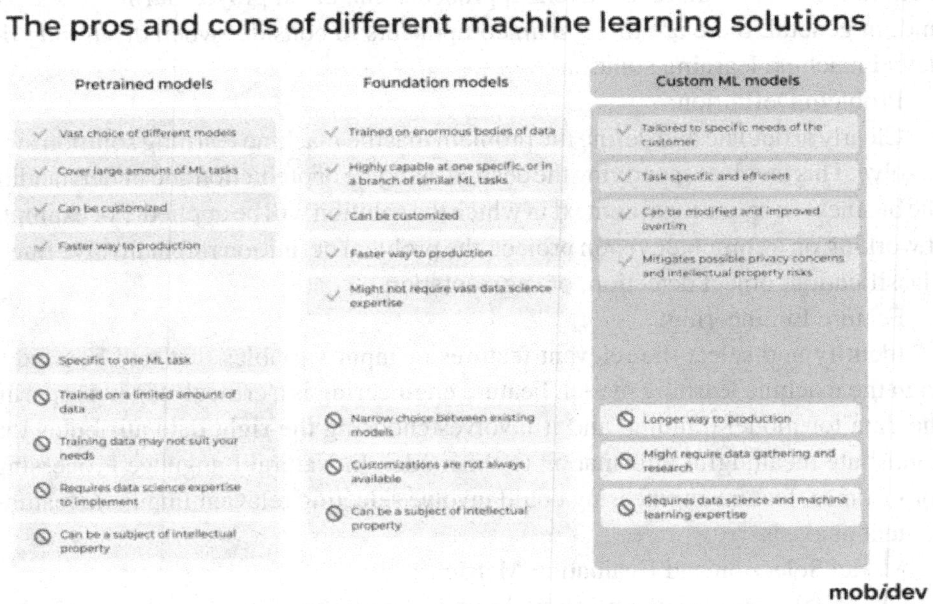

The pros and cons of different machine learning solutions

Pretrained models	Foundation models	Custom ML models
✓ Vast choice of different models	✓ Trained on enormous bodies of data	✓ Tailored to specific needs of the customer
✓ Cover large amount of ML tasks	✓ Highly capable at one specific, or in a branch of similar ML tasks	✓ Task specific and efficient
✓ Can be customized	✓ Can be customized	✓ Can be modified and improved over time
✓ Faster way to production	✓ Faster way to production	✓ Mitigates possible privacy concerns and intellectual property risks
	✓ Might not require vast data science expertise	
⊘ Specific to one ML task		
⊘ Trained on a limited amount of data		
⊘ Training data may not suit your needs	⊘ Narrow choice between existing models	⊘ Longer way to production
⊘ Requires data science expertise to implement	⊘ Customizations are not always available	⊘ Might require data gathering and research
⊘ Can be a subject of intellectual property	⊘ Can be a subject of intellectual property	⊘ Requires data science and machine learning expertise

mob/dev

APIs, providing customers with a potent machine learning module. While they may require additional training for domain-specific tasks, their integration demands less involvement from the data science team if extensive customizations are not needed. The Pros and Cons of different Artificial Intelligence and Machine Learning algorithms are shown in Figure 2.

Custom Machine Learning Models:

Custom machine learning models offer the utmost flexibility among the available options, allowing for the implementation of any functional variation and optimization according to the customer's specific tasks. However, the primary challenge in custom development lies in the availability of data. For small businesses or startups without a lengthy operational history, acquiring sufficient data can be a significant hurdle. Despite this challenge, custom models provide unparalleled adaptability and optimization for addressing unique business requirements.

QUALITY ASSURANCE

By harnessing the capabilities of neural networks, high optical resolution cameras, and robust GPUs, real-time video processing, coupled with machine learning and computer vision, excels in accomplishing visual inspection tasks with superior efficiency compared to human capabilities. This technological integration serves as a reliable means to verify the proper functioning of the "factory in a box" and guarantees the identification and removal of defective products from the system, enhancing overall quality control and operational efficiency (Azari et al., 2023; Wuest et al., 2014).

Historically, the application of machine learning in video analysis has faced criticism due to the variable quality of video inputs. The challenge arises from the potential blurriness and fluctuations in images from frame to frame, which can introduce errors in inspection algorithms. However, advancements in technology, such as the use of high-quality cameras and increased graphical processing power, have significantly improved the efficiency of neural networks in real-time defect detection without relying heavily on human intervention.

Moreover, machine learning proves invaluable in product testing without causing any damage, employing a range of Internet of Things (IoT) sensors. Algorithms can analyze real-time data, identifying patterns that align with defective versions of the product. This capability allows the system to promptly flag potentially undesirable products, contributing to enhanced quality control and defect detection processes (Wang et al., 2020).

NON- DESTRUCTIVE TESTING

Non-destructive testing offers an alternative approach to defect detection by assessing the stability and integrity of materials without causing harm. For instance, ultrasound machines can be employed to identify anomalies such as cracks in materials. While humans can manually analyze the data to spot outliers, advanced technologies, including outlier detection algorithms, object detection algorithms, and segmentation algorithms, bring automation to this process.

Machine learning significantly enhances efficiency by analyzing data for discernible patterns that may be challenging for humans to perceive. Importantly, machine learning applications in this context are less susceptible to errors compared to human assessments, offering a more accurate and consistent means of defect detection in materials (Zhang et al., 2020; Zhang et al., 2019).

PREDICTIVE MAINTENANCE

One of the core tenants of machine learning's role in manufacturing is predictive maintenance. PwC reported that predictive maintenance will be one of the largest

Figure 3. Implementation of AI and ML from 2018 to 2023

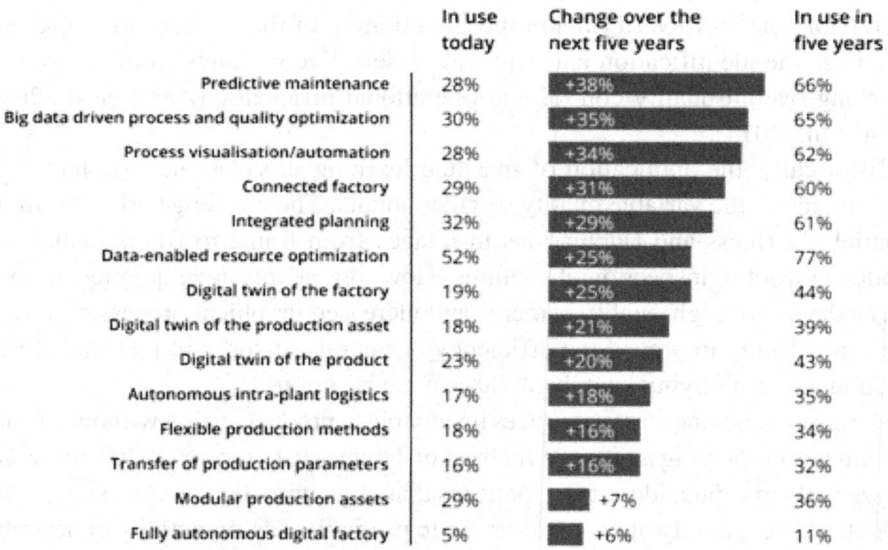

	In use today	Change over the next five years	In use in five years
Predictive maintenance	28%	+38%	66%
Big data driven process and quality optimization	30%	+35%	65%
Process visualisation/automation	28%	+34%	62%
Connected factory	29%	+31%	60%
Integrated planning	32%	+29%	61%
Data-enabled resource optimization	52%	+25%	77%
Digital twin of the factory	19%	+25%	44%
Digital twin of the production asset	18%	+21%	39%
Digital twin of the product	23%	+20%	43%
Autonomous intra-plant logistics	17%	+18%	35%
Flexible production methods	18%	+16%	34%
Transfer of production parameters	16%	+16%	32%
Modular production assets	29%	+7%	36%
Fully autonomous digital factory	5%	+6%	11%

growing machine learning technologies in manufacturing, having an increase of 38% in market value from 2018 to 2023 (Azari et al., 2023; Zonta et al., 2020) as shown in Figure 3.

Unscheduled maintenance can pose a significant financial burden on businesses, prompting the exploration of predictive maintenance as a proactive solution. This approach empowers factories to preemptively make adjustments and corrections before machinery succumbs to potentially more expensive failures (Yang et al., 2018). Ensuring maximum uptime and minimizing delays for the factory in a box becomes achievable through the implementation of predictive maintenance strategies.

IoT sensors play a pivotal role in enabling predictive maintenance by continuously collecting essential data about the operational conditions and status of machinery. This comprehensive data set may encompass parameters such as humidity, temperature, and other relevant factors. By leveraging this wealth of information, manufacturers can anticipate potential issues, schedule timely maintenance, and optimize the overall reliability and performance of the factory in a box (Bordeleau et al., 2018; Dai et al., 2019; Sisinni et al., 2018; Zonta et al., 2020). A machine learning algorithm can analyze patterns in data collected over time and reasonably predict when the machine may need maintenance. There are several approaches to achieve this goal:

Regression Models: These predict the Remaining Useful Life (RUL) of the equipment. This uses historical and static data and allows manufacturers to see how many days are left until the machine experiences a failure.

Classification Models: these models predict failures within a predefined time span.

Anomaly Detection Models: These flag devices upon detecting abnormal system behavior.

PROBLEM LOCALIZATION

Indeed, the integration of IoT sensors in predictive maintenance not only enables real-time monitoring but also facilitates machine learning analysis of data patterns. By scrutinizing the data, machine learning algorithms can identify specific components of the machine that require maintenance to prevent potential failures. In cases where recurring patterns hint at a trend of defects, it becomes possible to pinpoint hardware or software behaviors as potential causes (Su et al., 2015; Zhang & Yang, 2018).

This valuable insight empowers engineers to devise targeted solutions aimed at correcting the system and mitigating the risk of future defects. The iterative process of data analysis, problem identification, and solution implementation contributes to a continuous improvement cycle, ultimately reducing the margin of error in the factory in a box scenario. As a result, predictive maintenance, coupled with machine learning, becomes a powerful tool for enhancing reliability The hardware landscape for AI visual inspection in defect detection offers a diverse array of options to cater to various needs. Notably, there is a demand for portable devices like Raspberry Pi or Arduino computers, emphasizing the flexibility and accessibility required for visual inspection purposes. Simultaneously, practical applications often benefit from the compact yet powerful capabilities of Nvidia Jetson Nano or Nvidia Jetson Xavier, especially when deployed in pairs for efficient optical analysis.

VISUAL INSPECTION

A distinctive advantage of deep learning models in this context lies in their adaptability and refinement post-deployment. The iterative process of gathering new data and re-training the model allows for continuous improvement in accuracy. I This dynamic approach enables the deep learning model to evolve and enhance its defect detection capabilities over time. The ability to refine the model during operation by incrementally increasing the volume of training data contributes to the creation of a more sophisticated and perfected visual inspection model. It has been shown in Figure 4.

Figure 4. AI powered Iterative visual inspection model

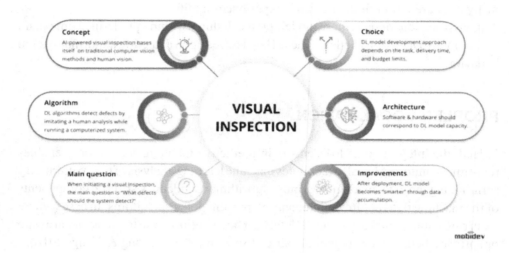

In essence, the hardware choices and the adaptive nature of deep learning models work in tandem to create a robust and flexible system for AI visual inspection in defect detection. This dynamic interplay ensures that manufacturers can choose the hardware configuration that best suits their specific requirements while benefiting from the ongoing refinement and improvement of the deployed deep learning models and efficiency in manufacturing processes.

MACHINE LEARNING MODELS FOR ENERGY CONSUMPTION FORECASTING

Efficient energy management is a crucial aspect of optimizing a factory, and utilizing sequential data measurements with machine learning algorithms proves instrumental in this endeavor. Data scientists leverage autoregressive models and deep neural networks to analyze power consumption patterns and enhance predictive capabilities (Kamble et al., 2018).

Autoregressive Models:

Strengths: Ideal for identifying trends, cyclicity, irregularities, and seasonality in power consumption.

Improvement Strategies: Data scientists can enhance accuracy by transforming raw data into features that provide valuable information for prediction algorithms.

Deep Neural Networks:

Strengths: Efficiently process large datasets to rapidly identify patterns in data consumption.

Feature Extraction: These networks can be trained to automatically extract features from input data, eliminating the need for explicit feature engineering as required in autoregressive models.

Neural Networks for Sequential Data:

RNN (Recurrent Neural Networks): Suitable for preserving information from previously inputted energy usage data using internal memory.

LSTM (Long Short-Term Memory)/GRU (Gated Recurrent Unit): These variations of RNNs offer improved capabilities in handling long-term dependencies and retaining information over extended sequences.

Attention-Based Neural Networks: Designed to focus on specific parts of the input sequence, allowing for more nuanced processing and learning.

By employing these machine learning techniques, the factory can optimize energy consumption, identify potential areas for improvement, and implement strategies to enhance energy efficiency, contributing to a more sustainable and cost-effective operation.

GENERATIVE DESIGN

The BMW iX Flow's innovative e-ink wrap, allowing the car to change its color or shade, showcases the application of generative design, where machine learning plays a pivotal role in optimizing product design. Generative design involves using machine learning algorithms to explore and refine the design of various products, including automobiles, electronic devices, toys, and more. The process involves cycling through numerous potential arrangements to identify the most optimal design based on specified criteria.

Key machine learning algorithms employed in generative design include:

Reinforcement Learning:

Process: Algorithms learn by interacting with the environment and receiving feedback in the form of rewards or penalties.

Application: Optimizing designs based on specific performance criteria, such as durability, weight, or cost.

Deep Learning:

Process: Neural networks with multiple layers process complex data to identify patterns and relationships.

Application: Analyzing and optimizing design parameters, including aesthetic considerations, by learning from vast datasets.

Genetic Algorithms:

Figure 5. Digital Twin model of manufacturing application

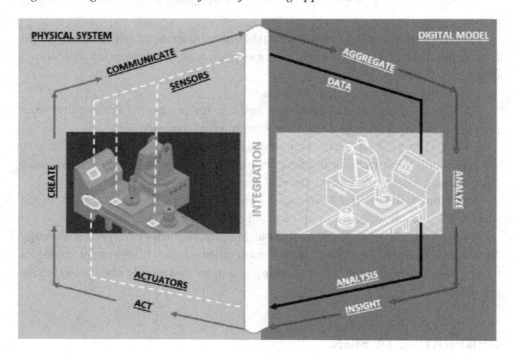

Process: Inspired by the principles of natural selection, genetic algorithms involve evolving a population of potential designs over successive generations.

Application: Finding optimal solutions by simulating the evolutionary process, often applied to parameters like shape, strength, and functionality.

These generative design processes enable machine learning algorithms to iteratively refine designs based on predefined objectives. By incorporating criteria such as weight, shape, durability, cost, strength, and aesthetic parameters, the outcome is a more efficient and tailored design for the intended purpose, as demonstrated by the dynamic color-changing feature of the BMW iX Flow.

DIGITAL TWIN

Digital Twin, a pivotal technology in Industry 4.0, combines Big Data, AI, Machine Learning (ML), and IoT. Various definitions of Digital Twin have been explored by researchers, with those provided by Grieves and NASA, being widely acknowledged. Currently, two globally accepted definitions are attributed to Grieves and NASA. According to NASA, a Digital Twin for a space vehicle is described as: "A Digital

Twin is an integrated multi-physics, multiscale, probabilistic simulation of an as-built vehicle or system that uses the best available physical models, sensor updates, fleet history, etc., to mirror the life of its corresponding flying twin". Numerous companies have embraced the Digital Twin concept, exemplified by Chevron, which achieves substantial savings in maintenance costs for its oil refineries and fields through its implementation. Siemens also leverages Digital Twin to minimize failures, reduce time to market, and explore innovative business avenues. A Digital Twin model of manufacturing application is presented in Figure 5.

Digital Twin (DT) serves as a means to enhance the human-machine connection, facilitating bi-directional communication between a digital model and the physical world. The simulation model relies on real-time sensor data for selected parameters to accurately replicate the performance and functioning of the system in question (Tao et al., 2018). The digital representation of physical systems is valuable for predictive analysis, health monitoring, refining business models, mitigating downtime, and optimizing product design at a reduced cost.

The significance and challenges of DT in personal healthcare are explored in Ahmadi-Assalemi et al. (2020), summarizing the necessary technologies and application requirements for implementing DT in this context. DT offers a novel approach to representing a physical system in a digital model, encompassing aspects such as position, shape, status, gesture, and motion. Leveraging real-time sensor data, along with AI, machine learning, and big data analytics, enables DT to be applied in diagnostics, monitoring, prognostics, and optimization (Cai et al., 2017; Zaccaria et al., 2018), thereby expanding its capabilities for diverse decision-making operations. Once DT models of facilities, environments, and individuals are developed, they can be utilized for training users, operators, maintenance workers, and service providers. Ultimately, DT proves to be a fruitful method for enhancing productivity and efficiency in various industries or companies.

DEMAND FORECASTING

The selection of machine learning models for demand forecasting is influenced by various factors such as business goals, data characteristics (type, amount, and quality), forecasting period, and more. Multiple approaches are often combined within a single forecasting system to enhance effectiveness. The following approaches have proven to be efficient for demand forecasting:

- ARIMA/SARIMA
- Exponential Smoothing
- Regression models

- Gradient Boosting
- Long Short-Term Memory (LSTM)
- Ensemble Models
- Transformer-based Models

The choice among these approaches depends on specific requirements. In the context of the SmartTab project, a time series approach was employed, combining Gradient Boosting and KNN models. The key components analyzed include trends, seasonality, irregularity, and cyclicity. This approach allows for predicting revenue with daily granularity for each venue and the entire venue chain over the upcoming year. It's important to note that the application of machine learning in analytical systems extends beyond retail cases.

AI IN PRODUCTION

The process of developing and deploying a forecasting model involves several key steps. Once the data is obtained, the following procedures are typically carried out iteratively until the model achieves the desired level of performance.

Training, Validation, and Improvement: The model undergoes training and validation processes, and adjustments are made iteratively to enhance its performance.

Utilization of Raw Predictions: The forecasting model can be used in its raw form, presenting predictions in a table or sending analytical reports via email for a specified period.

Development of Front-End Application: More commonly, a front-end application is developed to complement the forecasting model. This application often takes the form of a dashboard, offering a single interface that presents insights and visualizations. Users can query different reports, share them with stakeholders, customize visualization types, and more.

Testing and Approval: The developed front-end application is rigorously tested, and once approved, the entire product is ready for deployment.

Deployment into Production Environment: The product, including both the forecasting model and the front-end application, is deployed into the production environment for regular use.

It's crucial to note that to maintain the forecasting model's relevance and accuracy, regular updates with new data are necessary. This can include daily transactions, inventory turnover, or other relevant data sources. To streamline this process, setting up an automated pipeline is recommended. This pipeline periodically aggregates recent data and updates the machine learning model accordingly. This automated approach helps ensure the reliability of the developed demand forecasting model on an ongoing basis.

SUPPLY CHAIN MANAGEMENT

Effective management of the supply chain (SCM) is crucial for optimizing the movement of products and services, as well as streamlining overall business operations. From acquiring raw materials to overseeing dependable suppliers, automating warehouse processes, and optimizing shipping routes and delivery schedules, every stage in the supply chain must operate efficiently to enhance a company's profitability and maintain a competitive edge.

Numerous companies recognize the benefits of incorporating artificial intelligence (AI) to automate various supply chain tasks, encompassing back-office and warehouse logistics, quality checks, inventory management, and supplier relationship management. The implementation of AI in quality control within supplier management allows companies to automate time-consuming tasks, leading to increased accuracy, efficiency, and sustainability, fostering greener warehouse practices. Research indicates that 37 percent of businesses, including those in the supply chain sector, are already realizing the advantages of AI solutions, which are projected to contribute $15.7 trillion to the global economy by 2030.

When the supply chain is fully optimized through successful SCM practices, companies are likely to experience the following benefits:

Decreased operating costs: Companies can reduce operational costs by cutting purchasing and production expenses. For instance, direct procurement of fresh vegetables from a farmer eliminates the need for a third party, saving money and ensuring quicker availability of products in a grocery store. AI enhances data transparency, providing better visibility into the supply chain and yielding cost savings.

Better productivity and reduced labor costs: AI streamlines manual tasks, making them more efficient and reducing the need for human labor. Approximately 40 percent of the workload during the sales process can be automated through AI solutions, resulting in decreased operating costs.

Improved relationships with stakeholders: Effective relationship management among suppliers, manufacturers, retailers, and planners helps prevent overstock or out-of-stock scenarios.

Shorter delivery times and on-time delivery: AI analyzes data, identifies patterns, and tracks shipments, facilitating on-time delivery by helping managers make informed decisions. For example, Walmart uses AI to optimize inventory levels, reducing stockouts and ensuring faster delivery of fresh food, leading to improved customer satisfaction.

Improved transportation network and routes: AI determines the most efficient and cost-effective transportation methods by analyzing factors such as the number of trucks needed, fuel consumption, and travel time. Integrating AI into logistics operations, as seen with UPS, optimizes route planning, reduces fuel consumption, and enhances delivery accuracy.

Reduced risks: AI provides data on warehouse management systems, identifying weaknesses, gaps, and potential risks. Proactive measures can then be taken to create a safer work environment and a more efficient supply chain.

Enhanced decision-making capabilities: AI's rapid analysis of large datasets enables quicker, more accurate, and more precise decision-making. While AI doesn't replace human decision-makers, it provides increased data visibility and insights to support informed decisions.

CHALLENGES OF IMPLEMENTING AI IN MANUFACTURING

Empowering manufacturers to do more with less using Artificial Intelligence automation is the way to accelerate digital transformation, helping to reduce costs, improve efficiency and solve new problems. Manufacturing companies often face several barriers that impede their efforts in digital transformation and the adoption of AI initiatives. Some common challenges include:

Legacy Systems and Infrastructure:

Many manufacturing facilities still rely on legacy systems and outdated infrastructure. Integrating new technologies, such as AI, can be challenging when existing systems were not designed to accommodate modern digital solutions. Upgrading or replacing these legacy systems may be necessary but can be costly and time-consuming.

Data Quality and Availability:

Effective AI applications require high-quality, relevant data. Manufacturing processes may produce large volumes of data, but ensuring its accuracy, completeness, and accessibility can be a hurdle. Inconsistent data formats, siloed information, and poor data quality can hinder the training and performance of AI models.

Lack of Skilled Workforce:

Implementing AI initiatives requires a workforce with the necessary skills in data science, machine learning, and AI development. The shortage of skilled professionals in these areas can pose a significant barrier. Companies may need to invest in training programs or hire external experts to fill the skill gap.

Security Concerns:

The integration of AI and digital technologies introduces new cybersecurity risks. Protecting sensitive manufacturing data, intellectual property, and maintaining the security of interconnected devices become crucial. Developing robust cybersecurity measures is essential to prevent unauthorized access and data breaches.

Costs and Return on Investment (ROI) Uncertainty:

Implementing AI and digital transformation initiatives involves upfront costs for technology adoption, training, and system integration. Manufacturers may face challenges in accurately estimating the return on investment, especially when the benefits of these initiatives are not immediately apparent or quantifiable.

Regulatory Compliance:

Manufacturing industries are often subject to stringent regulations and compliance standards. Ensuring that AI solutions comply with industry-specific regulations, data protection laws, and ethical standards can be a complex task. Non-compliance may result in legal consequences and reputational damage.

Change Management and Organizational Resistance:

The introduction of new technologies can face resistance from employees accustomed to traditional methods. Implementing digital transformation requires effective change management strategies to address concerns, educate the workforce, and foster a culture that embraces innovation.

Interoperability Issues:

Integration challenges may arise when different systems, machines, or devices within a manufacturing environment need to work together seamlessly. Ensuring interoperability between various technologies is crucial for a cohesive and efficient digital transformation.

Overcoming these barriers requires a holistic approach, including strategic planning, investment in technology and talent, and a commitment to addressing organizational and cultural challenges associated with adopting AI and digital transformation in manufacturing.

CONCLUSION

In conclusion, the integration of artificial intelligence (AI) in manufacturing industries has ushered in a new era of innovation and efficiency, as evidenced by real-time applications across various facets of the production process. From automated assembly lines leveraging machine learning for quality assurance to predictive maintenance strategies powered by IoT sensors and advanced algorithms, AI technologies are reshaping the landscape of manufacturing. The implementation of neural networks, deep learning, and generative design techniques illustrates the adaptability and versatility of AI in addressing complex challenges. Machine learning algorithms, with their ability to analyze vast datasets and identify patterns, contribute significantly to visual inspections, defect detection, and energy optimization.

As showcased by the BMW iX Flow's e-ink wrap, which seamlessly changes the car's color through generative design processes, AI is not merely streamlining operations but also pushing the boundaries of creativity and customization in manufacturing. The amalgamation of reinforcement learning, deep learning, and genetic algorithms in generative design highlights the diverse approaches AI can employ to optimize product designs based on multifaceted criteria. In essence, the real-time applications of AI in manufacturing industries are enhancing productivity, reducing errors, and paving the way for smarter, more sustainable operations. As these technologies continue to evolve, the potential for further advancements in efficiency, cost-effectiveness, and product innovation appears boundless, promising a transformative impact on the future of manufacturing.

REFERENCES

Ahmadi-Assalemi, G., Al-Khateeb, H., Maple, C., Epiphaniou, G., Alhaboby, Z. A., & Alkaabi, S. (2020). Digital twins for precision healthcare. Cyber Defence in the Age of AI Smart Societies and Augmented Humanity. Springer.

Azari, M. S., Flammini, F., Santini, S., & Caporuscio, M. (2023). A Systematic Literature Review on Transfer Learning for Predictive Maintenance in Industry 4.0. *IEEE Access: Practical Innovations, Open Solutions, 11*, 12887–12910. doi:10.1109/ACCESS.2023.3239784

Bordeleau, F.-È., Mosconi, E., & Santa-Eulalia, L. A. (2018). Business intelligence in industry 4.0: State of the art and research opportunities. *Proc. 51st Hawaii Int. Conf. Syst. Sci.*, (pp. 1-10). IEEE. 10.24251/HICSS.2018.495

Cai, Y., Starly, B., Cohen, P., & Lee, Y.-S. (2017, January). Sensor data and information fusion to construct digital-twins virtual machine tools for cyber-physical manufacturing. *Procedia Manufacturing, 10*, 1031–1042. doi:10.1016/j.promfg.2017.07.094

Dai, W., Nishi, H., Vyatkin, V., Huang, V., Shi, Y., & Guan, X. (2019, December). Industrial edge computing: Enabling embedded intelligence. *IEEE Industrial Electronics Magazine, 13*(4), 48–56. doi:10.1109/MIE.2019.2943283

Dhanalakshmi, R., Kalpana, A. V., Umamageswaran, J., & Kumar, B. P. (2023). Health Information Broadcast Distributed Pattern Association based on Estimated Volume. *2023 Third International Conference on Artificial Intelligence and Smart Energy (ICAIS)*. 10.1109/ICAIS56108.2023.10073672

Kalpana, A. V., Venkataramanan, V., Charulatha, G., & Geetha, G. (2023). An Intelligent Voice-Recognition Wheelchair System for Disabled Persons. *2023 International Conference on Sustainable Computing and Smart Systems (ICSCSS)*. 10.1109/ICSCSS57650.2023.10169364

Kamble, S. S., Gunasekaran, A., & Gawankar, S. A. (2018, July). Sustainable industry 4.0 framework: A systematic literature review identifying the current trends and future perspectives. *Process Safety and Environmental Protection*, *117*, 408–425. doi:10.1016/j.psep.2018.05.009

Medida, L. H., & Renugadevi, R. (2023). Machine Learning Techniques for Predicting Pregnancy Complications. In D. Satishkumar & P. Maniarasan (Eds.), *Predicting Pregnancy Complications Through Artificial Intelligence and Machine Learning* (pp. 116–125). IGI Global. doi:10.4018/978-1-6684-8974-1.ch008

Praveen Kumar, B., kalpana, A. V., & Nalini, S. (2023, March 16). Gated Attention Based Deep Learning Model for Analyzing the Influence of Social Media on Education. *Journal of Experimental & Theoretical Artificial Intelligence*, 1–15. doi:10.1080/0952813X.2023.2188262

Renugadevi, R., & Sethukarasi, T. (2023). A Novel and Efficient Multi-Band Wireless Communication System for Healthcare Management System. *2023 International Conference on Intelligent and Innovative Technologies in Computing, Electrical and Electronics (IITCEE)*. 10.1109/IITCEE57236.2023.10090991

Shyam, M., & Amalasweena, M. (2023). Intellectual Design of Bomb Identification and Defusing Robot based on Logical Gesturing Mechanism. *2023 International Conference on Advances in Computing, Communication and Applied Informatics (ACCAI)*, Chennai, India. 10.1109/ACCAI58221.2023.10201034

Sisinni, E., Saifullah, A., Han, S., Jennehag, U., & Gidlund, M. (2018, November). Industrial Internet of Things: Challenges opportunities and directions. *IEEE Transactions on Industrial Informatics*, *14*(11), 4724–4734. doi:10.1109/TII.2018.2852491

Su, H., Maji, S., Kalogerakis, E., & Learned-Miller, E. (2015). Multi-view convolutional neural networks for 3D shape recognition. *Proc. IEEE Int. Conf. Comput. Vis. (ICCV)*, (pp. 945-953). IEEE. 10.1109/ICCV.2015.114

Tao, F., Cheng, J., Qi, Q., Zhang, M., Zhang, H., & Sui, F. (2018, February). Digital twin-driven product design manufacturing and service with big data. *International Journal of Advanced Manufacturing Technology*, *94*(9), 3563–3576. doi:10.1007/s00170-017-0233-1

Wang, F., Zhang, M., Wang, X., Ma, X., & Liu, J. (2020). Deep learning for edge computing applications: A state-of-the-art survey. *IEEE Access : Practical Innovations, Open Solutions, 8,* 58322–58336. doi:10.1109/ACCESS.2020.2982411

Wuest, T., Irgens, C., & Thoben, K.-D. (2014, October). An approach to monitoring quality in manufacturing using supervised machine learning on product state data. *Journal of Intelligent Manufacturing, 25*(5), 1167–1180. doi:10.1007/s10845-013-0761-y

Yang, Z.-X., Wang, X., & Wong, P. K. (2018, December). Single and simultaneous fault diagnosis with application to a multistage gearbox: A versatile dual-ELM network approach. *IEEE Transactions on Industrial Informatics, 14*(12), 5245–5255. doi:10.1109/TII.2018.2817201

Zaccaria, V., Stenfelt, M., Aslanidou, I., & Kyprianidis, K. G. (2018, August). Fleet monitoring and diagnostics framework based on digital twin of aero-engines. *Turbo Expo Power Land Sea Air, 51128,* 10. doi:10.1115/GT2018-76414

Zhang, P.-B., & Yang, Z.-X. (2018, January). A novel AdaBoost framework with robust threshold and structural optimization. *IEEE Transactions on Cybernetics, 48*(1), 64–76. doi:10.1109/TCYB.2016.2623900 PMID:27898387

Zhang, S., Zhang, S., Wang, B., & Habetler, T. G. (2020). Deep learning algorithms for bearing fault diagnostics—A comprehensive review. *IEEE Access : Practical Innovations, Open Solutions, 8,* 29857–29881. doi:10.1109/ACCESS.2020.2972859

Zhang, Z., Wang, X., Wang, X., Cui, F., & Cheng, H. (2019, March). A simulation-based approach for plant layout design and production planning. *Journal of Ambient Intelligence and Humanized Computing, 10*(3), 1217–1230. doi:10.1007/s12652-018-0687-5

Zonta, T., da Costa, C. A., da Rosa Righi, R., de Lima, M. J., da Trindade, E. S., & Li, G. P. (2020, December). Predictive maintenance in the industry 4.0: A systematic literature review. *Computers & Industrial Engineering, 150,* 106889. doi:10.1016/j.cie.2020.106889

Chapter 7

Harnessing the Power of Artificial Intelligence in Reinventing the Manufacturing Sector

Geetha Manoharan
(iD) https://orcid.org/0000-0002-8644-8871
SR University, India

Sunitha Purushottam Ashtikar
SR University, India

M. Nivedha
Robert Gordon University, UK

ABSTRACT

AI is revolutionising industry with unprecedented efficiency and innovation. AI's promise has transformed large manufacturing organizations. This chapter covers manufacturing AI applications from optimization to process automation. Image and video recognition, prescriptive modeling, smart automation, advanced simulation, complex analytics, and more employ AI. Machine learning and deep learning apply AI to manufacturing using neural networks and algorithms. AI improves computer vision and image identification for quality control, improving product inspections. AI helps supply chain management estimate demand, optimize stocks, improve logistics, and distribute procedures. Strong cyber security and staff upskilling are needed to protect sensitive production data and smoothly move to AI-driven processes. Manufacturing's widespread AI deployment raises ethical, legal, and social challenges that researchers, industry stakeholders, and policymakers must solve. AI-powered manufacturers have various obstacles in maximizing efficiency, creativity, and long-term success.

DOI: 10.4018/979-8-3693-2615-2.ch007

INTORUDUCTION TO THE EMERGENCE OF ARTIFICIAL INTELLIGENCE IN THE MANUFACTURING SECTOR

The proliferation of industrial IoT and smart factories in manufacturing has led to a wealth of data, creating numerous opportunities for the use of AI. AI in manufacturing uses machine learning and deep learning neural networks to improve data analysis and decision-making in manufacturing. Intelligent machines or vehicles that are capable of performing tasks one of the applications of artificial intelligence that is frequently discussed in the industrial sector is predictive maintenance. As a result of applying Artificial Intelligence to industrial data, businesses have the opportunity to improve their ability to precisely foresee and proactively minimize the effects of equipment faults. As a consequence of this, the industrial activities experience less costly times of inactivity. Artificial intelligence in production improves demand forecasting and reduces raw material waste. Due to the fact that industrial manufacturing environments generally require close collaboration between humans and robots, the relationship between artificial intelligence and manufacturing is inherently present. The concept of "Industry 4.0," which refers to the growing automation in production settings as well as the widespread gathering and transfer of data in these settings, is significantly influenced by Artificial Intelligence, which plays a significant part in the concept. In order for businesses to derive value from the large amounts of data that are produced by industrial machines, Artificial Intelligence and Machine Learning are essential technologies that should be utilized. Using artificial intelligence to integrate this data in the optimization of industrial processes can result in a plethora of benefits, including savings in costs, improvements in safety, and upgrades in supply-chain efficiency, amongst many others and independently of one another.

Artificial Intelligence in manufacturing automates complex tasks and uncovers new patterns in manufacturing workflows and processes. The application in manufacturing is a part of a larger push toward the complete automation of production processes. The introduction of "smart factories" has the potential to change the operational processes of businesses. Artificial intelligence technologies have the potential to enhance human capabilities, provide instantaneous insights, and streamline design and product creation, hence increasing overall efficiency. As a result of the transformative qualities that it possesses, artificial intelligence is causing a revolution in the industrial sector. Artificial Intelligence is being utilized by manufacturing companies to enhance the efficiency, precision, and productivity of a variety of procedures. Predictive maintenance, supply chain optimization, quality control, and demand forecasting are just some of the applications that can be brought about by the employment of artificial intelligence in the manufacturing industry.

In this article, we will investigate a number of use cases and present examples that illustrate how artificial intelligence might be applied in the industrial sector.

Figure 1. Examples of artificial intelligence in manufacturing industry

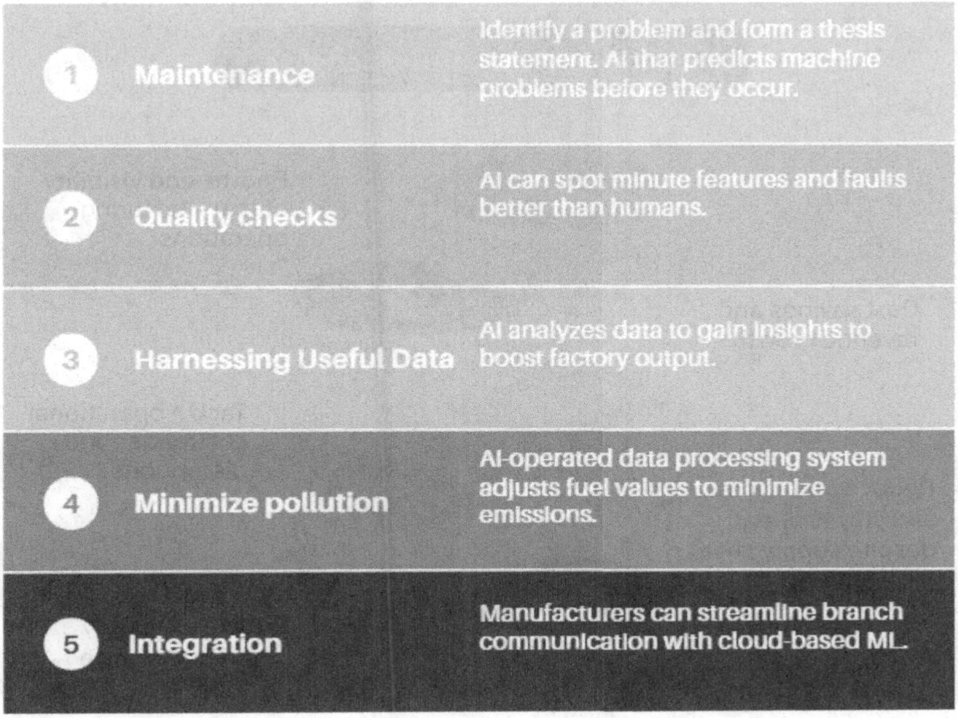

1	Maintenance	Identify a problem and form a thesis statement. AI that predicts machine problems before they occur.
2	Quality checks	AI can spot minute features and faults better than humans.
3	Harnessing Useful Data	AI analyzes data to gain insights to boost factory output.
4	Minimize pollution	AI-operated data processing system adjusts fuel values to minimize emissions.
5	Integration	Manufacturers can streamline branch communication with cloud-based ML.

The goal is to give manufacturing organizations with a range of applications of artificial intelligence in manufacturing and to assist them in speeding the growth of their companies. Artificial intelligence is ideal for manufacturing because it can manage massive amounts of data from the Internet of Things (IoT) and smart factories. Manufacturers analyse this data and make better decisions using AI (Shameem, A., Ramachandran, K. K., et al., 2023) including Machine Learning and deep learning neural networks.

In manufacturing, Artificial Intelligence is primarily used in the following ways:

Machine learning:

Programs are able to learn from data patterns without explicit programming.

Deep learning:

A version of machine learning more advanced and comprehends videos and images.

Autonomous objects:

Intelligent robots or vehicles that are capable of performing tasks on their own.

Increasing levels of innovation across the manufacturing sector (Durai, S., Krishnaveni, K., et al., 2022) has been facilitated by the incorporation of Artificial

Figure 2. AI and the future of manufacturing

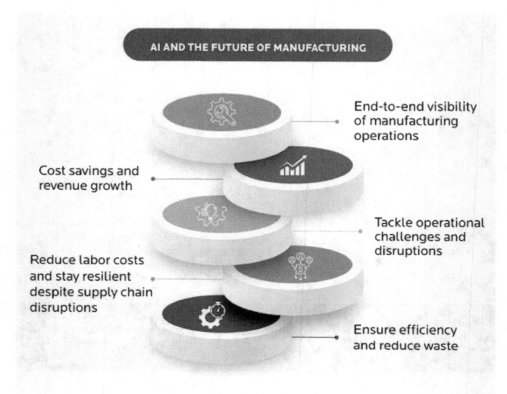

Intelligence into production processes, which has led to a substantial and significant change in industrial operations. In order to improve operations, make decisions based on data, and construct intelligent and flexible systems, factories and industries have been able to adopt artificial intelligence more easily as a result of the convergence.

Mypati, O., Mukherjee, A., et al (2023) indicate that the introduction of the fourth industrial revolution, has brought internet, artificial intelligence (AI), and machine learning (ML) in manufacturing. In manufacturing, understanding AI and ML capabilities and applications is crucial. This article presents a comprehensive analysis of the algorithms that are used in artificial intelligence and how they are applied in the industrial sector. The article identifies six distinct manufacturing verticals, which are as follows: casting, forming, machining, welding, additive manufacturing (AM), and supply chain management (SCM). The horizontal axis shows the progression of each process from mechanization to the present, as well as the process automation advances made through signal and image data processing, machine learning, and artificial intelligence algorithms.

The discourse also encompasses the progression of robotics and cloud-based technology. The critical study provides an objective perspective on the implementation of production automation and the advantages of artificial intelligence. In addition, the study explores various manufacturing scenarios in which AI and ML algorithms are implemented. As a prospective area of study, the concept of human-like intelligence is emphasized, emphasizing the importance of cognitive abilities in the field of manufacturing. Essentially, a reader may rationally elucidate the precise manner in which AI will comprehensively revolutionize the field of manufacturing, including the reasons, timing, and extent of its impact. In this study, Madrid, J.A. (2023) offers a succinct summary of the impact of Artificial Intelligence on automobile manufacturing and design. The author relies on both quantitative and qualitative data to produce an in-depth analysis. The results reveal an increasing utilization of AI technology in the sector, leading to significant cost reductions, enhanced design efficiency, and notable enhancements in production quality. Although these developments enhance competitiveness and sustainability, they also give rise to issues in worker adaptation and ethical considerations. The extensive analysis based on data highlights the significant influence of Artificial Intelligence on automotive manufacturing and design, establishing it as a crucial driving force for innovation and the restructuring of the industry's future landscape.

Miltiadou, D., Perakis, K., et al (2023) are the researchers who published these findings through their team of researchers. Nevertheless, manufacturers are cautious about embracing industrial Artificial Intelligence for smart factories in black-box AI models lacking transparency and the uncertainty around the reliability of the judgments made by these models. This is because black-box AI models are not very transparent. On top of that, there is a dearth of awareness among manufacturers regarding the locations and methods by which they might implement AI into their operational procedures and goods. XMANAI's Explainable AI platform is introduced in this article. This platform takes use of the most recent technological advancements and combines the most recent discoveries in the field of Explainable AI (XAI). Without sacrificing the efficiency of the AI, the platform intends to construct artificial intelligence models that are open to human scrutiny and easy to understand. Graph and hybrid artificial intelligence models make up the core of the platform's central component. These models are built, refined, and verified to be baseline artificial intelligence models that can be used for any manufacturing issue or trained AI models that are fine-tuned to handle specific manufacturing issues. These models are guaranteed to be reliable by the platform, which offers explanations that are founded on values and can be readily and efficiently comprehended by human beings. It is an essential component of educational institutions that serves to document the presence of each applicant in an examination system and track the particular number of response sheets for each applicant. This component is known as the record of answer books and attendance (RABA). It is a lengthy task, difficult to handle, and prone to human errors to manually record the RABA (Report of Academic and Behavioral Achievement) of pupils in the usual manner if manual recording is done. Among the potential solutions to these issues is the digital RABA system, which is a system powered by artificial intelligence that combines facial recognition with a fingerprint-based biometric system. A computerized system that makes use of biometrics for the purpose of identifying fingerprints and face features is presented in this paper. In addition, a camera module is utilized in order to scan the barcode that is attached to the response sheet in order to obtain RABA input. In the end, each and every one of these particulars is stored in a database, which the university possesses the ability to examine at any moment. A approach that is both more effective and more precise for preventing proxy attendance is presented in the research that was carried out by Kumaravel, D., Tomar, P.S., Tulanovna, K.D., Sharma, S., Pawar, K.P., and Patil, P.P. (2023 respectively).

It is essential to keep a close eye on the state of the production environment, as stated by Sundaram, S., and Zeid, A. (2023). This prevents unexpected repairs and shutdowns and detects defective products that could cost a lot of money. Through the utilization of data-driven approaches and advancements in sensor technology, in conjunction with the Internet of Things (IoT), the practical deployment of real-time tracking for a variety of systems has been made possible. Quality Control (QC) measures can be employed to consistently evaluate the product's health throughout its manufacturing lifespan. Quality inspection is a crucial procedure in which the product is assessed and either approved or denied based on its standards. A human operator uses their senses to inspect the product during visual inspection, or final inspection. Nevertheless, various factors influence the visual inspection process, leading to an industry-wide inspection accuracy of approximately 80%. In advanced manufacturing processes, manual visual inspection is a laborious and expensive process, hindering the achievement of 100% inspection. Computer Vision (CV) algorithms have aided in the automation of certain aspects of the visual inspection process, although there remain unresolved issues. This study introduces an Artificial Intelligence-driven methodology for visual assessment, employing Deep Learning (DL) techniques. The strategy uses a customized Convolutional Neural Network (CNN) for inspection and a shop floor computer program to simplify inspection. The proposed model achieves an inspection accuracy of 99.86% when applied to picture data of casting goods.

A vast quantity of data is collected and processed in the most recent semiconductor business. Prior yield prediction studies have focused on a single type of data or a dataset obtained from a specific process. According to Lee and Roh (2023), semiconductor device manufacture involves numerous procedures, and device yields are influenced by multiple factors. This study tackles this difficulty by employing a flexible input data-driven architecture that incorporates many aspects in the prediction process. It also uses explainable artificial intelligence (XAI) to read the model and change fabrication conditions. Following the completion of the data preprocessing step, the succeeding step entails optimizing and assessing different machine learning models in order to identify which one is the most effective for the dataset. In this particular instance, the model that has been chosen is a random forest (RF) regression, which has a root mean square error (RMSE) value of 0.648. The prediction improves production management, while the model's Shapley additive explanation (SHAP) values help explain yield factors. In this research, empirical evidence is presented by means of a case study of data pertaining to device production. The framework improves the accuracy of forecasts, and the utilization of SHAP values demonstrates the connections between yields and features. In addition, the framework boosts the precision of predictions. The method that has been suggested is able to investigate the broader areas of fabrication situations in order to provide an interpretation of the intricate components of semiconductor manufacturing.

Wire Arc Additive Manufacturing is an additive process that is referred to as Direct Energy Deposition, according to the findings of the research that was carried out by Mattera, G., Nele, L., and Paolella, D.A. (2023). Depositing layers of material and producing a finished component are both accomplished through the utilization of the wire welding principle. The manufacturing industry is showing a growing interest in this technology, mostly because to its affordability and the ability to construct large-scale components. Currently, the advancement in transitioning to smart manufacturing systems, along with the growing availability of computer resources, has facilitated the creation of intelligent applications for both on-site inspection and process parameter control in smart production systems. This paper aims to present a comprehensive overview of artificial intelligence applications in Wire Arc Additive Manufacturing. It specifically focuses on software modules for defect detection, feedback generation for control systems, and innovative control strategies such as reinforcement learning. These techniques are employed to address challenges related to model non-linearity and uncertainties. According to Wang, F. (2023), artificial intelligence is pushing traditional and emerging industries forward in the latest scientific and technological development. This is beneficial for improving the economic structure and achieving high-quality development. How can AI technology be effectively utilized in manufacturing firms to enhance product quality? This paper examines the current state of artificial intelligence implementation and proposes six key scenarios for its application in the manufacturing industry. It aims to advance the digital transformation of manufacturing enterprises by shifting from sales-oriented C2S level to a more profound manufacturing-oriented C2M level. Artificial intelligence is expected to upgrade the manufacturing industry chain, improve export quality, and boost China's manufacturing export competitiveness. This analysis offers specific solutions to China's AI challenges.

Currently, Akinsolu, M.O. (2023) disclosed that Artificial Intelligence is exerting significant influence in our modern society through a multitude of applications. Although AI applications have numerous benefits, the analytical frameworks that focus on the consequences of these applications are still developing. New technologies are constantly used in manufacturing and industrial production. Many engineering managers, who are crucial to the shift of manufacturing and industrial production toward Industry 4.0 and Industry 5.0, find AI and its applications unclear and uncertain. This paper examines AI's effects on manufacturing and industrial production. To give engineering managers valuable insights. AI in manufacturing and industrial production systems is examined using political, economic, social, and technical (PEST) factors. This article does not present a novel engineering management model. Here, we propose a debate that serves as a tool for engineering managers to appraise the consequences of the general applications of AI.

The essay authored by Heilala, J., and Singh, K. (2023) depicts the integration of Industry 6.0 (I6.0) with the metaverse, drawing upon a comprehensive analysis of relevant literature. This text briefly discusses the Key Enabling Technologies (KETs) that have evolved from the first industrial revolution (industry 1.0) with steam engines to the current information era (industry 4.0) that encompasses Automation and Robotics (AR) and Additive Manufacturing (AM). The objective is to understand the practical relationship between production, the core competence of business partners, the pattern of the Global Supply Chain (GSC), and the integration of Human Systems (HSI). For safety reasons, the manufacturing robots of level I5.0 must take into account Human Factors (HFs), as this level demands advanced systems. Furthermore, the design office must also take into account the well-being and welfare of individuals in terms of their health and safety. In order for humans to function in the production environment without limitations, HSI necessitates the presence of certain Robot Factors (RFs). This study explores the practical implementation of transferring Artificial Intelligence (AI) to a robot by developing a test setting that allows for the sharing of human behavior using high-frequency sampling devices. Create models that advance cognitive Augmented Reality (AR) to increase engagement in the early metaverse of the next-generation internet. The proposed clinical test setup programme looks at meta-level model construction methods and looks ahead to long-term advances in this study field. The significance and promising prospects of the recent study in this field suggest that the proposed technique will aid future research.

BENEFITS OF USING ARTIFICIAL INTELLIGENCE IN THE MANUFACTURING SECTOR

Artificial Intelligence offers numerous advantages in the manufacturing industry. Artificial intelligence plays a significant part in all processes within the industrial business. Artificial intelligence has the capability to receive various forms of data, including information from individuals and sensory equipment. This is employed with particularly determined algorithms for the optimization of operations in order to observe improvements in the outcomes. The application of artificial intelligence in manufacturing involves the use of various methods and techniques to enhance industrial productivity through marketing techniques (Manoharan, G., Durai, S., et al., 2024). Machine learning is a widely utilized form of artificial intelligence in the manufacturing industry, sometimes seen as a subset. Hence, the industrial sector can gain significant benefits and enhance profitability by leveraging artificial intelligence and machine learning. Here are the different benefits that can be harnessed by implementing artificial intelligence (Abdulwahid, A. H., Pattnaik, M., et al., 2023) in the industrial industry:

ENHANCED EXPEDIENCY IN MAKING DECISIONS

By incorporating IIOT (Industrial Internet of Things) with VR or AR and cloud computing, firms may effectively exchange simulations, discuss manufacturing activities, and transmit real-time critical or pertinent information regardless of their geographical location. Furthermore, the data obtained from sensors and beacons plays a crucial role in determining customer performance, allowing businesses to anticipate future needs, make prompt production decisions, and expedite transfers between manufacturers and suppliers.

CONTINUOUS PRODUCTION AROUND THE CLOCK

To ensure seamless and uninterrupted production, the working hours of human personnel are separated into three shifts. Conversely, there are AI-driven robots that operate continuously, without any periods or rests. They are specifically engineered to optimize efficiency and bolster profitability by effectively catering to the needs of a global consumer base.

Figure 3. Use of AI in manufacturing

SECURE SETTING

Human workers have the capacity to produce errors either intentionally or unintentionally. These errors have the potential to lead to significant blunders or accidents within the company. While Artificial Intelligence cannot entirely eradicate the risk variables, it can effectively mitigate or diminish the severity of errors. The presence of remote access controls obviates the necessity for human interventions. In addition, cutting-edge sensors combined with IIoT (Industrial Internet of Things) devices aid in the effective deployment of defense and security personnel.

POTENTIAL FOR HUMAN WORKERS

As artificial intelligence replaces the physical tasks performed by human workers, it relieves them from the burden of time-consuming, exhausting, and monotonous labor (. Human personnel may now allocate more time to concentrate on innovative and demanding duties that truly necessitate their intellectual capabilities to steer the organization in the correct trajectory.

REDUCED OPERATIONAL EXPENSES

Indeed, implementing AI in industrial and small and medium-sized enterprises (Jaichandran, R., Krishna, S. H., et al., 2023; Geetha Manoharan & Pavan Kumar Billa Sunitha Purushottam Ashtikar, 2022) necessitates a substantial financial outlay;

nevertheless, the potential return on investment (ROI) from AI is far greater. When intelligent robots begin to handle daily operations, firms will have the opportunity to significantly reduce operational expenses. The practical applications in manufacturing encompass:

ANTICIPATORY MAINTENANCE

Analyzing sensor data, manufacturers may identify potential problems and downtime, predict the point at which machines will cease functioning, and arrange for repairs in advance of any errors. This leads to enhanced efficiency as it eliminates the need to halt the functions and reduces the expenses associated with repairing and replacing malfunctioning machines.

CYBER SECURITY

This involves safeguarding computers, networks, and data from unauthorised access, damage, or theft. Given the high volume of data generated in security logs within the manufacturing industry, the process of identifying and removing suspicious entries during routine operations is a significant challenge. Artificial intelligence possesses the ability to autonomously detect fraud, infiltrators, malware, and other such dangers, allowing it to address contemporary cyber security concerns with more speed and accuracy compared to a human worker.

EDGE ANALYTICS

Edge analytics is a process that collects data from devices and sensors powered by artificial intelligence. This allows for the generation of decentralized and rapid statistical insights from the collected information. Edge collects data which may be analyzed and transformed into statistics that can be utilized to optimize operations. Additionally, edge analytics can be employed for the purposes of:

- Surveillance of employees' well-being and safety
- Improvement of production quality and results
- Detection of initial signs of performance decline and potential failure.

ROBOTICS

Robotics is the most efficient technology for automating company operations, reducing workload, increasing production, and minimizing errors. Robots alleviate the burden on employees and enable them to allocate more time towards significantly more intricate duties. By integrating Artificial Intelligence with robotics, machines are able to not only monitor their own behaviors, but also autonomously enhance their efficiency and performance on a daily basis.

QUALITY INSPECTIONS

This use case holds significant importance in the manufacturing sector, showcasing the pivotal role of artificial intelligence. Occasionally, identifying the internal defects of equipment can prove challenging, despite the expertise of professionals who are unable to closely monitor the functioning of the items and identify their deficiencies. However, this activity can be readily accomplished with the assistance of Artificial Intelligence and Machine Learning since Artificial Intelligence tools and apps can proficiently identify minor defects in machines. Hence, it is evident that artificial intelligence guarantees quality assurance and regulation in the production process. Intelligent AI systems monitor the efficiency and effectiveness of machines, improve productivity, identify malfunctions, and save maintenance expenses.

CLIENT RELATIONSHIP MANAGEMENT

Intelligent Artificial Intelligence Chatbots not only enhance sales, productivity, and performance, but also contribute to the enhancement of customer care. This enhances customer satisfaction by minimizing waiting time, providing immediate responses, fostering customer relationships using CRM technologies, and leveraging consumer data for informed decision-making.

TO MAKE A SUMMARY

In recent years, there has been substantial and noteworthy advancement in the fields of Artificial Intelligence and industrial automation. The advent of Artificial Intelligence as a superpower in the current period can be attributed to the development of advanced machine learning techniques, highly complex robotic technology, the exponential growth of computing power, and the utilization of cutting-edge sensors. AI's ability to employ deep learning and speech recognition facilitates the collection and extraction of data, pattern identification, learning, and continuous improvement. These considerations have rendered AI exceedingly versatile in the human environment. The integration of Artificial Intelligence in the manufacturing sector not only aids manufacturers but also facilitates technology developers in enhancing the caliber of their concepts, devising innovative digital technology solutions, and transforming them into tangible reality with the guidance of experts in various sectors like business and food industry (Ponduri, S. B., Ahmad, S. S., et al., 2024).

USE OF AI IN VARIED CASES IN MANUFACTURING SECTORS

1. Artificial intelligence in logistics

An ongoing issue in production is the inefficiencies resulting from excessive or insufficient inventory levels. Excessive inventory can frequently result in both product waste and reduced profit margins. Inadequate inventory levels can result in decreased sales, revenue, and client base. AI enables manufacturers to:

*   Monitor production floor operations
*   Enhance demand forecasting accuracy
*   Reduce inventory-related losses and
*   Streamline resource management

By utilizing advanced technologies like as 3D printing, manufacturers have the capability to generate many parts internally or at nearby facilities, thereby decreasing their dependence on distant, inexpensive production sites and improving their inventory management. Manufacturers can employ robots as substitutes for human couriers, particularly during pandemics, to guarantee continuous last-mile deliveries. Marble, a company specializing in last-mile logistics, employs LIDAR sensors to ensure the secure and cost-effective delivery of items.

2. Artificial Intelligence based robots

AI robots utilize machine learning algorithms to automate decision-making and repetitive operations in production facilities. As these algorithms are capable of self-learning, they continuously enhance their ability to manage their designated processes more effectively. In addition, AI robots do not require breaks and are less prone to errors compared to people. Manufacturers have the ability to readily expand their production capacity. Robots can perform the physically demanding duties in production facilities, allowing people to focus on more intricate responsibilities. This enhances workplace safety and overall production efficiency. According to varied huge companies, the use of robots that are capable of collaborating with humans and understanding their surroundings can enhance productivity by as much as 20% in work environments that need a lot of manual effort. Within the automobile sector, numerous manufacturers have already implemented robotic systems to manage car assembly lines. Robots offer a cost-effective, efficient, and highly accurate alternative to humans in the fields of e-commerce and packaging. Additional applications encompass:

- Joining materials through welding
- Applying paint to surfaces
- Creating holes through drilling
- Examining products for quality assurance
- Producing objects for die casting
- Shaping materials through grinding
  3.   Artificial intelligence in supply chain management

AI-powered solutions can assist manufacturers in evaluating many situations (including time, cost, and income) to enhance last-mile deliveries. Artificial intelligence has the capability to anticipate the most efficient routes for deliveries, monitor the performance of drivers in real-time, and analyze weather and traffic reports along with previous data to precisely estimate future delivery times. AI can enhance firms' ability to manage their supply chains by providing increased oversight and optimization of capacity planning, inventory tracking, and management. An effective approach would be to establish a supplier assessment and monitoring model that operates in real-time and predicts potential failures. This would enable prompt notifications in the event of a supplier failure and instant evaluation of the impact on the supply chain. An illustration can be seen in the case of the automobile manufacturer Rolls Royce. The company utilizes sophisticated machine learning algorithms and image recognition technology to operate its autonomous fleet of ships. This, in turn, enhances the efficiency of its supply chain and ensures the secure transportation of its goods.

4.   Artificial Intelligence autonomous vehicles

Autonomous vehicles employed in the manufacturing setting, like as those demonstrated by Porsche in a previous instance, have the capability to mechanize a wide range of tasks, including assembly lines and conveyor belts. Autonomous vehicles, such as self-driving cars and ships, have the ability to enhance the efficiency of deliveries, work continuously without any time constraints, and accelerate the total delivery procedure. Autonomous vehicles are experiencing a gradual increase in demand and are projected to account for 10-15% of worldwide automotive sales by 2030. Connected vehicles, outfitted with sensors, have the capability to track real-time information regarding traffic congestion, road conditions, accidents, and other relevant data. This enables the vehicles to optimize delivery routes, minimize accidents, and promptly notify authorities during emergencies. This enhances the effectiveness of transportation and enhances the level of safety on the roads.

5.   Artificial Intelligence for factory automation

Factory operators use intuition and expertise to manage multiple signals on multiple panels and manually change machine configurations.This method also imposes the responsibility of resolving issues, conducting testing, and doing other activities on the operators, so exacerbating their workload. Consequently, operators occasionally resort to expedient methods, inaccurately prioritize tasks, and may not necessarily concentrate on enhancing economic worth. There are two issues that occur with this approach:

- Human-intensive systems are susceptible to errors, equipment malfunctions, and decreased production efficiency.
- Relying on experience increases the difficulty of finding replacements for manufacturing operators. Furthermore, the departure of a proficient operator also leads to the depletion of contextual information regarding factory operations.

AI helps companies cut labour costs and boost factory productivity. Other uses include:

Implement automated processes to streamline multiple intricate operations within industrial facilities. Detect any irregularities promptly by ongoing surveillance and supervision of activities, and immediately notify the technicians. Create a centralized database to store all operational data, including relevant information, to facilitate smoother personnel migrations.

Minimize the resource consumption necessary for operating a factory:

Efficiently adjust production levels based on changes in demand and manufacturing approaches. Siemens is a notable illustration of manufacturing automation (Deviprasad, S., Madhumithaa, N., et al., 2023). Google has helped the company boost shop floor productivity with computer vision, cloud data, and AI algorithms.

6.   Artificial Intelligence for IT operations

Artificial Intelligence for IT operations, or AIOps, is essential for optimizing IT operations. Gartner defines AIOps as the integration of big data (Lourens, M., Sharma, S., et al., 2023) and machine learning to streamline and automate IT operations procedures.

- AIOps is mostly utilized for the automation of large data management. This would entail:
- Gathering and incorporating data from sensors and machinery within factories
- Continuously monitoring and tracking the production area in real-time, and evaluating its performance against certain criteria
- Employing predictive analytics to detect, forecast, and mitigate IT service problems, as well as to conduct precise capacity planning.
- Monitoring and improving cloud resource and infrastructure performance with big data (Tripathi, M. A., Tripathi, R., et al., 2023) analytics.

Additional applications encompass event correlation and analysis, performance analysis, anomaly detection, causality determination, and IT service management.

7.   Artificial Intelligence in design and manufacturing

AI-powered software can generate multiple product-optimal designs. Generative design software requires engineers to input characteristics like:

- Raw materials
- Dimensions and mass
- Production techniques
- Limitations in terms of cost and available resources

By utilizing these factors, the algorithm is capable of producing diverse design variants. The software enables engineers to evaluate several designs over a diverse set of production scenarios and settings in order to select the optimal conclusion. Nissan is utilizing Artificial Intelligence to rapidly create innovative automotive designs that have never been seen before. The procedure would require significant time and effort from human designers, potentially spanning several months or even years to reach completion. This program can also be utilized to select optimal recipes that result in minimal raw material and energy wastage.

8.   Artificial intelligence and IoT

The term IoT, or Internet of Things, pertains to intelligent and interconnected devices that are equipped with sensors capable of generating substantial amounts of operational data in real-time (Al-Safi, J. K. S., Bansal, A., et al., 2023). This phenomenon is commonly referred to as IIoT, which stands for the Industrial Internet of Things, in the manufacturing industry. AI, along with IIoT, can enhance manufacturing processes by enabling higher levels of accuracy and efficiency. Notable applications of Industrial Internet of Things (IIoT) encompass:

- Use smart glasses to access instructions hands-free and improve real-time awareness.
- Monitor equipment performance, energy consumption, environmental temperature, and hazardous gases to improve workplace safety. Utilize smart lighting and HVAC controls to save energy.
- Collect data from production floor edge devices for industrial analytics.
    9.   Artificial Intelligence in warehouse management

Artificial intelligence has the capability to automate multiple facets of warehouse operations. Manufacturers can enhance their logistics planning by continuously monitoring their warehouses (M Geetha P Murthi, K Poongodi, 2023).through real-time data collection. Utilizing demand forecasting enables manufacturers to proactively replenish their inventories and meet consumer demand while minimizing transportation expenses. Automated robots in warehouses possess the ability to monitor, elevate, transport, and categorize objects, hence delegating the more strategic responsibilities to human workers and diminishing the occurrence of occupational injuries. Automation of quality control and inventory management can reduce warehouse management costs, increase productivity, and reduce staff. As a consequence, manufacturers have the ability to increase their level of sales and revenue margins.

10.   Artificial Intelligence process automation

AI-powered process mining technologies can automatically find and fix manufacturing bottlenecks. These technologies also allow firms to evaluate factory operational efficiency across multiple regions. They can standardise and optimise operations to improve manufacturing processes. Another application is RPA, where robots perform repetitive shop floor tasks autonomously. Only exceptions or abnormalities require human interaction with robots. Computer vision allows robots to monitor and inspect operations without human intervention.

The user's text is a bullet point.

- Process automation has the potential to decrease cycle times,
- Enhance yields,
- Enhance accuracy,
- Improve workplace safety and
- Elevate staff morale and productivity.

According to major corporations, the utilization of Artificial Intelligence in the semiconductor sector for process automation can result in a significant improvement of up to 30% in yield, as well as a reduction in scrap rates and testing expenses.

11. Artificial Intelligence for predictive maintenance

According to a survey by major corporations, the primary benefit of Artificial Intelligence in the industrial industry is derived from predictive maintenance, which contributes to a global value of $0.5-$0.7 trillion. BCG identifies predictive maintenance as the primary focus of Industry 4.0, particularly for cement manufacturers.

- Artificial Intelligence -driven systems have the capability to collect and analyze large volumes of data, including audio, video, and GPS information, obtained from sensors located on the shop floor.
- Assist in identifying irregularities or equipment inefficiencies to avoid unexpected equipment failure. Such occurrences may manifest as an unusual noise emanating from a vehicle's engine or a fault in a manufacturing line.
- Enhance production efficiency and cost-effectiveness by mitigating unplanned equipment downtime.
- Additionally, repairing failures in separate components is more cost-effective than replacing the entire equipment.
  12. Artificial Intelligence -based product development

Before starting production, manufacturers can use AI to create several simulations and test them in AR and VR. Therefore, manufacturers can:

- Reduce trial-and-error costs
- Shorten market time
- Help their engineers anticipate and resolve issues before the product launches.
- Simplify maintenance and debugging
- AI-based product development helps manufacturers innovate faster and create new, innovative products before the competition.

With a continuous feedback loop, AI algorithms improve with each iteration and help build better products. Product developers can use AI to create multiple simulations and test AR and VR technologies before starting production. Benefits for manufacturers include:

- Minimize expenses associated with trial and error
- Accelerate the time it takes to bring a product to market
- Assist their engineers in anticipating and mitigating any issues before to the product's release to the market.
- Optimize the process of maintenance and troubleshooting by utilizing Artificial Intelligence -based product creation, firms may enhance and
- Expedite their innovation process, enabling them to create novel and more advanced products ahead of their competitors.

Furthermore, because to the presence of a continuous feedback loop in Artificial Intelligence algorithms, they progressively enhance their performance with each iteration, so contributing to the development of superior products.

13. Artificial Intelligence -based connected factory

The manufacturing industry should embrace the use of sensors and cloud technology to create connected factories or smart factories, as this is the most promising direction for future and sustainable development (Geetha Manoharan, Dr. Y. Saritha Kumari et al., 2022; Satpathy, A., Samal, A., et al., 2024). Smart factories facilitate the following:

- Offer immediate visibility into the work floor
- Track and analyze the utilization of assets • Implement remote, contactless systems
- Facilitate real-time interventions
- Establish a centralized and reliable repository for all production data
- Expand production capacity seamlessly and without significant interruptions

An example is GE's "Brilliant Factory". GE constructed a plant in Pune, India with the aim of enhancing production and minimizing operational interruptions. They noticed a 45%-60% boost in OEE in their networked equipment.

14. Artificial Intelligence -based visual inspections and quality control

Computer vision technology uses high-resolution cameras to monitor all production processes for AI-driven defect identification. This technology can detect issues that humans may miss and take automatic action. This reduces product recalls and waste. Real-time detection of abnormalities like poisonous gas releases reduces workplace risks and improves worker safety in industrial facilities. AR overlays, another AI-based method, compare physical assembly parts to supplier-supplied parts to detect quality differences. Augmented reality (AR) allows technicians from anywhere to guide site staff remotely.

15. Artificial Intelligence for purchasing price variance

Raw material price changes may affect manufacturers' profit margins. Estimating raw material costs and choosing suppliers is difficult. AI-powered dashboards can track resource characteristics (pitch, diameter, material type, finishing) and supplier dimensions for manufacturers (e.g., country, brand name, performance data).

AI-powered algorithms can also group product parts for manufacturing and predict a standard purchase price based on historical data and market trends. These tools also establish a baseline for supplier price comparisons. This simplifies monitoring components from multiple vendors and centralising procurement information.

16. Artificial Intelligence order management

Order management must be agile and adaptable to market, demand, consumer, and manufacturing strategy changes. Manufacturing can use AI-based systems or robots to:

- Automate tasks such as order entry and other repetitive tasks
- Utilize sensors to monitor inventory and generate buy requisitions automatically
- Effectively handle the intricacies of many sorts of orders across several channels
- Enhance the efficiency and clarity of inventory planning and order management.

17. Artificial Intelligence for cyber security

Studies indicate that manufacturers are the most vulnerable to cyber-attacks, as even a short interruption of the production line can result in significant financial losses. With the proliferation of IoT devices, the risks will inevitably escalate at an exponential rate. Cyber-attacks pose a significant risk to smart factories. Artificial intelligence-powered cyber security solutions and risk detection procedures can enhance the security of industrial facilities and minimize potential risks. Manufacturers may utilize self-learning Artificial Intelligence to swiftly detect and halt assaults on cloud services and IoT devices with exceptional accuracy. The system can additionally notify the appropriate teams to promptly take action in order to mitigate any additional harm. Implementing sandboxing, code signing, and other security measures can reduce IIoT cyber threats.

DISCUSSIONS AND CONCLUSION

Narrative artificial intelligence models, such as chat GPT, have emerged as highly influential technologies with widespread use across several industries. The problem-solving and communication skills, as well as the efficiency and productivity, have yielded significant results across numerous domains, surpassing traditional work approaches. Industry 5.0 integrates human intuition with technological technology, specifically artificial intelligence, to create a collaborative and safe working environment that significantly improves overall business productivity (S Gokula Krishnan & Geetha Manoharan, 2022). The rapid development of artificial intelligence has been driven by the progress in natural language processing and deep learning. It is evident that a higher level of customization consistently leads to increased customer satisfaction and provides useful insights into consumer behavior and market trends.

However, generative artificial intelligence also presents the potential for stimulating innovation by employing creative problem-solving strategies. Given the intricate structure of current industrial processes, they frequently encounter difficulties due to the dynamic and unpredictable character of traditional methods. Consequently, artificial intelligence has emerged as a solution across several industries like education (Manoharan, G., Durai, S., et al., 2023; Razak, A., Nayak, M. P., et al., 2023), business (Lourens, M., Raman, R., et al., 2022)on a global scale. The integration of artificial intelligence in manufacturing processes has consistently showcased the benefits of digitalization, resulting in higher levels of utilization in the manufacturing sectors (Keserwani, Hitesh; P. T., Rekha; P. R., et al., 2021). Currently, it is possible that jobs are undergoing a transformation and we may observe the implementation of human-machine collaborative methods in industrial facilities.

REFERENCES

Abdulwahid, A. H., Pattnaik, M., Palav, M. R., Babu, S. T., Manoharan, G., & Selvi, G. P. (2023, April). Library Management System Using Artificial Intelligence. In *2023 Eighth International Conference on Science Technology Engineering and Mathematics (ICONSTEM)* (pp. 1-7). IEEE.

Akinsolu, M. O. (2023). Applied Artificial Intelligence in Manufacturing and Industrial Production Systems: PEST Considerations for Engineering Managers. *IEEE Engineering Management Review, 51*(1), 52–62. doi:10.1109/EMR.2022.3209891

Al-Safi, J. K. S., Bansal, A., Aarif, M., Almahairah, M. S. Z., Manoharan, G., & Alotoum, F. J. (2023, January). Assessment Based On IoT For Efficient Information Surveillance Regarding Harmful Strikes Upon Financial Collection. In *2023 International Conference on Computer Communication and Informatics (ICCCI)* (pp. 1-5). IEEE. 10.1109/ICCCI56745.2023.10128500

Deviprasad, S., Madhumithaa, N., Vikas, I. W., Yadav, A., & Manoharan, G. (2023). The Machine Learning-Based Task Automation Framework for Human Resource Management in MNC Companies. *Engineering Proceedings, 59*(1), 63.

Durai, S., Krishnaveni, K., & Manoharan, G. (2022, May). Designing entrepreneurial performance metric (EPM) framework for entrepreneurs owning small and medium manufacturing units (SME) in Coimbatore. In AIP Conference Proceedings (Vol. 2418, No. 1). AIP Publishing.

Geetha, M., Murthi, P., & Poongodi, K. (2023). Inventory Management of Construction Project Through ABC Analysis: A Case Study. *National conference on Advances in Construction Materials and Management, Springer Nature Singapore*, (pp. 95-105). IEEE.

Geetha, M. (2022). *Covid-19: a special review of MSMES for sustaining entrepreneurship. Msmes and covid - 19 the opportunities, challenges and recovery measures, 2022*. Trueline Academic And Research Centre.

Heilala, J., & Singh, K. (2023). *Evaluation Planning for Artificial Intelligence-based Industry 6.0 Metaverse Integration*. 6th International Conference on Intelligent Human Systems Integration (IHSI 2023) Integrating People and Intelligent Systems, Venice, Italy.

Jaichandran, R., Krishna, S. H., Madhavi, G. M., Mohammed, S., Raj, K. B., & Manoharan, G. (2023, January). Fuzzy Evaluation Method on the Financing Efficiency of Small and Medium-Sized Enterprises. In *2023 International Conference on Artificial Intelligence and Knowledge Discovery in Concurrent Engineering (ICECONF)* (pp. 1-7). IEEE.

Keserwani, H. P. T., Rekha; P. R., Jyothi; Manoharan, Geetha; Mane, Pallavi; Gupta, Shashi Kant (2021). Effect Of Employee Empowerment On Job Satisfaction In Manufacturing Industry. Turkish Online *Journal of Qualitative Inquiry, 12*(3).

Kumaravel, D., Tomar, P. S., Tulanovna, K. D., Sharma, S., Pawar, K. P., & Patil, P. P. (2023). Advanced Manufacturing Process Development for Automation in the Industry using Artificial Intelligence-Based System. *2023 3rd International Conference on Advance Computing and Innovative Technologies in Engineering (ICACITE)*, (pp. 955-958). IEEE.

Lee, Y., & Roh, Y. H. (2023). An Expandable Yield Prediction Framework Using Explainable Artificial Intelligence for Semiconductor Manufacturing. *Applied Sciences (Basel, Switzerland), 13*(4), 2660. doi:10.3390/app13042660

Lourens, M., Raman, R., Vanitha, P., Singh, R., Manoharan, G., & Tiwari, M. (2022, December). Agile Technology and Artificial Intelligent Systems in Business Development. In *2022 5th International Conference on Contemporary Computing and Informatics (IC3I)* (pp. 1602-1607). IEEE. 10.1109/IC3I56241.2022.10073410

Lourens, M., Sharma, S., Pulugu, R., Gehlot, A., Manoharan, G., & Kapila, D. (2023, May). Machine learning-based predictive analytics and big data in the automotive sector. In *2023 3rd International Conference on Advance Computing and Innovative Technologies in Engineering (ICACITE)* (pp. 1043-1048). IEEE. 10.1109/ICACITE57410.2023.10182665

Madrid, J. A. (2023). The Role of Artificial Intelligence in Automotive Manufacturing and Design. *International Journal of Advanced Research in Science. Tongxin Jishu.*

Manoharan, G., Durai, S., Ashtikar, S. P., & Kumari, N. (2024). Artificial Intelligence in Marketing Applications. In Artificial Intelligence for Business (pp. 40-70). Productivity Press.

Manoharan, G., Durai, S., Rajesh, G. A., Razak, A., Rao, C. B., & Ashtikar, S. P. (2023). A study of postgraduate students' perceptions of key components in ICCC to be used in artificial intelligence-based smart cities. In *Artificial Intelligence and Machine Learning in Smart City Planning* (pp. 117–133). Elsevier. doi:10.1016/B978-0-323-99503-0.00003-X

Manoharan, G., Durai, S., Rajesh, G. A., Razak, A., Rao, C. B., & Ashtikar, S. P. (2023). A study on the perceptions of officials on their duties and responsibilities at various levels of the organizational structure in order to accomplish artificial intelligence-based smart city implementation. In *Artificial Intelligence and Machine Learning in Smart City Planning* (pp. 1–10). Elsevier. doi:10.1016/B978-0-323-99503-0.00007-7

Mattera, G., Nele, L., & Paolella, D. A. (2023). Monitoring and control the Wire Arc Additive Manufacturing process using artificial intelligence techniques: A review. *Journal of Intelligent Manufacturing*, 1–31.

Meenaakumari, M., Jayasuriya, P., Dhanraj, N., Sharma, S., Manoharan, G., & Tiwari, M. (2022, December). Loan Eligibility Prediction using Machine Learning based on Personal Information. *2022 5th International Conference on Contemporary Computing and Informatics (IC3I)* (pp. 1383-1387). IEEE. 10.1109/IC3I56241.2022.10073318

Miltiadou, D., Perakis, K., Sesana, M., Calabresi, M., Lampathaki, F., & Biliri, E. (2023). A novel Explainable Artificial Intelligence and secure Artificial Intelligence asset sharing platform for the manufacturing industry. *2023 IEEE International Conference on Engineering, Technology and Innovation (ICE/ITMC)*, (pp. 1-8). IEEE. 10.1109/ICE/ITMC58018.2023.10332346

Mypati, O., Mukherjee, A., Mishra, D., Pal, S. K., Chakrabarti, P. P., & Pal, A. (2023). A critical review on applications of artificial intelligence in manufacturing. *Artificial Intelligence Review*, 56(S1), 661–768. doi:10.1007/s10462-023-10535-y

Ponduri, S. B., Ahmad, S. S., Ravisankar, P., Thakur, D. J., Chawla, K., Chary, D. T., & Sharma, S. (2024). A Study on Recent Trends of Technology and its Impact on Business and Hotel Industry. *Migration Letters : An International Journal of Migration Studies*, 21(S1), 801–806.

Razak, A., Nayak, M. P., Manoharan, G., Durai, S., Rajesh, G. A., Rao, C. B., & Ashtikar, S. P. (2023). Reigniting the power of artificial intelligence in education sector for the educators and students competence. In *Artificial Intelligence and Machine Learning in Smart City Planning* (pp. 103–116). Elsevier. doi:10.1016/B978-0-323-99503-0.00009-0

Satpathy, A., Samal, A., Gupta, S., Kumar, S., Sharma, S., Manoharan, G., Karthikeyan, M., & Sharma, S. (2024). To Study the Sustainable Development Practices in Business and Food Industry. *Migration Letters : An International Journal of Migration Studies*, 21(S1), 743–747. doi:10.59670/ml.v21iS1.6400

Shameem, A., Ramachandran, K. K., Sharma, A., Singh, R., Selvaraj, F. J., & Manoharan, G. (2023, May). The rising importance of AI in boosting the efficiency of online advertising in developing countries. In *2023 3rd International Conference on Advance Computing and Innovative Technologies in Engineering (ICACITE)* (pp. 1762-1766). IEEE. 10.1109/ICACITE57410.2023.10182754

Sundaram, S., & Zeid, A. (2023). Artificial Intelligence-Based Smart Quality Inspection for Manufacturing. *Micromachines*, *14*(3), 14. doi:10.3390/mi14030570 PMID:36984977

Tripathi, M. A., Tripathi, R., Effendy, F., Manoharan, G., Paul, M. J., & Aarif, M. (2023, January). An In-Depth Analysis of the Role That ML and Big Data Play in Driving Digital Marketing's Paradigm Shift. In *2023 International Conference on Computer Communication and Informatics (ICCCI)* (pp. 1-6). IEEE. 10.1109/ICCCI56745.2023.10128357

Wang, F. (2023). Research on the application of artificial intelligence technology to promote the high-quality development path of manufacturing industry. *SHS Web of Conferences*. IEEE.

Chapter 8
Revolutionizing the Manufacturing Sector a Comprehensive Analysis of AI's Impact on Industry 4.0

Atharva Paymode
VIT Bhopal University, India

Janhvi Shukla
VIT Bhopal University, India

D. Lakshmi
iD https://orcid.org/0000-0003-4018-1208
VIT Bhopal University, India

ABSTRACT

The impact of AI on Industry 4.0 is discussed in this chapter, with a focus on how it works best with IoT to create autonomous factories. It is emphasized how important AI is for predictive maintenance, downtime reduction, and utilizing ML and deep learning for increased productivity. This study looks at how NLP and machine vision may revolutionize document processing and work in tandem with intelligent document processing to remove administrative bottlenecks. In addition to technical details, the chapter explores wider ramifications, including proactive field service and customized solutions. Supply chain optimization, cost reduction, and value extraction from datasets all depend on the integration of AI and data analytics. Designed with experts and policymakers in mind, this succinct story offers insightful information about AI's contribution to the evolution of manufacturing to a global audience ready to understand the rapid changes taking place.

DOI: 10.4018/979-8-3693-2615-2.ch008

INTRODUCTION

Artificial Intelligence (AI) integration has become a transformational force in the industrial industry, changing the traditional boundaries of day-to-day operations. This chapter sets out to explore the deep effects of AI on the industrial industry, following the technology's development from its initial stages of integration to the present, when Industry 4.0 was introduced.

Examining the history of AI integration in manufacturing makes it clear that, in this rapid sector, advancement and survival through technology have come to be associated with technical advancement. AI has been shown to be a driver of productivity, creativity, and competitiveness from the early days of automation to the current era of data-driven decision-making (Alenizi et al., 2023).

The subject matter then turns to Industry 4.0, an innovation that indicates the combination of digital and conventional industrial processes. This revolutionary stage introduces intelligent systems that are able to make decisions on their own, going beyond simple automation. Industry 4.0 is a break from traditional manufacturing methods, a new era of linked gadgets, data analytics, and AI that alters production as an entire process (Plathottam et al.,2023).

The symbiotic interaction between the IoT and AI in everyday life is a crucial component of this transition. The foundation of intelligent factories is this convergence, where networked devices exchange data in a smooth manner, creating massive data streams. AI then analyses this data to produce insightful insights and lead production processes to previously unattainable levels of responsiveness, efficiency, and adaptability (Stadnicka et al.,2022).

This chapter attempts to provide an extensive understanding of the technical fabric that shapes manufacturing's present and future as we traverse through the layers of Industry 4.0, AI integration, and the symbiosis of IoT and AI.

Background of AI Integration in Manufacturing

Recent years have seen a major advancement in the application of AI in manufacturing, with the idea of Industry 4.0 serving as a major catalyst for the development of technologies such as the IIoT. AI has a long history in manufacturing because of its capacity to facilitate human-machine collaboration, which increases productivity, safety, and efficiency. (Stadnicka et al.,2022)(Plathottam et al.,2023).

The capacity of AI to swiftly and effectively analyse large amounts of data is one of the technology's main advantages in manufacturing. Manufacturers may obtain insights from competitor analysis, industry trends, and consumer preferences by utilising ML algorithms. This allows them to make better decisions and run their businesses more efficiently. Synthetic intelligence technologies can now be used by industrial facilities to help optimise workflows and cut waste.

AI has been implemented in a number of manufacturing use cases, including product creation, warehouse management, and supply chain management. AI, for example, can be used to evaluate sensor data and forecast equipment failures, resulting in more effective maintenance plans and decreased downtime. Other than that, by automating monotonous jobs and enhancing safety in high-risk areas, AI can be utilised in conjunction with robots to free up human workers' attention for more productive aspects of the business (Javaid et al.,2022).

There is scepticism as well as enthusiasm over the use of AI in manufacturing. AI raises worries about job displacement and the possibility for increased complexity in manufacturing processes, even while it offers many benefits including better efficiency, improved safety, and lower prices. However, as more manufacturers become aware of the potential advantages and opportunities that AI offers, the technology's adoption in manufacturing is expanding, which bodes well for the technology's future in the sector.

Emergence of Industry 4.0

The start of Industry 4.0 signifies a noteworthy shift in the production process, propelled by the digitalization of manufacturing procedures. Building on the developments of earlier industrial periods, this shift—often referred to as the fourth industrial revolution—integrates new technology to produce production systems that are more productive, efficient, and networked.

The convergence of many technologies, such as cloud computing, AI, advanced robotics, and the IIoT, characterises Industry 4.0. With the use of these technologies, smart factories may be built, increasing productivity and efficiency by connecting and communicating amongst equipment to make autonomous decisions. Large-scale data analysis is made possible by the combination of AI and ML, which offers insightful information for improved decision-making and process optimization.

The transition from mass production to mass customization is another point of emphasis in the Industry 4.0 idea. Because smart manufacturing methods are flexible and adaptable, manufacturers can now quickly produce small batches of customised items for specific consumers. The smooth integration of production processes with open and effective supply chains—both crucial elements of Industry 4.0 strategies—makes this degree of customization possible.

Numerous industrial sectors, including mining, oil and gas, discrete and process manufacturing, and other industrial areas, are feeling the effects of Industry 4.0. During the Fourth Industrial Revolution, the adoption of these cutting-edge technologies is anticipated to change every industry, albeit to varying degrees. With a projected $3 trillion in value creation, Industry 4.0 offers manufacturers and suppliers significant opportunities.

Industry 4.0, which is being propelled by the incorporation of cutting-edge technology like robots, AI, and the industrial internet of things, signifies a major change in the manufacturing environment. The emergence of smart factories is being made possible by this transition. These factories will be able to satisfy the demands of mass customization with greater efficiency and productivity, as linked machines and systems collaborate to make autonomous decisions. All industrial sectors are predicted to be affected by Industry 4.0, which has the potential to significantly increase value for suppliers and manufacturers.

The Convergence of IoT and AI in Daily Operations

Many sectors are undergoing a revolution as a result of the everyday integration of IoT and AI, which offers increased productivity, better decision-making, and new business opportunities. These technologies work together to create more intelligent and efficient systems by enabling real-time analysis and decision-making. IoT devices collect data, which AI then analyses to provide more insightful analysis and wiser choices.

Predictive maintenance is one important area where IoT and AI are combining. AI systems examine sensor data and past equipment performance to forecast when and how equipment might break, enabling companies to plan maintenance in advance and minimise downtime. This strategy guarantees continuous output, lowers expenses, and increases efficiency (Plathottam et al., 2023).

The application of an IoT to smart agriculture is another illustration. In order to improve farming operations and increase efficiency, AI provides automated resource and irrigation management, livestock monitoring, and health management. IoT and AI are utilised in the energy and utility sectors to optimise and manage smart grids and perform predictive maintenance on utility infrastructure, which guarantees reliable and effective service.

The retail sector is likewise changing as a result of the confluence of AI and IoT. Retail is being revolutionised by IoT and AI, which streamline supply chains, improve inventory management, and offer personalised customer experiences. Retailers may increase overall efficiency, customise offers, and gain a deeper understanding of their customers with the use of these technologies.

The convergence of AI and IoT is accelerating innovation, opening up new commercial opportunities, and propelling the creation of disruptive applications in the communication infrastructure space. IoT is essential to smart cities, industrial applications, connectedness, and advances in healthcare, and AI makes intelligent data analysis, autonomous networks, predictive maintenance, and user-friendly interfaces possible.

INDUSTRY 4.0 FRAMEWORK

Intelligent Factories and Autonomous Decision-Making

Smart factories, or intelligent factories, are one of the main results of Industry 4.0, the fourth industrial revolution. These cutting-edge production sites are distinguished by their high degrees of autonomy, connectedness, and digitization. To work more productively and efficiently, they make use of a variety of cutting-edge technology, including AI, robotics, big data, the IoT, and sophisticated analytics. The objective of intelligent factories is to save labour costs and operating expenses while fostering a more responsive and flexible manufacturing environment. Figure 1 shows the Adoption Rates of Industry Technology in intelligent Factories from 2016 to 2022.

The capacity to make decisions on their own is one of the characteristics that set intelligent factories apart. This is made feasible by the factory's ability to analyse data, spot trends, and make choices without the need for human intervention In the smart factory, autonomous decision-making is used for a variety of purposes, such as supply chain management, monitoring of quality, predictive maintenance, and production planning.

Intelligent factories depend on a network of interlinked devices and sensors that gather real-time data on the factory's operations in order to achieve a high degree of autonomy. After that, this data is examined to find areas that could use modification and enhancement. Predictive maintenance systems, for instance, can schedule maintenance proactively by using sensor data to determine when a machine is likely to break (Zhong et al.,2017).

The employment of collaborative robots, or "bots," which are made to work alongside human workers to do a range of tasks, is another essential component of intelligent factories. With their sophisticated sensors and AI algorithms, these robots can securely collaborate with people in a variety of settings. This increases workplace safety while also allowing the production to be more adaptable and responsive to variations in demand.

Role of AI in Processing Manufacturing Data

The industrial sector is undergoing a transformation because of AI, which gives businesses the ability to analyse massive volumes of data and make wise decisions. AI is being applied in manufacturing for several purposes, such as supply chain

Figure 1. Adoption rates of industry technology in intelligent factories (2016-2022)

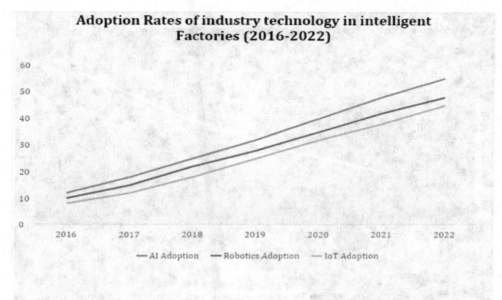

management, product creation, quality control, and predictive maintenance. The market for AI in manufacturing is anticipated to reach $16.7 billion by 2026, and the application of AI in manufacturing is anticipated to dramatically increase in the upcoming years. Figure 2 shows the Anticipated Growth of AI in Manufacturing Market

Predictive maintenance is one of the main uses of AI in manufacturing. To schedule maintenance proactively, predictive maintenance uses data from sensors and other sources to forecast when equipment is likely to break. This can enhance the overall efficiency of the manufacturing process and help save maintenance expenses and downtime (Kim el at.,2017).

Quality control is yet another significant area in which AI is being used in manufacturing. AI can analyse sensor and camera data to find product flaws and pinpoint areas that need development. This might reduce waste and raise the calibre of the products.

AI is also being used in manufacturing to enhance supply chain management. AI can assist manufacturers in making better decisions about where and when to acquire resources, how much to produce, and when to ship products by analysing data on inventory levels, production schedules and consumer demand. AI is also being used in manufacturing to speed up product development in addition to these other applications. AI can assist firms in identifying areas for innovation and streamlining the product development process by analysing data from production and trial operations.

Figure 2. Anticipated growth of AI in manufacturing market

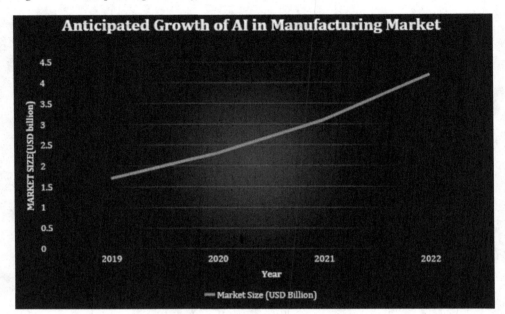

The use of AI in manufacturing has several benefits, but there are also drawbacks. A major obstacle is the absence of common industry data. Reliable AI model building is challenging in manufacturing because manufacturing data are frequently localised or customised to a certain industry area or a company's activities. Smaller manufacturers may find it challenging to implement AI because of the lack of specialised AI personnel.

In terms of analysing production data and helping manufacturers make wise decisions, AI is becoming increasingly significant. Manufacturers may lower costs, boost innovation, and enhance the effectiveness and quality of their operations by utilising AI technologies such as ML and Deep Learning. Although there are obstacles in the way of AI implementation in the industrial sector, the field is predicted to expand quickly in the upcoming years due to the technology's enormous potential benefits (Kim et al.,2017).

Predictive Maintenance as a Cornerstone Application

The aim of predictive maintenance, which is a proactive maintenance approach, is to predict the likelihood of equipment failure so that maintenance tasks can be planned accordingly. It has been applied in the industrial sector since the 1990s and is also referred to as condition-based maintenance. Predictive maintenance

Figure 3. Impact of predictive maintenance on key metrics

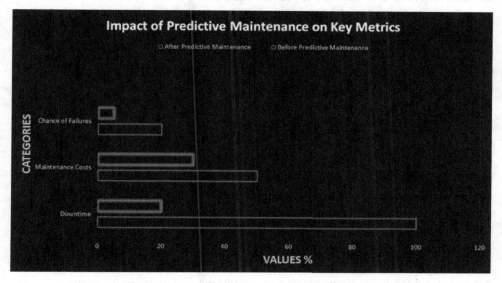

aims to first identify potential equipment failure points based on specific criteria, then use routinely planned and corrective maintenance to keep the equipment from breaking down. Using this method necessitates collecting asset data and extracting information that helps determine when maintenance is required. In sectors with intricate and specialised processes like manufacturing, aircraft, and chemicals, predictive maintenance is very helpful. By lowering downtime, maintenance costs, and the chance of failures, it can have a significant positive impact on corporate value.

On the other hand, predictive maintenance implementation calls for a large financial, human, and educational commitment. Predictive maintenance is essential to a successful maintenance programme because, in spite of the difficulties, the return on investment greatly exceeds the initial expenditures. Figure 3 shows the impact of predictive maintenance on key metrics (Plathottam et al.,2023) (Alexopoulos et al.,2021).

The primary stages to the predictive maintenance procedure. Initially, sensors and devices actively gather information about the state and functionality of equipment during normal operations. This collected data is then archived and examined closely in order to spot patterns and trends that might indicate future problems. In order to find patterns suggestive of imminent equipment breakdowns, sophisticated tools and techniques, including ML algorithms, are utilised during the second phase of data analysis. With the use of this predictive data, maintenance requirements may be predicted, allowing for strategic scheduling of repairs and replacements at the most advantageous periods. The last stage is maintenance scheduling, which is an essential step to save downtime and prevent equipment failures. This schedule includes proactive maintenance, which involves repairing or replacing components before they break down, as well as reactive maintenance, which deals with problems after they arise. All things considered, the predictive maintenance procedure smoothly combines data gathering, analysis, and planned scheduling to improve operational effectiveness and guarantee equipment longevity.

Benefits of Predictive Maintenance

Businesses can minimise operational disturbance by anticipating any problems ahead of time and scheduling maintenance and repairs during the most opportune times. By minimising unplanned downtime and overhauls and enabling organisations to schedule repairs and replacements at the most favourable times, predictive maintenance lowers maintenance costs. Organisations can lower costs and boost system and process performance by anticipating possible problems and planning maintenance appropriately. Predictive maintenance finds and fixes possible problems before they get worse, which helps firms maximise operations and increase efficiency.

Predictive maintenance is essential to contemporary industrial operations since it can save costs, increase productivity, and minimise downtime, among other advantages. Businesses may minimise downtime and maximise productivity by anticipating equipment breakdowns and scheduling maintenance tasks appropriately by utilising data analytic tools, ML, and the IoT.

MACHINE LEARNING IN MANUFACTURING

Significance of Machine Learning and Deep Learning

Two branches of AI, ML and deep learning, are revolutionising data processing and analysis. While deep learning teaches computers to think using structures modelled after the human brain, ML aims to make computers more capable of thinking and acting without human intervention. While deep learning is more complex to set up and requires less intervention once it is operational, ML requires more continuous human interaction to achieve outcomes. ML struggles to analyze photos, movies, and unstructured data in the same manner as deep learning (Kim et al., 2022).

The primary benefit of deep learning algorithms is their gradual approach to extracting high-level features from data. This implies that massive volumes of data may be used to train deep learning algorithms, which will eventually increase their accuracy. Deep learning algorithms are especially helpful for applications such as NLP, audio recognition, and image recognition.

A significant distinction between deep learning and ML is the volume of data required to train the algorithms. While deep learning algorithms can learn from unstructured data and require less labelled data, ML algorithms usually require a significant quantity of labelled data to be effective. Therefore, deep learning is especially helpful for applications in which a large amount of unstructured data is available, such as speech and image recognition.

Automating feature extraction is another benefit of deep learning. Typically, feature extraction in classical ML is performed manually by human professionals. Using algorithms, deep learning can automatically extract features from data, thereby minimising the need for human interaction and increasing process efficiency.

Support and customer service are also being enhanced through the application of deep learning. Deep learning algorithms can help firms enhance their customer care and support procedures by identifying patterns and trends in client data. Deep learning algorithms, for instance, can be used to examine client feedback and pinpoint areas in need of development (Kim et al., 2022).

Supervised Machine Learning: Predicting Equipment Failure

Supervised ML for predictive maintenance is an important application that has drawn much interest from various sectors. To train ML models that can anticipate when a failure is likely to occur, historical data from the equipment is used. By taking a proactive approach to maintenance, companies can lower maintenance expenses, prevent expensive downtime, and increase overall operational effectiveness (Susto et al., 2014).

In the AI subfield of Supervised ML, models are trained on labelled data to provide predictions or judgments. Supervised learning algorithms can be trained on historical data, which includes details about the equipment's performance before a breakdown, to forecast equipment failure. Along with the time to failure, these data usually have characteristics such as vibration, temperature, pressure, and other sensor measurements. Supervised ML models can be trained to recognize patterns and correlations that point to an imminent equipment failure by examining past data. After training, the model can be used to forecast fresh data, enabling maintenance staff to proactively avoid malfunctions.

Benefits of Equipment Failure Prediction

Businesses across a range of industries can greatly benefit from the ability to predict equipment failure. One of the primary benefits is that organisations may reduce the impact on operations by planning maintenance during planned downtime and predicting when an issue is likely to arise. Proactively addressing potential equipment issues can help businesses avoid costly emergency repairs and reduce overall maintenance costs. Businesses with the ability to predict equipment failures can reduce the risk of accidents and injuries by identifying and fixing potential safety issues ahead of time. Predictive maintenance can assist businesses in optimising their maintenance schedules and ensuring that resources are spent efficiently.

ML techniques can be applied to predictive maintenance. Regression models are a useful tool for estimating the lifespan of equipment by using past data as a foundation. Classification models are helpful resources for determining how likely it is that a piece of machinery may malfunction. Time series analysis techniques can be utilised to detect patterns and trends in equipment performance data that point towards impending failure. Using anomaly detection methods, one might find unusual patterns in equipment performance data that may point to a potential failure (Çınar et al.,2020).

Obstacles and Things to Consider

While there are many benefits to using supervised ML for predictive maintenance, there are certain challenges and things to keep in mind. Predictive maintenance requires high-quality labelled data. If the ML models are not trained using trustworthy and realistic data, the strategy will not be successful. Among the ML models that can be difficult to comprehend are deep learning models. Ensuring that maintenance staff can understand and rely on model forecasts is crucial. As the amount of data produced by equipment continues to rise, businesses will need to make sure that their predictive maintenance systems are scalable and able to handle huge volumes of data. Predictive maintenance systems need to be connected with the current maintenance and asset management systems to ensure that model predictions may be put into practice.

Supervised ML in conjunction with predictive maintenance is a crucial application that could revolutionise how firms handle equipment maintenance. Businesses may anticipate when equipment problems are likely to occur by utilising past data to train ML models. This enables them to take proactive measures to prevent downtime, lower costs, and increase safety. Predictive maintenance has several advantages, and even though there are certain drawbacks and things to keep in mind, its significance is certain to grow in the years to come.

Unsupervised Machine Learning: Identifying Anomalies in Production Lines

Unsupervised ML is a valuable method for detecting anomalies in production processes. Unsupervised ML algorithms can identify trends, outliers, and uncommon occurrences that substantially deviate from the typical behaviour of the dataset by examining data from the production line. Producers can reduce expenses, improve product quality, and simplify operations by doing so.

Producing Anomaly Identification

Anomaly detection is an essential part of production because it helps identify unusual events, patterns, and outliers that could affect output and product quality. With deeper insight into their processes, manufacturers can react to variations in their processes quicker and more effectively.

Unsupervised ML is particularly useful for identifying industrial anomalies because it can examine massive volumes of data produced from production lines. These data may comprise sensor readings, process data, and other features that indicate regular activities. Any predicted departure from these patterns is considered abnormal.

Unsupervised Machine Learning Techniques for Anomaly Detection

Manufacturing processes can use a variety of Unsupervised ML approaches to identify anomalies. In addition to identifying dissimilar data points, clustering algorithms can be used to group similar data points together. Data points that do not fit into any cluster are known as outliers. An unsupervised learning method that can identify outliers in the data automatically is called an isolation forest. To find outliers in the data, it randomly selects a feature at first, and then it randomly selects a split value for that feature. One kind of unsupervised learning model that can infer the underlying structure of the data is called an auto coder. The model may be able to recognize regular patterns in the data and classify anomalies as data points that differ from these patterns by being trained to replicate the input data.

Unsupervised Machine Learning Advantages for Recognizing Anomalies

The capacity to ascend Because Unsupervised ML algorithms can analyse big datasets, they are appropriate for usage in large-scale industrial contexts. Unsupervised learning algorithms can identify both known and unknown anomalies since they can adjust to shifting data patterns. Unsupervised ML can drastically cut the labor expenses related to human monitoring and analysis by automating the anomaly detection process. Real-time anomaly detection is possible with unsupervised ML models, enabling prompt and efficient remedial action.

Challenges and Considerations

When it comes to identifying irregularities in industrial processes, unsupervised ML offers several advantages, but there are also disadvantages that must be considered. A crucial factor in the efficacy of unsupervised ML models is the quality of the training data. Missing or tainted data might cause a model to perform poorly and generate more false positives or negatives. Two examples of Unsupervised ML methods that may be difficult to comprehend are isolation forests and autoencoders. Ensuring the accuracy and utility of the model's predictions is crucial. To get the best outcomes, hyper parameters for unsupervised ML models are continuously enhanced and modified. This is a difficult and somewhat convoluted process.

Unsupervised ML is a useful technique for identifying irregularities in production lines. By analysing massive amounts of data generated by manufacturing lines, unsupervised ML algorithms can identify trends, anomalies and unusual occurrences that may affect production procedures and product quality. While unsupervised ML presents significant benefits for anomaly identification in manufacturing processes, there are also considerations and limitations. This means that this approach will likely become more popular in the future.

ADVANCED TECHNOLOGIES IN ACTION

Natural Language Processing (NLP) in Manufacturing

Using the implementation of Natural Language Processing (NLP) technology, computers can now comprehend and process spoken or written human language. NLP can be used in the manufacturing sector to evaluate unstructured data, including emails, social media posts, product evaluations, online surveys, and customer support issues, to obtain insightful information that can improve decision-making procedures. With the use of NLP tools, equipment performance may be tracked, allowing for real-time monitoring of machinery and the implementation of major preventative steps in the case of equipment failure. Manufacturing sectors may efficiently create a digitally sound environment by eliminating the middleman with NLP. NLP can also be used to track trends in production line data that might indicate problems with quality control. Manufacturers can obtain vital business information from the internet by using NLP capabilities to perform web scraping, data extraction, data scanning, or data mining. By analysing many shipment documents, NLP can provide manufacturers with enhanced supply chain insights, allowing them to optimise their operations by upgrading specific process steps or altering logistics. NLP is an effective tool that increases productivity and efficiency in the manufacturing sector.

However, putting NLP into practice in the manufacturing sector can be difficult and requires careful consideration of which data to use and how to use it, as well as a lot of training and the right interface. NLP is still a promising area of AI despite these drawbacks, and its prospective uses in manufacturing are only expected to grow in the future (Alenizi et al., 2023) (May et al., 2022) (Kim et al., 2022).

Applications of NLP

Natural Language Processing (NLP) has been applied in many different sectors and has had a big influence on industrial procedures. Manufacturers use NLP to examine unstructured data in market reports, supplier notes, and email correspondence as it relates to supply chain optimization. This allows for a thorough assessment of supplier performance, a sophisticated comprehension of market trends, and the capacity to anticipate any supply chain disruptions. Manufacturers may manage their supply networks more effectively and strategically by utilising the capability of NLP. Another crucial area where NLP is essential is quality assurance. NLP algorithms analyse textual data from customer evaluations, maintenance logs, and inspection reports to find patterns linked to product defects. This enables producers to reduce flaws, improve overall product quality, and take proactive measures to resolve quality-related problems. The capacity to derive meaningful conclusions from textual data guarantees an iterative process of improvement in line with changing client demands and quality requirements. By facilitating natural language interactions between workers and machines, NLP promotes human-machine collaboration in the manufacturing setting. A safer and more efficient production floor is facilitated by voice-activated instructions, hands-free machine operation, and higher productivity. This incorporation of NLP into routine operations demonstrates the growing synergy between computer capabilities and human skills, improving overall efficiency and safety at work.

NLP also has transformational effects in the areas of knowledge management and documentation. Reports, specs, and manuals are among the many documents produced by manufacturers. NLP helps with the organisation and extraction of important information from these papers, making it easier for employees to access. Employees may easily obtain relevant information to improve their comprehension and performance, which leads to better knowledge management and well-informed decision-making.

Sustainable manufacturing processes depend heavily on energy administration, to which NLP lends a hand by evaluating textual information on production schedules, energy usage, and environmental factors. NLP helps factories to save costs and lessen their environmental effect by finding chances for energy optimization, which is in line with more general sustainability objectives.

Challenges in Implementing NLP in Manufacturing

There are benefits and drawbacks to using Natural Language Processing (NLP) in the manufacturing industry. Unstructured data, which includes written documents, images, and sensor readings, presents a major obstacle to the successful implementation of NLP solutions by guaranteeing data quality and relevance. Because industrial facilities frequently use equipment and technologies that are not inherently compatible with cutting-edge NLP, integrating NLP with older systems creates compatibility and integration challenges that impede general adoption. With the sensitive nature of industrial data, including trade secrets and intellectual property, security and privacy concerns become paramount. Strong security protocols are essential to protect against possible hacks.

Developing, implementing, and maintaining NLP systems in manufacturing effectively requires addressing the skills gap in the workforce and offering sufficient training. To fully profit from NLP, skilled professionals must be able to traverse the intricacies of these cutting-edge technologies. The cost of implementation is another issue; there are significant up-front expenditures for gear, software, and training. It can be especially difficult for small- and medium-sized firms to set aside money for these expenses. As is the case with any AI technology, ethical issues also become paramount. The application of NLP in manufacturing brings up moral concerns about algorithmic bias, decision-making transparency, and accountability for AI-driven activities. The positive effects of NLP in the manufacturing sector require careful thought and adherence to appropriate and ethical application.

Potential Impact on Manufacturing

Due to NLP, manufacturers may extract actionable insights from vast amounts of unstructured data. As a result, this enhances decision-making in many areas, such as daily operations and strategic planning. Natural language interfaces and language-related tasks can be automated with NLP. This can reduce the amount of physical labour required, improve workflow, and raise production levels overall on the manufacturing floor. Manufacturers can utilise NLP algorithms to minimise unscheduled downtime and facilitate scheduled maintenance by employing predictive maintenance to anticipate equipment issues. As such, it improves operational continuity and reduces maintenance costs. NLP may analyse market trends, customer feedback, and competitor data to give firms valuable insights for innovation and new product creation. This might help in keeping a competitive edge in the market. Nonverbal Learning can analyse textual data from several sources to assist supply chain management become more adaptable and responsive. Manufacturers are able to react swiftly to changes in demand, disruptions in the supply chain, and changes in market dynamics. In production environments, NLP can enhance communication and teamwork. By facilitating easier technological interaction, natural language interfaces can improve employee accessibility and promote a safer, more productive workplace.

Machine Vision and Its Role in Quality Assurance

Machine Vision is used in the industrial sector to ensure product quality by identifying impurities, faults, and other anomalies in manufactured goods. Machine Vision inspection can be used to check for defects in medication tablets, displays to confirm the presence of icons or pixels, or touch screens to gauge backlight contrast. In the food and pharmaceutical industries, Machine Vision can verify that a product matches its package and that safety seals, tops, and rings on bottles are in place. Inspection-specific Machine Vision systems monitor the material's exterior appearance. By applying statistical analysis, the system automatically detects possible surface flaws in the material and groups the flaws according to their similarity in terms of contrast, texture, and/or geometry. The majority of Machine Vision systems come with a software toolkit that allows you to perform various inspections and combine multiple inspections from the collected images (Penumuru et al., 2020).

One crucial area in the industry where automation has a significant impact on quality assurance is Machine Vision inspection. The reputation and financial performance of a manufacturer are greatly enhanced by quality management. A successful program for improving quality can lower the expenses associated with subpar work in a growing business, directly increasing earnings. By eliminating the human factor from the equation, Machine Vision can improve operational effectiveness. It consistently yields results and is operational around the clock, seven days a week. Every product on the line can be examined using Machine Vision, which also allows for the early detection of slight differences and deviations. This makes it possible to take action before the quality falls short of the necessary level or issues arise later.

In addition to detecting defects in products down to submillimeter sizes, Machine Vision inspection can also be used to decipher unclear and blurry text, measure and position with submillimeter precision, detect, recognize, and classify objects and patterns, and identify production process issues or predictive maintenance. Compact and able to fit into tight spaces, Machine Vision inspection offers high resolution, from hundreds to millions of measurement points, is fast, traceable, and capable of enabling the tracking and storing of individual product results. It can also be used to inspect any product because of its low cycle duration (Baygin et al., 2017).

Implementing Machine vision inspection can be difficult and requires careful consideration of which data to use and how, training, and a suitable interface.

However, as processing power has increased, Machine Vision—one of the most promising areas of AI has seen a plethora of practical applications. Manufacturers can obtain vital business information from the Internet using Machine Vision techniques such as site scraping, data extraction, data scanning, and data mining. To increase production and manufacturing efficiency, Machine Vision is a powerful tool. Future manufacturing applications for it are likely to grow.

Intelligent Document Processing (IDP) Revolutionising Data Capture

Data capture and document processing are being revolutionised by a new technology called IDP, which combines ML and AI. IDP transforms modern enterprises by automating the extraction of important data, capturing context, and interpreting nuances hidden in documents.

A number of important advantages of IDP greatly increase organisational effectiveness. The fact that IDP streamlines data extraction and automates the procedure to do away with the necessity for human data entry is one of its main benefits. In addition to saving time, this lowers the possibility of mistakes, freeing teams to concentrate on higher-level organisational duties. IDP prioritises industry best practices to comply with legal obligations while ensuring compliance with data governance and legislation. One noteworthy aspect of IDP is its continuous learning capabilities, which allows the system to become more intelligent and effective over time by processing more documents and increasing its accuracy and knowledge base. Another significant advantage is contextual processing, which guarantees that documents are processed in the proper context and improves process accuracy overall. IDP streamlines operations and improves document workflows by providing additional support for automated document flow. IDP easily handles massive numbers of documents, especially those in sensitive fields like medical and educational records, because of its sophisticated document parsing and text-to-speech algorithms. All things considered, IDP proves to be a revolutionary instrument that promotes accuracy, efficiency, and compliance in document-centric processes.

IDP extracts important information, gathers context, and deciphers subtleties from documents using sophisticated algorithms. IDP uses intelligent document identification, automated data extraction, and automated document processing to extract useful information from documents without requiring human data entry. The precision and accuracy of data capture are integrated into data extraction procedures using IDP technologies such as NLP and Optical Character Recognition (OCR) (bin Abdullah et al., 2022).

AI-based document processing is transforming data collection in many ways, including the handling of structured, semi-structured, and unstructured documents, contextual understanding, and adaptive learning. IDPs improvements in operational performance, customer satisfaction, and efficiency are revolutionising many industries. IDP, for instance, can be used to handle medical records in the healthcare sector, guaranteeing fast and accurate information for patient treatment. IDP can be used to handle loan applications in the financial service sector, thereby speeding up decision-making and reducing processing times (Baviskar et al., 2017).

BEYOND EFFICIENCY: BROADER IMPLICATIONS OF AI IN INDUSTRY 4.0

Personalized Customer Outcomes Through AI

AI integration has become a revolutionary force in the quickly changing business landscape, especially when it comes to improving consumer experiences. AI is transforming traditional corporate strategies, delivering personalised consumer outcomes, and opening up new growth opportunities. Businesses are using AI as technology develops to evaluate enormous volumes of data, comprehend consumer behaviour, and customise goods and services to specific desires.

In customer outcomes, personalization is the capacity to provide individualised experiences according to each person's demands, choices, and behaviours. By processing and analysing a variety of data sets, such as customer interactions, past purchases, and internet activity, AI plays a critical part in doing this. The intention is to make every customer feel personally cherished and understood by going beyond standard methods to marketing and service delivery.

With AI, companies can tailor their products to the unique needs of each customer. Businesses may employ AI algorithms to assess what each client is likely to value, from targeted promotions to personalised product recommendations. This increases revenue and fosters client loyalty in addition to improving consumer satisfaction (Ameen et al.,2021).

AI-powered Chabot's and virtual assistants play a major role in providing individualised client experiences. These smart technologies can interact with clients in real time, offering prompt support, responding to questions, and even predicting their needs. As a result, customers receive a smooth, customised, round-the-clock experience that raises satisfaction levels.

Beyond meeting immediate demands, AI's predictive capabilities help organisations predict customer preferences and behaviour in the future. AI algorithms can predict what goods or services a client would be interested in next by examining data patterns. This helps firms keep ahead of changing consumer preferences.

Figure 4. AI applications in customer personalization (2023)

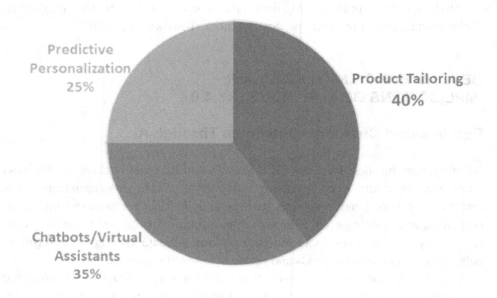

AI APPLICATIONS IN CUSTOMER PERSONALIZATION (2023)

Predictive Personalization 25%

Product Tailoring 40%

Chatbots/Virtual Assistants 35%

AI's capacity to produce in-depth behavioural insights about customers is one of its primary contributors to personalised customer outcomes. Businesses may anticipate client requirements by using ML algorithms to analyse past data and detect patterns, preferences, and trends. By taking a proactive stance, customer contacts are changed from reactive to proactive, strengthening the bond between the company and its customers. Although AI-driven personalization has many advantages, it also presents difficulties and ethical questions. Critical issues include permission, data privacy, and responsible customer data use. Establishing and preserving client trust requires finding the ideal balance between personalization and privacy. In figure 5 we can see AI applications in customer personalization.

The potential for customised client outcomes with AI is increasingly larger as technology develops. Technological advancements like Augmented Reality (AR) and Virtual Reality (VR) have the potential to improve customer experience personalization by facilitating customised and immersive interactions.

Proactive Field Service Delivery

IoT sensors and AI are used in conjunction for condition monitoring in predictive maintenance, which is an essential component of contemporary industrial processes. By analysing equipment data in real-time, possible problems are predicted before they arise. Predictive analytics also helps with proactive maintenance plans, which reduce downtime and extend the life of equipment. AI solutions enable dynamic routing and scheduling that optimises field service plans dynamically based on variables such as technician proficiency, service urgency, and geography. Precise arrival timings are guaranteed by this real-time adjustment, which also takes into account external factors like traffic and weather. Inventory procedures are streamlined by mobile inventory management and automated stock monitoring, which also give technicians access to real-time data for on-site repairs and automatically refill inventory.

AI-driven technologies such as Augmented Reality (AR) allow personnel to provide remote support for complex repairs or inspections in the field of remote help and analysis. AI-powered remote diagnostics examine equipment data in advance, increasing the first-time fix rate. AI-powered automation is used by customer self-service portals to enable consumers to submit service requests, track their progress, and get real-time updates. Clients can resolve minor difficulties on their own with the help of knowledge sources that are embedded into these portals. AI-powered communication platforms for clients and mobile collaboration tools for field professionals provide instantaneous communication, guaranteeing pertinent information and timely updates.

AI-powered KPIs and performance analytics offer insights into important performance metrics including response times, customer satisfaction, and resolution rates. Through the identification of trends and opportunities for improvement, this data-driven research facilitates ongoing field service operations improvement. AI is used to monitor security and regulatory compliance, ensuring that industry standards and safety laws are followed during field service operations. Based on AI findings, proactive risk mitigation strategies are put into place to lower the possibility of mishaps or noncompliance. Field service operations are transformed by this thorough integration of AI technologies, which increases productivity, decreases downtime, and ultimately improves customer happiness.

Mission-Critical Applications Driven by AI and Data Analytics

Due to their direct impact on operational efficacy, safety, and efficiency, mission-critical applications are essential to several sectors. As AI and data analytics continue to advance, these applications are becoming more sophisticated and reliant on accurate, timely, and actionable data (Lee et al., 2020).

Industries Dependent on AI Applications With Critical Missions

In many different sectors, mission-critical AI applications are crucial. Because it streamlines operations, enhances safety procedures, and reduces downtime particularly in rail systems, AI is vital to the transportation industry. By empowering healthcare professionals to make educated decisions that improve patient outcomes while effectively reducing costs, mission-critical AI has a significant beneficial influence on the healthcare industry. For example, early illness detection and precise treatment planning may be achieved with predictive analytics. In another crucial area, predictive maintenance uses AI-driven analytics to find and fix potential issues before they worsen, boosting overall efficiency and cutting down on maintenance expenses. These applications demonstrate how mission-critical AI is advancing productivity, security, and cost while revolutionising a number of sectors.

Important Distinctions in AI Applications for Mission-Critical

AI is an indispensable cornerstone, possessing three features that are important to the success of mission-critical applications. The most important of these is accuracy because it is the key to ensuring optimal performance and safety. AI systems are rigorously trained on copious volumes of high-quality data in order to do this. Furthermore, because scalability enables AI applications to handle vast and varied datasets from several sources and accommodate the expanding operating sizes of businesses, scalability is essential to AI applications. Strict adherence to Service-Level Agreements (SLAs) is also necessary to meet the demands of mission-critical situations. SLAs dictate prompt detection, prompt response, and precise alerting and warning signals. To ensure that essential information is only accessed by those who are authorised and to prevent undesired access, strict processes are in place. When combined, these components give mission-critical AI systems a solid base and make them more reliable and effective in a range of high-stakes scenarios.

   In the area of mission-critical AI applications, there are similarities between many industries and scenarios. Criticality levels are often used to categories these applications, albeit they might vary depending on the sector and environment in question. Ensuring these applications operate reliably and effectively within their designated domains is the primary objective, regardless of whether they are categorised as mission-critical, time-critical, life-critical, or safety-critical. Large-scale, high-quality training data availability is a basic need for mission-critical AI systems. The accuracy and dependability of the outcomes are determined by the type and volume of data used to train these AI systems. Another characteristic that emerges as frequent is real-time decision-making, underscoring the necessity for these applications to identify problems promptly, respond appropriately, and give stakeholders trustworthy alerts and cautions. These parallels demonstrate how important real-time skills and high-quality data are for developing and deploying mission-critical AI applications across several sectors.

Opportunities and Challenges for Mission-Critical AI Applications

Mission-critical applications are starting to rely more and more on AI, but there are a variety of potential and challenges that need to be carefully evaluated. Since mission-critical apps often handle sensitive data, data security and privacy are key objectives, necessitating strong, enterprise-level protection. The processing and storing of such data raises legitimate concerns due to potential hazards and repercussions for privacy. Another challenge is that, despite AI's advancements, human involvement is still required in some circumstances. Since AI technology is still in its early stages of development and may not be able to fully replace human decision-making in all circumstances, human operators are usually required to ensure optimal functioning in mission-critical AI systems (Ranasinghe et al.,2022).

Integrating AI into existing systems is another challenge. This process needs to be carefully planned and carried out to guarantee a seamless transition without causing any disruptions. Consideration must be given to compatibility issues, bottlenecks, and the need for comprehensive staff training when incorporating AI into current procedures and systems. These problems must be overcome in order to properly employ AI in mission-critical applications. This will ensure that creativity and the realistic worries that come with working in complex contexts are balanced.

CHALLENGES AND CONSIDERATIONS

Ethical Considerations in AI Integration

It is critical to address important issues that protect people, communities, and society at large in the goal of ethical AI integration. As essential values, accountability and transparency force AI engineers to dissect the intricate inner workings of their systems. Developers need to make sure that user interfaces, thorough documentation, and visualisations clearly communicate the decision-making processes in order to maintain responsibility and build trust (Felzmann et al.,2020).

The ethical integration of AI is predicated on two fundamental pillars: transparency and accountability. The complex internal mechanisms of sizable AI systems make it difficult to comprehend how decisions are made, which breeds opacity that erodes trust and prevents accountability. To promote transparency, it is imperative for AI engineers to clearly explain the logic and principles underlying system choices via user interfaces, thorough documentation, and visualisations.

It is important to balance human control with autonomy in AI systems. AI should not replace human judgement completely; rather, it should enhance human capability and offer instruments for educated decision-making. Maintaining human supervision is necessary to avoid self-governing behaviour and potentially dangerous decisions. The appropriate and moral application of AI systems is ensured by explicit supervision protocols.

The public, researchers, policymakers, and AI developers must all be involved in the joint endeavour to address ethical challenges in AI integration. AI may benefit society without undermining fundamental rights or inflicting harm if open communication is promoted, moral growth methods are supported, and strong ethical norms are established (Wang et al.,2018).

Overcoming Challenges in Implementation

AI and other Industry 4.0 technologies are being rapidly adopted, profoundly changing the manufacturing sector. This thorough examination explores the revolutionary effects of AI on Industry 4.0 and the essential actions required overcoming deployment roadblocks.

AI's Transformational Impact on Industry 4.0

AI has the potential to completely change the manufacturing sector in a number of ways. The productivity and efficiency of the industry are greatly increased by the application of AI algorithms. Overall productivity can be significantly increased by using these algorithms to optimise industrial processes, automate repetitive tasks, and detect equipment problems. AI-driven Machine Vision systems significantly enhance quality control by instantly identifying anomalies and defects, guaranteeing constant product quality and reducing the need for thorough human inspections.

AI has an impact on supply chain management as well, where it can save costs and increase efficiency. AI enhances the efficiency and cost-effectiveness of supply chains by anticipating variations in demand, streamlining logistical processes, and optimising inventory levels. AI has the ability to analyse customer data and customise goods and services to the tastes of individual users, increasing customer satisfaction and loyalty. AI systems are able to anticipate possible issues by analysing sensor data from machinery. Predictive maintenance is made possible by this proactive strategy, which eventually minimises downtime and maximises the effectiveness of production processes (Tien et al.,2020).

Overcoming Implementation Challenges

An effective AI approach must follow a few essential steps. Businesses must, first and foremost, develop a clear AI strategy that outlines particular use cases to address pressing business issues and complements overarching organisational goals. Investing in knowledge and data infrastructure is essential to achieving these goals. This entails learning the requisite data science skills and building a solid infrastructure for the gathering, organising, and analysing of data required for AI algorithms.

The skills gap must be closed in order to implement AI effectively. To guarantee the efficient deployment and upkeep of AI systems, businesses should think about retraining current employees or recruiting certified AI specialists. Promoting the use of AI and overcoming any resistance require equally as much attention to detail as it does innovation and adaptability.

Future Trends and Evolving Technologies

A wide variety of inventions that are reshaping the globe are included in the future trends and advancing technologies. The primary topics for emerging technologies, according to the 2022 Gartner Hype Cycle for Emerging Technologies, are optimising technologist delivery, accelerating AI automation, and evolving/expanding immersive experiences. Over the next two to ten years, these technologies are anticipated to have a big impact on society and business by enabling digital business transformation and giving people more control over their experiences (Ganz et al.,2023).

The World Economic Forum also lists a number of other ways that technology might alter the world by 2027, including the development of cutting-edge technologies like Web3 and quantum, the management of flexible grids, manufacturing on demand, and the democratisation of financial wellbeing and growth through the use of AI and ML advisors (Attarha et al.,2020).

Global development, health, agriculture, and energy advancements are all supported by technological evolution. Emerging technologies can be dangerous even though they have the potential to improve people's lives. The general good impact of these advances is contingent upon their price and accessibility.

Educational technology, information technology, nanotechnology, biotechnology, robotics, and AI are just a few of the domains that fall under the umbrella of emerging technologies. These technologies, which might be disruptive or gradual, are examples of progressive developments. New technical domains emerge as a result of the confluence of many systems.

Future trends and developing technologies include a broad range of innovations that are anticipated to have a substantial impact on business, society, and the environment, among other elements of life. These technologies, which have the potential to promote positive transformation and sustainable development, vary from AI and immersive experiences to the fusion of digital, biotech, and nanotech areas.

CASE STUDIES

Real-World Applications of AI in Manufacturing

The integration of AI is driving a fundamental transition in the manufacturing industry. This technology revolution is transforming the industry in many ways, increasing production, and improving quality control. One crucial use is predictive maintenance, where AI-powered systems examine sensor data from machines to anticipate possible malfunctions and allow for proactive maintenance scheduling. Siemens provides a good example of this, as they were able to predict wind turbine component failures and reduce downtime by 20%. Quality control is another area that is heavily influenced. As demonstrated by General Electric's 30% increase in aircraft engine quality, AI algorithms examine sensor data for abnormalities and vision systems driven by AI may discover minute faults. AI has also transformed supply chain optimization, bringing with it better demand forecasting, industrial scheduling, and inventory control. Amazon, for example, has effectively utilised AI-driven demand forecasting to lower inventory costs by 10%. The amalgamation of AI and robotics is augmenting automation in perilous or intricate operations. Toyota's 20% productivity surge in automotive bodywork welding is a prime example of this. Using data from engineering testing, market research, and customer input, AI is also helpful in product creation. AI is used by Airbus to build aero plane wings, which produces lighter and stronger wings. While AI is still in its infancy, it has already shown that it has the ability to revolutionise the manufacturing industry. Going forward, this might lead to significant breakthroughs in productivity, cost reduction, and product quality (Moyne et al.,2017).

Success Stories and Lessons Learned

There are many examples of success and valuable insights to be gained from the AI in manufacturing content. Predictive maintenance using AI is one of the biggest success stories. AI is used by manufacturers to evaluate sensor data, forecast malfunctions and accidents, and lessen the impact of large-scale equipment downtime. This has resulted in improved quality, lower costs, more efficiency, and less downtime. The application of AI-powered robots with computer vision and ML algorithms to accomplish difficult jobs with accuracy and flexibility is another success story. These robots are capable of doing complex assembly tasks, performing quality control checks, and even working seamlessly in tandem with human workers. Figure 6 shows the evolution of AI applications in manufacturing.

AI integration in the industrial sector has resulted in creative solutions and optimised workflows that are transforming how businesses develop and launch new goods. Among the most important uses of AI in manufacturing are supply chain optimization, inventory control, and predictive maintenance. In order to precisely forecast demand patterns, AI systems can examine historical sales data, present stock levels, and market trends. This helps warehouses to maintain product availability while cutting carrying costs by optimising inventory levels.

Significant advancements in quality control have also resulted from the application of AI in manufacturing. AI-driven NLP and Machine Vision have revolutionised quality assurance and document processing. One rising star in AI technology is IDP, which has expedited data collection processes and decreased paperwork inefficiencies.

Industry 4.0's broader implications of AI are as important. As industrial organisations increasingly use sophisticated analytical tools, AI and data analytics are key to unlocking value in massive datasets, leading to cost reductions, safety improvements, and increased supply-chain efficiencies. Additionally, personalised client outcomes, proactive field service delivery, and the development of vital apps have all benefited from AI.

There are many examples of success and useful insights to be gained from the AI in manufacturing content. AI integration in the industrial sector has resulted in creative solutions and optimised workflows that are transforming how businesses develop and launch new goods. Among the most important uses of AI in manufacturing are supply chain optimization, inventory control, and predictive maintenance.

AI-driven NLP and Machine Vision have revolutionised quality assurance and document processing. The wider ramifications of AI in Industry 4.0 are noteworthy as well, since they support customised client outcomes, proactive field service delivery, and the development of vital apps. Figure 6 shows the Evolution of AI applications in manufacturing over the past 5 years.

CONCLUSION

Summarising the Transformative Impact of AI

AI is revolutionising a wide range of industries and reshaping the nature of work in the future. When it comes to efficiency and automation, AI has become a change agent, greatly lowering the need for human intervention, increasing productivity, and automating repetitive work in a variety of industries. Businesses use AI to optimise operations, simplifying procedures to reduce errors and increase productivity. AI

Figure 5. Evolution of AI applications in manufacturing

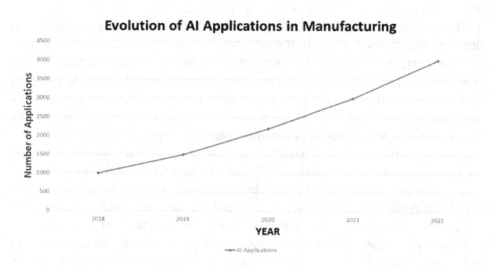

systems are pushing the envelope of human imagination in the fields of invention and creativity, producing art, music, and literature through creative processes. Furthermore, AI fosters creativity by offering fresh solutions to challenging issues via advanced algorithms and data analysis.

Particularly noteworthy areas impacted by AI include personalization and enhanced user experience. AI provides individualised experiences in industries like e-commerce, entertainment, and healthcare by customising offers to each user's preferences. Due to their responsive and user-friendly interfaces, Chabot's and virtual assistants improve user engagements. AI developments in healthcare are critical to diagnosis, treatment planning, and personalised medicine, improving patient outcomes by personalising care based on individual patient data.

Ethical issues become more pressing as AI continues to transform sectors. The need for ethical frameworks is highlighted by concerns about fairness and bias in algorithms, as well as ongoing discussions about privacy issues and the responsible use of personal data. Concerns about job displacement stemming from automation's possible influence on employment have prompted efforts to retrain workers for changing duties. But AI also help to create new employment, such as those in ethical supervision, ethical AI creation, and ethical AI maintenance.

Global collaboration is facilitated by AI as specialists and scholars work together to solve problems and exchange knowledge. However, questions have been raised concerning how the advantages of AI should be distributed fairly, emphasising the need for action to guarantee accessibility for everyone and stop the escalation of already-existing inequities. A sustainable and fair future necessitates striking a balance between innovation, ethical considerations, and inclusive accessibility when traversing the complicated terrain of AI.

Looking Ahead: The Future of AI in Manufacturing

A new era of linked systems and AI-powered intelligent manufacturing processes has been brought in by Industry 4.0 and the emergence of smart factories. AI-driven machine and system integration makes industrial processes more responsive and nimble by enabling real-time communication. AI systems facilitate predictive maintenance, which reduces downtime by predicting equipment problems and streamlining maintenance plans. Collaborative robots, or Cabot's, are a type of robotic automation that uses AI to power autonomous systems that can work with humans and adjust to changing conditions.

AI-enhanced computer vision systems revolutionise quality assurance by enabling real-time problem detection, minimising manual inspections, and guaranteeing higher-quality products. With predictive quality analytics, production data is further analysed to foresee and avert quality problems. Supply chain optimization enables enterprises to react quickly to changes in demand and logistical obstacles using AI-driven demand forecasting and dynamic supply chain management.

AI enables mass customization and flexible manufacturing lines that quickly reorganise to suit demand or accept changes in product design. Personalization and adaptive production are made possible by AI. AI algorithms that optimise energy usage and locate environmentally friendly materials for sustainable manufacturing methods address sustainability and energy efficiency.

AI is driving AR and VR technologies that improve human-machine coordination and facilitate maintenance, training, and collaboration between workers and machines. AI is essential for skill augmentation as well, freeing up workers to concentrate on more difficult jobs while computers take care of monotonous chores. AI is critical to the development of strong cybersecurity defences against cyberattacks and data breaches in the manufacturing sector. Anomaly detection algorithms scan network traffic for any peculiar patterns that might point to a security breach in manufacturing systems. The manufacturing landscape is changing as a result of the integration of AI across several Industry 4.0 components, which are bringing about previously unheard-of levels of efficiency, flexibility, quality, and sustainability.

REFERENCES

Alenizi, F. A., Abbasi, S., Mohammed, A. H., & Rahmani, A. M. (2023). The Artificial Intelligence Technologies in Industry 4.0: A Taxonomy, Approaches, and Future Directions. *Computers & Industrial Engineering, 185,* 109662. doi:10.1016/j.cie.2023.109662

Alexopoulos, K., Hribrenik, K., Surico, M., Nikolakis, N., Al-Najjar, B., Keraron, Y., & Makris, S. (2021). *Predictive maintenance technologies for production systems: A roadmap to development and implementation.*

Ameen, N., Tarhini, A., Reppel, A., & Anand, A. (2021). Customer experiences in the age of artificial intelligence. *Computers in Human Behavior, 114,* 106548. doi:10.1016/j.chb.2020.106548 PMID:32905175

Attarha, S., Narayan, A., Hage Hassan, B., Krüger, C., Castro, F., Babazadeh, D., & Lehnhoff, S. (2020). Virtualization management concept for flexible and fault-tolerant smart grid service provision. *Energies, 13*(9), 2196. doi:10.3390/en13092196

Baviskar, D., Ahirrao, S., Potdar, V., & Kotecha, K. (2021). Efficient automated processing of the unstructured documents using artificial intelligence: A systematic literature review and future directions. *IEEE Access : Practical Innovations, Open Solutions, 9,* 72894–72936. doi:10.1109/ACCESS.2021.3072900

Baygin, M., Karakose, M., Sarimaden, A., & Erhan, A. K. I. N. (2017, September). Machine vision based defect detection approach using image processing. In 2017 international artificial intelligence and data processing symposium (IDAP) (pp. 1-5). IEEE. doi:10.1109/IDAP.2017.8090292

bin Abdullah, M. R., & Iqbal, K. (2022). A Review of Intelligent Document Processing Applications Across Diverse Industries. *Journal of Artificial Intelligence and Machine Learning in Management, 6*(2), 29-42.

Çınar, Z. M., Abdussalam Nuhu, A., Zeeshan, Q., Korhan, O., Asmael, M., & Safaei, B. (2020). Machine learning in predictive maintenance towards sustainable smart manufacturing in industry 4.0. *Sustainability (Basel), 12*(19), 8211. doi:10.3390/su12198211

Felzmann, H., Fosch-Villaronga, E., Lutz, C., & Tamò-Larrieux, A. (2020). Towards transparency by design for artificial intelligence. *Science and Engineering Ethics, 26*(6), 3333–3361. doi:10.1007/s11948-020-00276-4 PMID:33196975

Ganz, C., & Isaksson, A. J. (2023). Trends in Automation. In *Springer Handbook of Automation* (pp. 103–117). Springer International Publishing. doi:10.1007/978-3-030-96729-1_5

Javaid, M., Haleem, A., Singh, R. P., & Suman, R. (2022). Artificial intelligence applications for industry 4.0: A literature-based study. *Journal of Industrial Integration and Management*, 7(01), 83–111. doi:10.1142/S2424862221300040

Kim, S. W., Kong, J. H., Lee, S. W., & Lee, S. (2022). Recent advances of artificial intelligence in manufacturing industrial sectors: A review. *International Journal of Precision Engineering and Manufacturing*, 23(1), 1–19. doi:10.1007/s12541-021-00600-3

Lee, M. S., Grabowski, M. M., Habboub, G., & Mroz, T. E. (2020). The impact of artificial intelligence on quality and safety. *Global Spine Journal*, 10(1, suppl), 99S–103S. doi:10.1177/2192568219878133 PMID:31934528

May, M. C., Neidhöfer, J., Körner, T., Schäfer, L., & Lanza, G. (2022). Applying Natural Language Processing in Manufacturing. *Procedia CIRP*, 115, 184–189. doi:10.1016/j.procir.2022.10.071

Moyne, J., & Iskandar, J. (2017). Big data analytics for smart manufacturing: Case studies in semiconductor manufacturing. *Processes, 5*(3), 39., J., & Iskandar, J. (2017). Big data analytics for smart manufacturing: Case studies in semiconductor manufacturing. *Processes (Basel, Switzerland), 5*(3), 39. doi:10.3390/pr5030039

Penumuru, D. P., Muthuswamy, S., & Karumbu, P. (2020). Identification and classification of materials using machine vision and machine learning in the context of industry 4.0. *Journal of Intelligent Manufacturing, 31*(5), 1229–1241. doi:10.1007/s10845-019-01508-6

Plathottam, S. J., Rzonca, A., Lakhnori, R., & Iloeje, C. O. (2023). A review of artificial intelligence applications in manufacturing operations. *Journal of Advanced Manufacturing and Processing*, 5(3), 10159. doi:10.1002/amp2.10159

Plathottam, S. J., Rzonca, A., Lakhnori, R., & Iloeje, C. O. (2023). A review of artificial intelligence applications in manufacturing operations. *Journal of Advanced Manufacturing and Processing*, 5(3), 10159. doi:10.1002/amp2.10159

Ranasinghe, K., Sabatini, R., Gardi, A., Bijjahalli, S., Kapoor, R., Fahey, T., & Thangavel, K. (2022). Advances in Integrated System Health Management for mission-essential and safety-critical aerospace applications. *Progress in Aerospace Sciences*, 128, 100758. doi:10.1016/j.paerosci.2021.100758

Stadnicka, D., Sęp, J., Amadio, R., Mazzei, D., Tyrovolas, M., Stylios, C., Carreras-Coch, A., Merino, J. A., Żabiński, T., & Navarro, J. (2022). Industrial needs in the fields of artificial intelligence, Internet of Things and edge computing. *Sensors (Basel)*, *22*(12), 4501. doi:10.3390/s22124501 PMID:35746287

Susto, G. A., Schirru, A., Pampuri, S., McLoone, S., & Beghi, A. (2014). Machine learning for predictive maintenance: A multiple classifier approach. *IEEE Transactions on Industrial Informatics*, *11*(3), 812–820. doi:10.1109/TII.2014.2349359

Tien, J. M. (2020). Toward the fourth industrial revolution on real-time customization. *Journal of Systems Science and Systems Engineering*, *29*(2), 127–142. doi:10.1007/s11518-019-5433-9

Wang, W., & Siau, K. (2018). *Ethical and moral issues with AI*.

Zhong, R. Y., Xu, X., Klotz, E., & Newman, S. T. (2017). Intelligent manufacturing in the context of industry 4.0: A review. *Engineering (Beijing)*, *3*(5), 616–630. doi:10.1016/J.ENG.2017.05.015

Chapter 9
Industrial Revolution (Industry 5.0) and Artificial Intelligence Technology

P. Vijayakumar
Nehru Institute of Technology (Autonomous), Coimbatore, India

S. Satheesh Kumar
Nehru Institute of Technology (Autonomous), Coimbatore, India

B. S. Navaneeth
Nehru Institute of Technology (Autonomous), Coimbatore, India

R. Anand
Nehru Institute of Technology (Autonomous), Coimbatore, India

ABSTRACT

Industry 4.0, the fourth industrial revolution, brought about a revolutionary shift in manufacturing by utilizing digital technology to improve productivity and communication. This chapter examines the deep integration of artificial intelligence (AI) technologies inside the industrial landscape as we approach Industry 5.0. Industry 5.0 represents a turn toward human-machine harmony, with artificial intelligence (AI) emerging as the key to coordinating this mutually beneficial partnership. It examines the transition from automation to cooperation, emphasizing AI technologies' role in enabling this peaceful cohabitation. This chapter explains how enhancing human capabilities with AI technology can empower workers. To ensure that the workforce continues to be a key factor in Industry 5.0, measures such as upskilling initiatives and the development of new positions that use AI are investigated. This chapter thoroughly reviews the relationship between AI and Industry 5.0, laying the basis for future research and discussion in the domains of ethics, industry, and technology.

DOI: 10.4018/979-8-3693-2615-2.ch009

INTRODUCTION

Contrary to the previous industrial revolutions, the fifth industrial revolution focused on symbiosis between humans and machines. This is because of the growth of cognitive technologies that permit greater collaboration between humans and machines. They are made up of cyber-physical cognitive systems and robots that are adaptable. These technologies in Industry 5.0 are vital to enable the digital revolution to happen without taking workers off their jobs. These tools were designed to collect, monitor, store, and analyze data from various sources in real time. The data is used to improve processes, make better decisions, and correct errors. This technology can also be used to automatize work and increase the efficiency of employees. However, their benefits are beyond productivity. They also improve health and safety security, reduce the amount of information available, and improve decisions and overall efficiency.

The latest generation of intelligent machines can communicate with each other and adapt to production requirements. This technology allows businesses to offer customers customized products and services. Furthermore, it permits firms to modify their manufacturing processes to change the needs of their customers and the changing market dynamics. Again, the intelligent robots can also be connected to other smart devices and programs to carry out complex tasks. This improves the effectiveness of manufacturing and reduces the risk of sustaining accidents. This helps speed up production and decrease costs. This can lead to an increase in the production of more goods quickly.

Furthermore, it permits the incorporation of renewable energy in manufacturing. This helps reduce the requirement for fossil fuels, as well as environmental pollution. This will also help to stop the depletion of Earth's precious resources. It is essential for sustainable development to remove the linear take-make-waste system and encourage circular economies and sustainable innovation, in addition to renewable energy. A new generation of intelligent robots may change the game in manufacturing. The goal is to transform manufacturing's future by enabling companies to offer customers a personalized experience. It can also be used to improve the quality of products and services. It can even help create safer and healthier working environments for employees. This will help achieve the highest-quality results and boost employees' morale and efficiency. In addition, it helps eliminate manual and repetitive work, allowing employees to focus on more valuable tasks.

Additionally, it allows for greater efficiency and higher quality work. A new generation of intelligent robots will likely revolutionize manufacturing by helping manufacturers reach their goals faster and more efficiently. They also hope to improve customer satisfaction and retention, which could result in more revenue. Companies must embrace the latest technology to compete against their competitors. It will

Figure 1. Overview of Industry 5.0

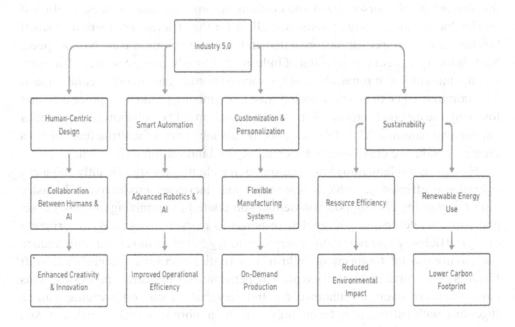

enable them to stay relevant in a constantly changing global market. Furthermore, it will assist them in attracting and keeping talent, among the most challenging issues businesses face.

The chapter provides background by discussing the technological advances altering our daily lives. It focuses on the rise of concepts such as Smart Society 5.0, Healthcare 5.0, and Agriculture 5.0, indicating a significant change in how we live and our societal activities. The emphasis is on the Industrial Revolution 5.0 (IR 5.0), regarded as the next logical step in industrialization, incorporating the most cutting-edge technologies, such as Explainable Artificial Intelligence (XAI), into everyday life to reach greater levels of wealth and intelligence.

Overview of Industry 5.0 is shown in Figure 1. It encapsulates the evolution towards a more human-centric, sustainable, and customized approach in industry and manufacturing, highlighting the integration of human needs and capabilities at the forefront of technological and industrial advancements. At its core, Industry 5.0 prioritizes human-centric design, fostering a synergistic collaboration between human intelligence and artificial intelligence (AI) to leverage the strengths of both for enhanced creativity, innovation, and decision-making. This era introduces smart automation by incorporating advanced robotics and AI, aiming to boost productivity and flexibility in manufacturing processes, enabling the execution of more complex and adaptive tasks. A hallmark of Industry 5.0 is its capacity for offering highly personalized and customized products and services, facilitated by flexible manufacturing systems that allow for the efficient production of small batches tailored to specific customer demands without compromising cost or speed. Sustainability forms a crucial pillar of Industry 5.0, emphasizing resource efficiency and the integration of renewable energy sources to mitigate environmental impact. This commitment extends to reducing waste, optimizing production processes, and lowering the carbon footprint of industrial operations. The collaborative dynamics between humans and AI and the adoption of advanced manufacturing technologies create a conducive environment for creativity and innovation to flourish.

Moreover, implementing smart automation technologies significantly enhances operational efficiency, reduces downtime, and increases productivity. Industry 5.0's adaptability also supports on-demand production, minimizing inventory costs and waste while enabling a quicker response to market changes. By prioritizing energy efficiency and renewable energy, Industry 5.0 aims to substantially reduce the environmental footprint of industrial activities, marking a significant shift towards more sustainable, efficient, and personalized industrial processes. This comprehensive overview underscores Industry 5.0's focus on merging human ingenuity with cutting-edge technology to foster a more sustainable, efficient, and customer-centric industrial landscape.

The main goal of this research is to examine the technology that is essential for Industry 5.0. It aims to identify key research areas that require greater concentration to achieve the full potential of this new revolution. The research also examines the importance of Explainable Artificial Intelligence (AI) in analyzing and implementing appropriate tools for a data-connected and Industry 5.0-compatible society.

This chapter covers a broad spectrum of topics that relate to the transformation of Industry 5.0:

- There has been a shift from traditional mass-production techniques to mass-personalization of manufacturing.
- The anticipated integration and dependence upon Cyber-Physical Systems (CPS) as well as digital twins for industrial processes.
- The advancement in intelligent factories and the effects on productivity, with a focus on the human-machine interaction.
- Opportunities and challenges in data transmission, interoperability privacy, and security issues in industry 5.0.
- The possibilities exist for Explainable Artificial Intelligence to bridge the gap between technology and human demands.

The scope of the chapter includes a look at the technological, operational, and social aspects associated with Industry 5.0, providing an in-depth overview of the current trends and the future direction.

THE EVOLUTION OF INDUSTRIAL REVOLUTIONS

The development of the Industrial Revolution, which ranged from Industry 1.0 to 4.0, was a significant change in the economic and technological environment.

1. **Industry 1.0**: This era started in the 18th century and transitioned from manual and animal-based jobs to mechanization. Industry 1.0 was marked by the development of mechanical machines and steam power, which were crucial in automating manufacturing processes and establishing the factory system era (Colombo et al., 2021).
2. **Industrial 2.0**: Starting towards the close of the late 19th century, this period saw significant technological advances, including the mass production of goods, electrification, and new transport modes. Electricity increased efficiency in manufacturing and distribution, radically changing how industries functioned (Colombo and Co. 2021).
3. **Industrial 3.0**: Emerging in the 1970s, this era was characterized by the advent of electronics, telecommunications, and computers, and this period brought complete automation and robotics in manufacturing, significantly increasing productivity and efficiency (Colombo and Co. 2021).
4. **Industrial 4.0**: In the early 21st century, industry 4.0 incorporates digital technology into manufacturing. This refers to technology such as the Internet of Things (IoT), cyber-physical systems, big data analytics, and cloud computing, transforming how industries work and interact with their surroundings (Colombo et al. 2021).

The transformation between Industry 1.0 and 4.0 shows the ever-growing integration of technology advancements into industrial processes, resulting in incredible levels of productivity, efficiency, and connectivity in manufacturing.

EMERGENCE OF INDUSTRY 5.0

Built on the foundations of Industry 4.0, Industry 5.0 is expected to enable more cooperative interaction between machines and humans. It will focus on integrating

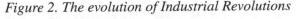

Figure 2. The evolution of Industrial Revolutions

human intelligence and AI with a focus on flexibility and sustainability and developing human innovation and creativity throughout manufacturing.

Industry 5.0 seeks to establish an equilibrium between human-centric and automated approaches, focusing on sustainable development, individualization, and social accountability. It is more than just a technological advance but also a change towards a more sustainable, human-centered industrial society.

Figure 2 above depicts the progression from Industry 1.0 to the envisioned Industry 5.0, showcasing how each industrial revolution builds upon the previous, leading to an increasingly integrated and advanced manufacturing landscape.

Industry 5.0 represents a significant change in the industry field that emphasizes human imagination and machine efficiency. What is essential to this time will be collaborative robots (cobots) that work securely alongside humans, improving productivity and ergonomics. Advanced AI in Industry 5.0 goes beyond simple automation and focuses on systems that can learn from and adjust to human input to create more efficient manufacturing. Digital twins are essential in creating dynamic digital copies of physical systems, which combine real-time data with human insight to enhance decision-making. The primary focus is sustainable production incorporating the environment and circular economic concepts. Industry 5.0 also emphasizes personalization and customization, employing techniques such as 3D printing to satisfy customer needs. The increased integration of cyber-physical systems results in more responsive manufacturing environments. Edge computing plays a crucial role in efficient data handling within intelligent factories. IoT connectivity enhances data sharing and collection, creating connected manufacturing ecosystems. Blockchain technology provides an open, secure supply chain management, ensuring data integrity and traceability. Extended Reality (XR) includes virtual, augmented, and mixed reality, which can be used to train, simulate, and improve manufacturing processes. Together, these technological breakthroughs open the way to an increasingly human-centric, durable, resilient, and sustainable industrial future in conformity with the goals of Industrial 5.0.

FUNDAMENTALS OF ARTIFICIAL INTELLIGENCE IN INDUSTRY

Artificial intelligence (AI) is revolutionizing the business world, profoundly altering how companies operate in decision-making and communicate with customers. AI is a multidisciplinary field. AI covers many methods and technologies that allow machines to mimic human intelligence, carry out tasks that previously required human brains, and gain knowledge from past experiences.

In the industry sector, AI is increasingly being integrated into various industries like manufacturing, healthcare and finance, logistics, and much more. AI in the sector presents unimaginable opportunities for efficiency, creativity, and growth. This chapter aims to give an overview of AI basics in the industrial context.

Artificial Intelligence (AI) in the industry field is shown in Figure 3. AI is a significant technological breakthrough affecting various industries with its ingenious applications.

1. **Automation and precision in production** Artificially-driven robotic arms are changing production lines by increasing efficiency and accuracy. They can perform complicated tasks with precision and minimal errors, as El-Namaki pointed out in 2019.
2. **Big Data and Analytics** AI's role in analyzing massive data sets is crucial for identifying patterns and providing information for making decisions across various industries such as retail and finance. The ability to analyze and process large amounts of data, as outlined by Anonymous in the year 2019, is essential.

Figure 3. Artificial intelligence in industry

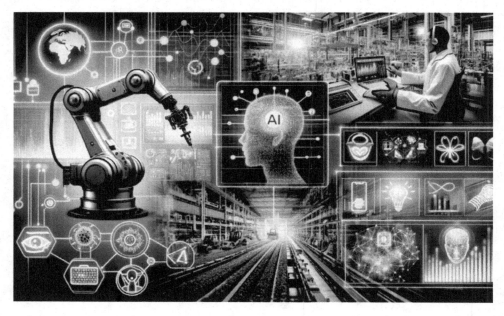

3. **AI in medicine**: AI significantly aids healthcare with diagnostics, personalized medical procedures, and surgical procedures, improving results and efficiency. Patel et al. 2022 stressed AI's capacity to analyze medical images, anticipate disease progress, and help during surgeries.

4. **AI in Agriculture**: Smart farming machines powered by AI are changing how we farm, from soil analyses to managing crops. Behgounia & Zohuri, in 2020, discussed how this will lead to more efficient farming, better yield predictions, and more sustainable farming methods.

5. **Customer Services** Chatbots and digital assistants that AI powers improve customer service by offering personalized assistance and handling customer queries, usually independently, as noted in a study by Lee & In-Seok in 2018.

The implications of AI in the field of industry show its transformational impact, increasing efficiency, precision, and decision-making across different sectors.

Artificial Intelligence in Various Sectors is shown in Figure 4. Artificial Intelligence (AI) is an evolving and multifaceted field encompassing many technologies and applications. AI is usually defined as the capacity of computers or machines to carry out tasks that generally require human intelligence, such as reasoning, learning, problem-solving perception, and understanding of language. Essential aspects of AI, as outlined by El-Namaki's research in 2019, include neural networks, machine

Figure 4. Artificial intelligence in various sectors

learning robotics, natural language processing, and computer vision, all making AI able to imitate human behaviour and intelligence. One of the most critical aspects, like the one that Dhamija & Bag highlighted in 2020, is AI's capability to learn from data through unsupervised, supervised, or reinforcement learning and its capacity to change and improve. Its interaction with AI and humans, as highlighted in the work of Nikitas et al. in 2020, is essential, particularly in how AI aids, enhances, or automates human-related tasks. Additionally, ethical issues like privacy, fairness, security, and transparency are becoming increasingly important in the development of AI and its application. The constant advancement of AI will ensure that its definitions and critical concepts are constantly improved as new technology and applications emerge.

AI TECHNOLOGIES IN MANUFACTURING

Artificial Intelligence (AI) technologies have revolutionized the manufacturing industry by transforming various applications—artificially driven robot arms and self-contained machines, such as those explained by Lee and Co. (2020). They are utilized in factories for packaging, assembly, and material handling tasks. They improve accuracy, efficiency, and speed while minimizing human error—quality control and predictive maintenance, as described by Li and Co. (2017) include AI algorithms that analyze data from machinery to anticipate maintenance requirements and spot defects, thus ensuring that breakdowns are avoided and providing better quality products—integrating sensors with intelligent technology and IoT devices as Javaid and co. (2021) mention allows instantaneous data analysis and collection within manufacturing equipment. This is essential for monitoring process operations, optimization of processes, and safety improvement. AI-driven analytics that optimizes production as described by Ding and colleagues. (2020) analyzes vast quantities of data to boost manufacturing productivity, resource management, and the optimization of supply chains. In addition, AI's contribution to product design, as outlined in Xingyu & Chuntang (2020) and Chuntang (2020), includes predictive modeling simulation and design optimization. This helps designers in developing more efficient and creative products quickly. These AI technologies are fundamental changes in the field of manufacturing, which will result in higher-end, more efficient, and genuine manufacturing processes.

Figure 5. AI technologies in manufacturing

Figure 5 above visually represents the diverse applications of AI in modern manufacturing, showcasing how AI technologies are pivotal in advancing the industry.

AI'S ROLE IN INDUSTRY TRANSFORMATION

Artificial intelligence (AI) has fundamentally changed the face of many industries due to its numerous robust applications. The process begins with data collection, in which AI systems collect massive amounts of information from various sources, laying the foundation for sophisticated analysis and understanding. The data is processed using AI algorithms to generate significant patterns and insights for educated decisions. For the optimization of processes, AI algorithms can be employed to improve and refine industrial processes, resulting in increased efficiency and efficiency. One of the most essential functions that AI plays in AI in this paradigm shift is predictive maintenance. This involves anticipating the requirements for maintenance of equipment, thereby reducing the time it takes to repair and prolonging its life. The ability to anticipate and prevent problems transforms into better decision-making, which is where AI-driven analysis provides companies with crucial insights to aid in strategic planning and implementation. When it comes to the development of products, AI contributes significantly by helping in the creation and design of new products, stimulating creativity as well as technological innovation. The final result of AI applications can be seen in the enhanced customer experience, where AI-generated insight allows the customization of services and products to meet customers' requirements and preferences, thus completing the loop of AI's transformative effects across all industries.

APPLICATIONS OF AI IN INDUSTRY 5.0

Integration of Artificial Intelligence (AI) within Industry 5.0 ushers in a new age of industrial transformation, marked by the seamless integration of digital technology and advanced manufacturing processes. AI applications within Industry 5.0 are pivotal in shaping the future of intelligent manufacturing processes, predictive maintenance, and optimizing logistics management. With the capability to process massive amounts of data and make educated decisions, AI brings efficiency, accuracy, and flexibility to industrial processes. This presentation sets the stage for a deeper exploration of the diverse uses possible with AI within the context of Industry 5.0, where the interaction between humans and machines and technological innovation lead to breakthroughs that rewrite the way we work.

Table 1. Area of expertise in smart manufacturing

Company Name	Area of Expertise in Smart Manufacturing
Siemens	Advanced automation systems, digitalization, and industrial software
General Electric (GE)	Predix platform for industrial IoT, digital twins, and analytics
Bosch	IoT sensors, automation technology, and Industry 4.0 solutions
Rockwell Automation	Integrated control and information solutions, industrial automation
Schneider Electric	Energy management and automation solutions
Honeywell	Industrial automation, control systems, and performance materials
ABB	Robotics, power, heavy electrical equipment, and automation technology
Emerson	Automation solutions for industrial, commercial, and residential markets
Mitsubishi Electric	Automation products and solutions, including PLCs and robotics
IBM	AI solutions, cloud computing, and IoT for smart manufacturing

Area of Expertise in Smart Manufacturing

Based on the information available, I can provide a table showcasing companies that are pioneers or significant contributors to intelligent manufacturing and production. However, please note that the specific companies mentioned in the smart manufacturing and production research papers are not always explicitly stated. Therefore, the table below includes well-known companies in the field of smart manufacturing based on general industry knowledge:

Table 1 represents a snapshot of key players in the smart manufacturing domain, each contributing various ways to advancing manufacturing technologies, IoT, automation, and AI-driven solutions.

Application of AI in Supply Chain Management

Each company leverages AI technology uniquely to improve efficiency, reduce costs, predict market trends, and enhance overall supply chain performance. This table is a snapshot of the diverse applications of AI in modern supply chain management.

AI Application in Quality Control and Maintenance

Based on the research available, here is a table showcasing various companies and their applications of Artificial Intelligence (AI) in Quality Control and Maintenance. Table 3 includes the specific AI applications utilized by these companies to enhance their quality control and maintenance processes:

Table 2. Application of AI in supply chain management

Company Name	Application of AI in Supply Chain Management
DHL	AI is used for predictive analytics in logistics and route optimization.
Amazon	Utilizes AI for inventory management, demand forecasting, and delivery optimization.
Maersk	Employs AI for container logistics optimization and predictive maintenance.
Walmart	Uses AI for inventory management, demand forecasting, and replenishment.
FedEx	Implements AI for route planning, logistics optimization, and customer service enhancements.
UPS	Applies AI for package sorting, optimizing delivery routes, and forecasting volumes.
Coca-Cola	Leverages AI for demand forecasting, supply chain optimization, and marketing analysis.
Nike	Utilizes AI for inventory management, personalized customer experiences, and predictive analytics in manufacturing.
Apple	Employs AI for supply chain optimization, demand forecasting, and product customization.
Procter & Gamble	Uses AI for demand planning, inventory optimization, and market trend analysis.

Table 3. AI application in quality control and maintenance

Company Name	AI Application in Quality Control and Maintenance
Central Textiles (HK) Ltd.	Integrated AI-based maintenance management system (Tu & Yeung, 1997).
Bosch	Expert systems for production planning, quality control, and preventive maintenance (Meyer & Isenberg, 1990).
Siemens	Intelligent decision support system for equipment diagnosis and maintenance management (Zhang, Tu, & Yeung, 1997).
Samsung	AI-ML image recognition for Quality Control in manufacturing (Maurya, Gaikawad, & Salvi, 2022).
General Electric (GE)	Machine learning for predictive maintenance, process control, and optimization in manufacturing (Fahle, Prinz, & Kuhlenkötter, 2020).

Each company in the table utilizes AI technologies to enhance its quality control and maintenance operations, thereby improving efficiency, reducing downtime, and ensuring higher product quality.

CHALLENGES AND OPPORTUNITIES

The risks and challenges associated with Industry 5.0 and Artificial Intelligence (AI) technology are numerous and complex, including technological, social, and cybersecurity aspects. Industry 5.0 is the introduction of disruptive technology that can have significant social and economic consequences, such as psychological effects and a possible reduction in creativity capabilities, dependence on information privacy concerns, heightened risk of data breaches, and the threat of losing control by humans over cyber systems as described by Melnyk and others. (2020). Carayannis et al. (2021) expose the difficulties of AI-driven and viral disruptions, which affect society's physical and mental aspects. They also raise worries about environmental impacts, displaced populations, virus pandemics, and changes in the governance structure. As outlined by Garg and co, the interconnectedness of industry and the dependence on AI increases cybersecurity risks and could lead to network failures because of hacking, virus threats, and security concerns of interconnected systems. (2022). 5.0's human-centric approach comes with the same problems highlighted by Nahavandi (2019) and others, such as creating jobs that complement robotics and AI and ensuring that technology does not replace human capabilities. The complex nature of AI systems calls for a rational AI (XAI) as well as also raises privacy issues, which are essential for confidence and understanding among users, significantly when AI decisions impact the lives of people, as discussed by Wang and colleagues. (2021). Lastly, Jagatheesaperumal et al. (2022) discuss data-related issues that include the management of data, availability of data bias, interpretability integrity of data, and defense against adversarial attacks. These multiple challenges emphasize the necessity of a balanced and considerate approach when implementing AI for Industry 5.0 and aligning technological advances with the needs of humans and social values.

ETHICAL CONSIDERATIONS AND WORKFORCE TRANSFORMATION

The transformation of the Industrial Revolution into Industry 5.0, combined with Artificial Intelligence advancements, is causing significant changes to the workforce and raising the issue of ethics. There's been a noticeable shift in the required skills in the workforce, brought on by the advancements in Industry 5.0 and AI technologies that are resulting in the automation of specific tasks and the creation of new positions that require advanced technical expertise and innovative problem-solving. This time is marked by a radical interaction with humans and AI that improves productivity and creativity and the increasing importance of lifelong education and learning to adapt to the ever-changing technological environment. The technological shift raises ethical issues regarding job displacement and loss, especially in sectors that depend on repetitive work, requiring ethical methods to reduce unemployment and social disruption. AI systems could develop biases based on training data, which poses issues in ensuring fairness and transparency in making decisions. Security and privacy of data are crucial, given the increasing use of data-driven technology requiring strong protection against abuse and data breaches. Furthermore, the rise of AI and Industry 5.0 calls for an effective regulatory system and governance that balances technology with ethical concerns and protects worker's rights. This convergence between the industries of Industry 5.0 and AI is not just a technological change but a societal one, which requires an equilibration approach that balances technological advances with ethical responsibilities and the overall well-being of our workforce.

CASE STUDIES ON INDUSTRY 5.0 AND ARTIFICIAL INTELLIGENCE TECHNOLOGY

Case studies provide valuable insights into the practical applications and implications of Industry 5.0 and Artificial Intelligence technology. This topic explores a few case studies that highlight different aspects of this technological evolution:

SMART MANUFACTURING: A GERMAN AUTOMOTIVE COMPANY

A well-known German automobile manufacturer has made advancements in smart manufacturing by using AI and robotics in their manufacturing lines. In this project, the most modern robots with AI capabilities were employed to complete tasks like welding and assembly, working with humans. Furthermore, AI algorithms are essential in predicting maintenance, quality control, and enhancing your supply chain. The result of this integration is positively resonant, leading to greater efficiency, reduced time to repair, and an increase in the quality of the product. This technological advancement has also enabled human workers to concentrate on more challenging and creative work and create an environment for greater creativity and innovation within the business.

AI IN HEALTHCARE: A U.S. HOSPITAL'S JOURNEY

A renowned U.S. hospital undertook a revolutionizing journey by integrating the use of artificial intelligence (AI) into its daily operations to improve the quality of care for patients and improve efficiency. The process involved the implementation of AI algorithms that provide medical support, analyze patient information, and simplify administrative tasks. In addition, robotic systems were utilized to aid during

Figure 6. Smart manufacturing: A German automotive company

Figure 7. AI in Healthcare: A U.S. hospital's journey

surgeries and logistical procedures. The results of this program were characterized by significant advances in the field of diagnostic accuracy, such as improved diagnosis, the development of customized treatment plans, and better resource management. The ethical considerations were carefully negotiated, and a commitment was made to ensure the transparent and accountable use of patient data during the use of AI technologies in healthcare.

SUSTAINABLE AGRICULTURE: SMART FARMING IN THE NETHERLANDS

In the Netherlands, a shift to sustainable agriculture has been pushed by Dutch farmers who adopted smart farming techniques aided by Artificial Intelligence (AI). This innovative approach entailed using AI with Internet of Things (IoT) devices for comprehensive crop monitoring and soil analysis and applying predictive analytics to anticipate weather conditions and crop diseases. The results of this program have been impressive, as shown by increased crop yields, reduced resource consumption,

Figure 8. Sustainable agriculture: Smart farming in the Netherlands

and a significant decrease in environmental impacts. This case study provides an enthralling example of the vital importance of AI in encouraging sustainable agriculture practices. It highlights the positive impact that technology can have on improving productivity and reducing environmental impact.

FUTURE DIRECTIONS AND RESEARCH TRENDS

Over the last ten years, it has seen an incredible increase in publications in the field of citations, publications, and research in the field of Industry 5.0 and Industry 5.0, which prompted this article to carry out an extensive study of research trends as well as the future direction of this rapidly growing area through the analysis of bibliographic coupling. The study uncovers six distinct research themes, each showing an understanding of crucial factors essential to Industry 5.0. The first, focusing on sustainability and technology, emphasizes the vital role played by technology in creating sustainable, human-centered, and resilient societies and industries by focusing on integrating and enhancing technology to meet these objectives. The second group delved into collaboration between humans and machines, highlighting the necessity to incorporate human capabilities with advanced technology to improve industrial production. This includes leveraging human-robot collaboration within manufacturing processes to enhance productivity, product quality, and customer satisfaction while reducing operating expenses and encouraging innovation. The third cluster focused on intelligent manufacturing, combining machines, smart robots, AI, and machine learning to improve efficiency in operations by automating processes, decreasing labour costs, ensuring workers' safety, and increasing product quality within a constantly changing and real-time manufacturing system. In addition, the fourth cluster focuses on the intersection with Industry 5.0 and the health industry, especially concerning the COVID-19 pandemic. It focuses on the potential of technology to provide individualized treatment and therapy processes and efficient data delivery for healthcare providers, thereby contributing to better patient outcomes and prevention of transmission of viruses. The fifth group examines the broader impact of emerging manufacturing technologies that stimulate economic growth, boost productivity, and extend applications in diverse sectors like healthcare, supply chain management, and cloud production. The authors acknowledge worries about shortages in labor security threats and privacy issues. To tackle these concerns, they suggest increasing understanding of the impact of human capital as well as the ability to innovate and establishing policies to ensure the use of technology is ethical and encouraging scholars and industrialists to think about the human aspects that the Fourth Industrial Revolution brings when conducting their research and development efforts. This holistic approach is designed to ensure that technology is developed and used to benefit all of society and to prevent the development of an artificial divide that might exclude vulnerable populations from the benefits derived from the latest technological advances.**Top of Form**

GLOBAL DIGITAL TRANSFORMATION MARKET TRENDS (UP TO 2023) – AN ANALYSIS

Digital transformation has been a key driver of change across industries worldwide, significantly impacting how businesses operate, innovate, and deliver value to customers. This transformation encompasses integrating digital technology into all company areas, fundamentally changing how they work and providing customer value. It's also a cultural change that requires organizations to continually challenge the status quo, experiment, and get comfortable with failure.

- **Rapid Growth**: The digital transformation market has multiplied due to the increasing adoption of IoT, AI, cloud computing, and big data analytics across various sectors such as healthcare, banking, manufacturing, and retail.
- **Impact of COVID-19**: The pandemic accelerated digital transformation efforts across the globe as businesses adapted to remote work, digital customer interactions, and online commerce.
- **Regional Variations**: North America, particularly the United States, has led the digital transformation market thanks to its robust tech industry, significant investments in R&D, and quick adoption of innovative technologies. Europe and Asia-Pacific have also shown strong growth, driven by government initiatives, the growing tech startup ecosystem, and increasing digital literacy among the population.

Future Projections (2023–2030):

- **Continued Expansion**: The market is expected to continue expanding as more businesses undergo digital transformation to improve efficiency, agility, and customer service.
- **Emerging Technologies**: Technologies like 5G, AI, machine learning, and edge computing are expected to play a significant role in shaping the future of digital transformation, offering new opportunities for innovation and growth.
- **Sector-Specific Growth**: While all sectors are affected by digital transformation, industries such as healthcare, finance, and manufacturing might see exceptionally high growth rates due to the increasing need for digitalization in response to global challenges and consumer demands.

Figure 9. Digital transformation market size (USD billion) by region, 2018–2030 (Polaris Market Research)

- **Regional Growth**: Asia-Pacific is expected to grow significantly due to the rapid digitalization of economies in countries like China, India, and Southeast Asia. Europe's growth will likely be driven by GDPR and other regulations that push for digital security and privacy, while North America continues to innovate at the forefront of technology.

RESEARCH GAPS AND OPPORTUNITIES

Examining the interaction between Industrial Revolution (Industry 5.0) and Artificial Intelligence (AI) technology provides a rich potential landscape but is also characterized by notable research gaps and promising possibilities. As the industry 5.0 continues to develop with the introduction of AI, it is an empowering force, enabling advances in manufacturing and industrial processes. But, identifying and fixing the current research gaps is essential to fully harness the potential this technology can bring. The need for thorough studies concerning the ethical consequences associated with AI adoption in the workplace, the effects on the employment dynamic, and the creation of robust regulatory frameworks is evident. Furthermore, there is an immense research opportunity to make it easier to integrate AI technologies with the existing infrastructures in industrial settings that ensure seamless integration while eliminating disruptions. Investigating the possible biases in AI algorithms in industrial settings and devising strategies to counter these biases is an additional avenue for further research. Furthermore, there is a pressing necessity to examine the long-term durability of Industry 5.0, augmented by AI, and analyze energy use and waste production. While Industry 5.0 and AI continue to influence our industrial environment, experts can add to the corpus of knowledge that addresses current issues and opens the way to an ethical, efficient, and sustainable future for the industrial sector.

CONCLUSION

This convergence in Industry 5.0 and Artificial Intelligence (AI) technology signals an essential era in industrial development, characterized by the shift to more human-centric approaches that place a high value on sustainability, customization, and cooperation between machines and humans. The role that AI plays in the transformation that AI plays in AI in this revolution can't be overemphasized since it can process massive datasets, enhance the process of making decisions, and accelerate technological innovation that has led to significant technological advancements in smart manufacturing, automated maintenance, and improved quality of supply chain management. The implications for the economy and society are immense. From a business perspective, the current revolution is the shift towards more efficiency in the workplace, greater flexibility, and more customization that will enable businesses to streamline processes, cut costs, and create more customized goods and solutions. In the social sphere, there are a lot of issues to consider, such as the possibility of a transformation of work, with AI performing routine tasks, and the creation of new jobs based on machines and data analysis. This is also raising concerns about job displacement, which requires the need to reskill.

The investigation of the complex tapestry created by the fusion between industry 5.0 with Artificial Intelligence (AI) technology provides a landscape that is brimming over with potential for transformation, ethical considerations, and the potential for an environmentally sustainable human-centric, efficient, and sustainable future. We are at the threshold of a new era in industrial technology. It is clear that AI is more than an automation instrument but an engine for redefining the relationship between machines and humans in fostering innovation and transforming industrial and manufacturing practices towards more personalization, sustainability, and inclusion.

The journey into the world of smart manufacturing and supply chain optimization, the control of quality, more illustrated by case studies and theoretical insights, highlights the crucial function of AI in improving the efficiency of operations in decision-making and creating value that was previously not imagined. However, this path is not without its issues and ethical dilemmas, from the loss of jobs and security concerns to the importance of ensuring that everyone has equal gain from this technological revolution. These challenges require a balanced approach in which technological advances are matched with ethical stewardship, the transformation of the workforce, and the protection of humanity's dignity.

The research and future direction developments discussed in this discussion insist on the necessity of continuous advancement, ethical reflection, and collaboration across industries, disciplines, and borders. As we explore the complexity of infusing AI into Industry 5.0, The focus should be on harnessing the power of AI to benefit the larger good and ensuring it functions as a tool for empowerment of people, preserving our environment, and helping to create an equitable and sustainable world.

In conclusion, the convergence of industry 5.0 and AI technology offers an unbeatable chance to transform the landscape of industrial production and make it more responsive to the needs of humans, adaptable to environmental demands, and a favorable environment for development and innovation. But, realizing this technology's full potential requires a concerted effort to address its challenges, especially issues concerning ethics, the development of workforces, and social equality. As we enter this new age, it is the responsibility of researchers and policymakers, as well as industry leaders and the entire society, to be engaged in continuous discussion with research, action, and dialogue to make sure that our future business is not just technologically advanced, but as well ethically-based and human-centered.

REFERENCES

Anonymous. (2019). The Revolution of Artificial Intelligence and the Significance of Adopting AI in Different Industries. Inter*national Journal of Recent Technology and Engineering*.

Behgounia, F., & Zohuri, B. (2020). Artificial intelligence integration with nanotechnology. *Journal of Nanosciences Research & Reports, 117.*

Carayannis, E. G., Christodoulou, K., Christodoulou, P., Chatzichristofis, S. A., & Zinonos, Z. (2021). Known unknowns in an era of technological and viral disruptions—Implications for theory, policy, and practice. *Journal of the Knowledge Economy*, 1–24.

Colombo, A. W., Karnouskos, S., Yu, X., Kaynak, O., Luo, R. C., Shi, Y., Leitao, P., Ribeiro, L., & Haase, J. (2021). A 70-year industrial electronics society evolution through industrial revolutions: The rise and flourishing of information and communication technologies. *IEEE Industrial Electronics Magazine, 15*(1), 115–126. doi:10.1109/MIE.2020.3028058

Dhamija, P., & Bag, S. (2020). Role of artificial intelligence in operations environment: A review and bibliometric analysis. *The TQM Journal, 32*(4), 869–896. doi:10.1108/TQM-10-2019-0243

Ding, H., Gao, R. X., Isaksson, A. J., Landers, R. G., Parisini, T., & Yuan, Y. (2020). State of AI-based monitoring in smart manufacturing and introduction to focused section. *IEEE/ASME Transactions on Mechatronics, 25*(5), 2143–2154. doi:10.1109/TMECH.2020.3022983

El Namaki, M. S. S. (2016). *How companies are applying AI to the business strategy formulation*. The Conversation.

Fahle, S., Prinz, C., & Kuhlenkötter, B. (2020). Systematic review on machine learning (ML) methods for manufacturing processes–Identifying artificial intelligence (AI) methods for field application. *Procedia CIRP, 93*, 413–418. doi:10.1016/j.procir.2020.04.109

Garg, S., Mahajan, N., & Ghosh, J. (2022). Artificial Intelligence as an emerging technology in Global Trade: the challenges and Possibilities. In *Handbook of Research on Innovative Management Using AI in Industry 5.0* (pp. 98–117). IGI Global. doi:10.4018/978-1-7998-8497-2.ch007

Jagatheesaperumal, S. K., Rahouti, M., Ahmad, K., Al-Fuqaha, A., & Guizani, M. (2021). The duo of artificial intelligence and big data for industry 4.0: Applications, techniques, challenges, and future research directions. *IEEE Internet of Things Journal, 9*(15).

Javaid, M., Haleem, A., Singh, R. P., & Suman, R. (2022). Artificial intelligence applications for industry 4.0: A literature-based study. *Journal of Industrial Integration and Management, 7*(01), 83–111. doi:10.1142/S2424862221300040

Johnson, M., & Anderson, R. (2021). AI in Healthcare: A U.S. Hospital's Journey. *Journal of Healthcare Innovation, 5*(3), 123–140.

Lee, B. R., & Kim, I. S. (2018). The role and collaboration model of human and artificial intelligence considering human factor in financial security. *Journal of the Korea Institute of Information Security & Cryptology, 28*(6), 1563–1583.

Lee, E. S., Bae, H. C., Kim, H. J., Han, H. N., Lee, Y. K., & Son, J. Y. (2020). Trends in AI technology for smart manufacturing in the future. *Electronics and telecommunications trends, 35*(1), 60-70.

Li, B. H., Hou, B. C., Yu, W. T., Lu, X. B., & Yang, C. W. (2017). Applications of artificial intelligence in intelligent manufacturing: A review. *Frontiers of Information Technology & Electronic Engineering, 18*(1), 86–96. doi:10.1631/FITEE.1601885

Maurya, P., Gaikawad, C., & Salvi, S. (2022). Visual Inspection for Industries. International Journal of Advanced Research in Science. *Tongxin Jishu.*

Melnyk, L. H., Dehtyarova, I. B., Dehtiarova, I. B., Kubatko, O. V., & Kharchenko, M. O. (2019). *Economic and social challenges of disruptive technologies in conditions of industries 4.0 and 5.0: the EU Experience.* Research Gate.

Meyer, W., & Isenberg, R. (1990). Knowledge-based factory supervision: EP 932 results. *International Journal of Computer Integrated Manufacturing, 3*(3-4), 206–233. doi:10.1080/09511929008944450

Nahavandi, S. (2019). Industry 5.0—A human-centric solution. *Sustainability (Basel), 11*(16), 4371. doi:10.3390/su11164371

Nikitas, A., Michalakopoulou, K., Njoya, E. T., & Karampatzakis, D. (2020). Artificial intelligence, transport and the smart city: Definitions and dimensions of a new mobility era. *Sustainability (Basel), 12*(7), 2789. doi:10.3390/su12072789

Patel, A. I., Khunti, P. K., Vyas, A. J., & Patel, A. B. (2022). Explicating artificial intelligence: Applications in medicine and pharmacy. *Asian Journal of Pharmacy and Technology, 12*(4), 401–406. doi:10.52711/2231-5713.2022.00061

Tu, Y., & Yeung, E. H. (1997). Integrated maintenance management system in a textile company. *International Journal of Advanced Manufacturing Technology, 13*(6), 453–461. doi:10.1007/BF01179041

Van der Burg, S., Bogaardt, M. J., & Wolfert, S. (2019). Ethics of smart farming: Current questions and directions for responsible innovation towards the future. *NJAS Wageningen Journal of Life Sciences, 90*(1), 100289. doi:10.1016/j.njas.2019.01.001

Wang, S., Atif Qureshi, M., Miralles-Pechuan, L., Reddy Gadekallu, T., & Liyanage, M. (2021). Explainable AI for B5G/6G: technical aspects, use cases, and research challenges. *arXiv e-prints.*

Xingyu, C., & Chuntang, C. (2020, November). The Development of Machinery Manufacturing And the Application Analysis of Artificial Intelligence. *Journal of Physics: Conference Series, 1684*(1), 012017. doi:10.1088/1742-6596/1684/1/012017

Zhang, J., Tu, Y., & Yeung, E. H. H. (1997, July). Intelligent decision support system for equipment diagnosis and maintenance management. In Innovation in Technology Management. The Key to Global Leadership. PICMET'97 (p. 733). IEEE. doi:10.1109/PICMET.1997.653599

Chapter 10
Unleashing the Power of Industry 5.0

A. Gobinath
Velammal College of Engineering and Technology, India

P. Rajeswari
Velammal College of Engineering and Technology, India

N. Suresh Kumar
Velammal College of Engineering and Technology, India

M. Anandan
Vel Tech Rangarajan Dr. Sagunthala R&D Institute of Science and Technology, India

R. Pavithra Devi
Velammal College of Engineering and Technology, India

ABSTRACT

This chapter discusses the convergence of the Fifth Industrial Revolution (Industry 5.0) with artificial intelligence (AI) technologies, emphasizing the revolutionary influence on industries, economies, and societal paradigms. Industry 5.0 is the most recent stage in the progression of industrial revolutions, and it is distinguished by the incorporation of smart technology, networking, and human-machine collaboration. The seamless integration of physical and digital systems lies at the heart of this revolution, resulting in a comprehensive and interconnected industrial environment. Artificial intelligence technology emerges as a key enabler of Industry 5.0, enabling intelligent automation, predictive analytics, and decision-making skills. This abstract investigates the deep implications of artificial intelligence in industrial environments, such as increased efficiency, optimal resource usage, and the development of adaptive, self-learning systems.

DOI: 10.4018/979-8-3693-2615-2.ch010

INTRODUCTION

The Industrial Revolution, a transformative period that began in the late 18th century and extended into the 19th century, marked a pivotal shift in human history, reshaping economies, societies, and lifestyles. This era unfolded against the backdrop of agrarian economies, where manual labor and traditional craftsmanship were the norm. The catalyst for change emerged in Britain, where a confluence of factors, including technological innovations, access to raw materials, and a burgeoning population, set the stage for unprecedented advancements. The transition from agrarian societies to industrialized ones was characterized by the mechanization of production processes, resulting in increased efficiency and the birth of a new socioeconomic order (Gervais, J. 2019).

Technological innovation played a central role in the Industrial Revolution, with key inventions such as the steam engine, spinning jenny, and power loom dramatically altering the landscape of production. The steam engine, pioneered by figures like James Watt, replaced traditional sources of power, enabling factories to operate at unprecedented scales. The mechanization of textile production, exemplified by inventions like the spinning jenny and power loom, revolutionized the manufacturing of fabrics. These technological advancements not only increased output but also paved the way for the factory system, a departure from cottage industries to centralized manufacturing hubs.

The shift towards industrialization was not confined to the textile sector; it permeated various industries, including coal mining, iron and steel production, and transportation. Coal became a primary energy source, powering steam engines and fueling the furnaces of burgeoning factories. The iron and steel industry experienced a surge in demand, providing the materials necessary for constructing machinery, railways, and infrastructure. The development of efficient transportation, epitomized by the steam locomotive, facilitated the movement of goods and people, connecting regions and expanding markets. These interconnected developments laid the foundation for a new economic paradigm characterized by mass production, urbanization, and the rise of a working class (Lueth, K. L, 2020).

Societal changes accompanying the Industrial Revolution were profound and multifaceted. Rural populations migrated to urban centers in search of employment opportunities in factories, leading to the rapid growth of cities. The nature of work underwent a radical transformation, as traditional artisanal skills gave way to specialized, repetitive tasks on factory assembly lines. While the Industrial Revolution fueled economic growth and technological progress, it also brought about social challenges, including poor working conditions, long hours, and exploitation of labor.

The Industrial Revolution was a watershed moment in history that reshaped the foundations of human societies. It was characterized by the widespread adoption of

Figure 1. Industry 5.0

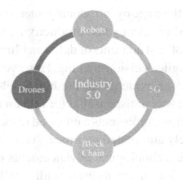

mechanized production processes, technological innovations, and the rise of industrial capitalism. This era marked the transition from agrarian economies to industrialized ones, setting the stage for the modern industrial age. The impacts of the Industrial Revolution, both positive and negative, continue to shape the trajectory of global societies and economies to this day.

Introduction to Industry 4.0 and the Evolution Towards Industry 5.0

The advent of Industry 4.0 represents a new chapter in the ongoing narrative of industrial evolution, seamlessly merging the physical and digital realms. Building upon the foundations laid by its predecessors, Industry 4.0 is characterized by the integration of advanced technologies, such as the Internet of Things (IoT), artificial intelligence (AI), machine learning, and big data analytics, into manufacturing processes. This convergence of digital technologies has ushered in an era of smart factories and connected ecosystems, where machines communicate and collaborate with each other in real-time. Industry 4.0 is not merely a technological upgrade; it signifies a paradigm shift in the way industries operate, emphasizing automation, data-driven decision-making, and the optimization of production processes.

As Industry 4.0 continues to reshape industrial landscapes, a new concept emerges on the horizon—Industry 5.0. While Industry 4.0 laid the groundwork for automation and digitization, Industry 5.0 takes a distinctive turn by placing human collaboration at its core. Unlike its predecessors, Industry 5.0 recognizes the irreplaceable role of human intuition, creativity, and emotional intelligence in the manufacturing process. This evolution signifies a move beyond pure automation, emphasizing the harmonious coexistence of human workers and intelligent machines. Industry 5.0 envisions a future where technology enhances human capabilities, fostering a symbiotic relationship that leverages the strengths of both (Kumar et.al.2018).

The journey from Industry 4.0 to Industry 5.0 reflects a maturation of industrial paradigms. Industry 4.0 set the stage by introducing interconnected systems and smart technologies that optimized efficiency and productivity. It laid the groundwork for the collection and analysis of vast amounts of data, enabling predictive maintenance, real-time monitoring, and agile decision-making. However, as industries progressed into this digital era, it became evident that the human element, with its cognitive and emotional dimensions, was indispensable for certain aspects of the production process. Industry 5.0, therefore, represents a nuanced response to the limitations and challenges posed by a purely automated approach. The evolution towards Industry 5.0 is not a rejection of the technological advancements of Industry 4.0 but rather a strategic integration of human capabilities with intelligent technologies. This evolution acknowledges that while machines excel in precision, speed, and data processing, humans bring intuition, creativity, and adaptability to the table. Industry 5.0 envisions a future where technology is harnessed to empower workers, enhance job satisfaction, and create a more inclusive and dynamic work environment. It signifies a departure from the fear of job displacement towards a collaborative model that leverages the unique strengths of both humans and machines (Díaz, M et.al, 2016).

The transition from Industry 4.0 to Industry 5.0 represents a natural progression in the ongoing saga of industrial transformation. From the digitalization and automation of Industry 4.0, we move towards a more human-centric approach in Industry 5.0, where the collaboration between humans and machines becomes the cornerstone of industrial innovation. This evolution reflects a commitment to harnessing the full spectrum of human and technological capabilities, ensuring a future where industries are not just efficient and productive but also enriching and sustainable.

Role of Artificial Intelligence in Shaping the Future of Industries

Artificial Intelligence (AI) stands at the forefront of revolutionizing industries across the globe, holding the potential to reshape the future in profound ways. Its role extends beyond mere automation, transcending traditional boundaries and permeating various sectors. At its core, AI is a catalyst for unprecedented innovation, efficiency, and decision-making capabilities. In industries, AI is becoming the driving force behind smart systems, predictive analytics, and autonomous technologies, heralding a future where machines collaborate seamlessly with human intelligence.

One pivotal aspect of AI's impact on industries lies in data analysis and decision support. As industries accumulate vast amounts of data, AI algorithms sift through this information with unparalleled speed and precision, extracting valuable insights that would be impractical for humans to discern. Machine learning algorithms, a subset of AI, enable systems to learn and adapt from data patterns, enhancing their

Figure 2. Role of AI in shaping the future of Industries

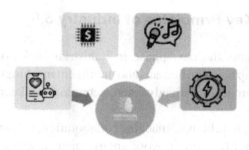

ability to make predictions and optimize processes over time. This capability is particularly potent in fields like finance, healthcare, and manufacturing, where data-driven decision-making is crucial. For instance, in finance, AI algorithms analyze market trends and predict investment opportunities, while in healthcare, AI aids in disease diagnosis and treatment planning by processing vast medical datasets. In manufacturing, predictive maintenance powered by AI ensures optimal equipment performance, reducing downtime and enhancing overall efficiency. AI, therefore, emerges as a transformative force, amplifying the capacity of industries to make informed decisions and navigate complex scenarios (Dinculeana et.al, 2019).

Moreover, AI is driving the automation revolution, transcending routine tasks and freeing human resources for more creative and strategic endeavors. In manufacturing, for example, robotic systems equipped with AI can handle intricate assembly processes with precision, speed, and adaptability. This not only enhances production efficiency but also reduces the risk of errors. AI-powered automation extends beyond the factory floor, reaching into logistics, customer service, and even creative industries. Chatbots and virtual assistants, driven by AI, provide efficient and personalized customer support, enhancing user experiences. In creative fields like content creation and design, AI algorithms assist in generating ideas, improving workflows, and even creating art. By automating routine and time-consuming tasks, AI allows human workers to focus on higher-level thinking, problem-solving, and innovation. In this way, AI is not a replacement for human labor but a collaborator, augmenting human capabilities and paving the way for a more dynamic and productive future across diverse industries.

INDUSTRY 5.0: THE HUMAN-CENTRIC REVOLUTION

Definition and Key Principles of Industry 5.0

Industry 5.0 represents the next phase in industrial evolution, emphasizing the symbiotic relationship between humans and intelligent machines. It seeks to integrate human-centered approaches with advanced technologies like artificial intelligence and automation.

Key principles include human-machine collaboration, acknowledging the unique strengths of both, and fostering a work environment where technology enhances human skills. Industry 5.0 aims to optimize processes by combining the creativity, intuition, and emotional intelligence of humans with the precision and efficiency of machines, creating a more inclusive, adaptive, and productive industrial landscape.

Importance of Human Skills, Creativity, and Intuition in the Industrial Process

In the context of Industry 5.0, the importance of human skills, creativity, and intuition in the industrial process cannot be overstated. While technology and automation have brought efficiency and precision to industries, the unique qualities of human workers play a vital role in driving innovation and ensuring the success of complex processes. Human skills encompass a range of attributes, including critical thinking, problem-solving, and effective communication. These skills are essential for decision-making, adapting to unforeseen challenges, and collaborating in dynamic work environments. The ability of humans to navigate ambiguity and exercise judgment adds a layer of adaptability that machines, despite their capabilities, currently struggle to achieve (Shivraj et.al, 2015).

Creativity is another indispensable aspect that humans bring to the industrial table. Innovations, design improvements, and novel solutions often arise from the human capacity to think outside the confines of programmed algorithms. Creative thinking becomes particularly crucial in product design, process optimization, and problem-solving, fostering a culture of continuous improvement.

Intuition, often associated with gut feelings or tacit knowledge, is a valuable asset in Industry 5.0. It allows humans to make swift decisions based on experience and contextual understanding, especially in situations where data might be incomplete or ambiguous. This intuitive capacity becomes a complement to the analytical capabilities of machines, creating a more holistic decision-making process (Sembroiz, 2018).

In essence, Industry 5.0 recognizes that the human touch is irreplaceable. By embracing and leveraging human skills, creativity, and intuition, industries can achieve a harmonious collaboration between humans and machines, resulting in a more resilient, adaptive, and innovative industrial landscape. The coexistence of these human qualities with advanced technologies ensures a balanced and sustainable approach to industrial processes in the era of Industry 5.0.

AI TECHNOLOGIES IN INDUSTRY 5.0

Industry 5.0 is witnessing a transformative impact from the latest Artificial Intelligence (AI) technologies, which are redefining the dynamics of human-machine collaboration and optimizing industrial processes. Machine Learning (ML) algorithms, a subset of AI, are central to Industry 5.0, enabling systems to learn from data and make predictions. In manufacturing, ML contributes to predictive maintenance, identifying potential equipment failures before they occur, thus minimizing downtime and optimizing production efficiency. This predictive capability extends to supply chain management, where AI-driven algorithms analyze data to anticipate demand fluctuations, enabling more agile and responsive operations.

Furthermore, AI technologies like Natural Language Processing (NLP) and computer vision are enhancing human-machine interfaces in Industry 5.0. NLP facilitates communication between humans and machines through language, enabling intuitive interactions. This is particularly valuable in collaborative settings where humans and machines need to work seamlessly. Computer vision, on the other hand, empowers machines with the ability to interpret and understand visual information. In manufacturing, this translates to quality control through image recognition, and in logistics, it aids in autonomous navigation for robots. The integration of these advanced AI technologies into Industry 5.0 not only improves efficiency and productivity but also facilitates a more intuitive and natural collaboration between humans and intelligent systems (Lee et.al, 2017).

Role of Machine Learning, Deep Learning, and Neural Networks

In the background of Industry 5.0, Machine Learning (ML), Deep Learning, and Neural Networks play pivotal roles, contributing to the evolution of smart and adaptive industrial processes. Machine Learning, a subset of artificial intelligence, involves the development of algorithms that enable systems to learn and make predictions or decisions without explicit programming. In Industry 5.0, ML is instrumental in predictive maintenance, where it analyzes historical data to predict when machinery might fail, helping companies implement proactive measures to prevent downtime. Moreover, ML algorithms contribute to quality control by identifying patterns and anomalies in manufacturing processes, ensuring the production of high-quality goods. The adaptive nature of ML allows systems to continuously learn and improve, making them well-suited for dynamic and complex industrial environments.

Deep Learning, a specialized form of machine learning, has gained prominence in Industry 5.0 for its ability to process vast amounts of unstructured data. Deep Learning models, such as neural networks with multiple layers, excel at recognizing patterns and features in complex datasets. In manufacturing, these technologies find application in image and speech recognition, enabling machines to interpret visual or auditory information. For instance, in quality control, deep learning algorithms can analyze images from production lines to detect defects or irregularities with high accuracy. Similarly, in natural language processing, deep learning enhances human-machine communication, facilitating more intuitive interactions on the factory floor. The depth and complexity of deep learning models make them adept at handling intricate tasks that were once challenging for machines to master (Zhao et.al, 2011).

Neural Networks, a fundamental component of both Machine Learning and Deep Learning, emulate the structure and function of the human brain. They consist of interconnected nodes or artificial neurons arranged in layers, with each layer contributing to the extraction of specific features from data. In Industry 5.0, neural networks are employed for tasks ranging from process optimization to autonomous decision-making. For example, in supply chain management, neural networks can analyze historical data to predict demand fluctuations, enabling more efficient inventory management. In collaborative robotics, neural networks empower machines to adapt and learn from human guidance, fostering a flexible and responsive industrial environment. The ability of neural networks to model complex relationships and adapt to changing conditions aligns with the dynamic nature of Industry 5.0, where flexibility and adaptability are paramount.

The triumvirate of Machine Learning, Deep Learning, and Neural Networks stands as a cornerstone in the edifice of Industry 5.0. These technologies bring intelligence, adaptability, and efficiency to industrial processes, fostering a collaborative ecosystem where human workers and machines operate in synergy. From predictive maintenance to quality control, from image recognition to demand forecasting, the applications of these technologies are diverse, promising a future where industries are not just automated but also intelligent and responsive to the dynamic challenges of the modern era.

SMART FACTORIES AND ADVANCED MANUFACTURING

Artificial Intelligence (AI) is catalyzing a profound transformation in traditional factories, ushering in an era where these manufacturing environments evolve into smart, connected ecosystems. This paradigm shift, often referred to as the fourth industrial revolution or Industry 4.0, leverages AI technologies to enhance efficiency, productivity, and responsiveness within the manufacturing landscape.

One key aspect of this transformation is the implementation of AI-driven predictive maintenance. Traditional factories often grapple with unexpected equipment failures leading to downtime and production losses. AI changes this by analyzing vast amounts of data generated by sensors on machinery. Machine Learning algorithms process this data to identify patterns indicative of potential failures. By predicting when equipment is likely to malfunction, proactive maintenance can be scheduled, minimizing unplanned downtime, reducing maintenance costs, and optimizing overall operational efficiency (Jiang et.al, 2016).

Another notable impact of AI in smart factories is the facilitation of real-time monitoring and control. Through the deployment of IoT (Internet of Things) devices and sensors, AI systems continuously gather data from various points in the manufacturing process. This real-time data allows for precise monitoring of production metrics, quality parameters, and equipment conditions. AI algorithms can instantly detect anomalies or deviations from set benchmarks, triggering immediate responses such as adjustments to the production process or alerts for human intervention. This level of real-time control enhances the adaptability of factories, making them more responsive to dynamic production requirements.

Moreover, AI is instrumental in automating and optimizing complex manufacturing processes. Robotic systems, guided by AI algorithms, perform tasks with precision and speed, reducing the need for manual labor in repetitive or hazardous processes. Machine Learning algorithms in robotics enable adaptive learning, allowing machines to refine their performance over time based on feedback from the production environment. This adaptability is particularly valuable in environments where product variations or customization are prevalent, as AI-driven systems can seamlessly adjust to changing requirements.

Collaborative robots, or cobots, represent another facet of AI's impact on smart factories. These robots work alongside human operators, assisting in tasks that require a combination of precision and flexibility. AI algorithms enable cobots to adapt to human movements, ensuring safe and efficient collaboration. This not only enhances productivity but also augments the overall work environment, creating a harmonious balance between human skills and machine efficiency. Furthermore, AI contributes to supply chain optimization in smart factories. By analyzing historical data and market trends, AI algorithms can forecast demand and optimize inventory levels. This predictive capability ensures that materials and components are procured efficiently, minimizing excess inventory costs and reducing the risk of shortages (Haripriya et.al, 2016).

AI is fundamentally transforming traditional factories into intelligent, interconnected ecosystems. The integration of predictive maintenance, real-time monitoring, process automation, and collaborative robotics empowers smart factories to operate with unprecedented efficiency, adaptability, and responsiveness. As industries continue to embrace these AI-driven innovations, the vision of a fully interconnected, intelligent manufacturing landscape becomes increasingly tangible, promising a future where traditional factories evolve into dynamic, smart ecosystems at the forefront of industrial progress.

Examples of AI-Driven Automation and Optimization in Manufacturing Processes

AI-driven automation and optimization are revolutionizing manufacturing processes, enhancing efficiency, reducing costs, and improving overall productivity. Here are some examples of how AI is making a significant impact in manufacturing:

Predictive Maintenance:

Application: Predictive maintenance powered by AI algorithms.

Description: AI analyzes data from sensors on machinery to predict potential equipment failures before they occur. This allows for scheduled maintenance, minimizing downtime and preventing unexpected breakdowns. For instance, General Electric uses AI for predictive maintenance in their jet engines, optimizing performance and reducing maintenance costs.

Quality Control:

Application: AI-powered image recognition for quality assurance.

Description: Machine Learning algorithms can analyze images or visual data from the production line to identify defects or irregularities in real-time. This ensures that only high-quality products reach the market. Automotive manufacturers like BMW utilize AI for visual inspection in their assembly lines, improving the accuracy and efficiency of quality control.

Process Optimization:

Application: AI-driven process optimization in chemical manufacturing.

Description: AI algorithms analyze various parameters such as temperature, pressure, and ingredient ratios in chemical processes. This optimization enhances yield, reduces waste, and ensures adherence to quality standards. Companies like BASF leverage AI to optimize chemical production processes, improving efficiency and sustainability.

Supply Chain Management:

Application: AI-driven demand forecasting and inventory management.

Description: AI algorithms analyze historical data, market trends, and external factors to predict demand fluctuations. This helps in optimizing inventory levels, preventing stockouts, and minimizing excess inventory costs. Procter & Gamble

(P&G) uses AI to optimize their supply chain, ensuring timely production and delivery of consumer goods.

Robotics and Automation:

Application: AI-guided robotic systems for complex tasks.

Description: Robots equipped with AI algorithms can perform intricate tasks with precision and adaptability. For example, in the electronics industry, robots with computer vision capabilities handle delicate tasks like soldering and assembly. Companies like Foxconn use AI-driven robots for assembling electronics components.

Energy Management:

Application: AI for energy consumption optimization.

Description: AI algorithms analyze energy usage patterns in manufacturing plants and optimize consumption based on production demand. This helps in reducing energy costs and environmental impact. Schneider Electric employs AI for energy management in their smart factories, ensuring efficient use of resources.

Collaborative Robots (Cobots):

Application: AI-driven collaborative robots working alongside humans.

Description: Cobots are equipped with AI algorithms that enable them to adapt to human movements and work collaboratively on tasks. This enhances efficiency and safety in the manufacturing process. Companies like Rethink Robotics produce collaborative robots that can be easily programmed and integrated into various manufacturing tasks.

These examples demonstrate the versatility of AI in transforming various facets of manufacturing, from maintenance and quality control to supply chain management and collaborative robotics. As technology continues to advance, AI-

Figure 3. Concept of cobots

driven innovations are likely to play an even more prominent role in shaping the future of manufacturing processes.

Increased Efficiency, Reduced Downtime, and Improved Quality Through AI

Artificial Intelligence (AI) is a transformative force in manufacturing, delivering substantial benefits in terms of increased efficiency, reduced downtime, and improved product quality. One of the key contributions of AI lies in predictive maintenance. By leveraging machine learning algorithms to analyze data from sensors on machinery, manufacturers can anticipate potential equipment failures before they occur. This proactive approach minimizes unplanned downtime, as maintenance activities can be scheduled precisely when needed. As a result, the overall efficiency of manufacturing processes is significantly enhanced, and production continuity is optimized (Kim et.al, 2019).

Furthermore, AI's impact on quality control is profound. Through AI-powered image recognition and data analysis, manufacturers can ensure stringent quality standards. AI algorithms can swiftly identify defects or irregularities in real-time, allowing for immediate corrective actions. This not only reduces the likelihood of producing faulty products but also contributes to the overall improvement of product quality. The precision and speed with which AI can analyze vast datasets make it an invaluable tool in maintaining consistently high-quality standards throughout the manufacturing process.

In essence, the integration of AI in manufacturing translates into a more efficient and reliable production environment. The predictive capabilities of AI reduce the disruptions caused by equipment failures, leading to increased operational efficiency. Simultaneously, AI's role in quality control ensures that each product meets stringent standards, contributing to enhanced overall product quality. Together, these advancements underscore the transformative impact of AI on the manufacturing landscape, ushering in an era of optimized processes, reduced downtime, and elevated product quality (BizIntellia, 2018).

HUMAN-MACHINE COLLABORATION

The significance of human-machine collaboration in Industry 5.0 is pivotal, marking a departure from the purely automated approaches of previous industrial revolutions. In this era, the integration of advanced technologies, such as artificial intelligence and robotics, is designed to complement and amplify human capabilities rather than replace them. Industry 5.0 recognizes the unique strengths that each brings to the table, emphasizing a symbiotic relationship where humans and machines work together seamlessly.

Human-machine collaboration is essential for tasks that require emotional intelligence, creativity, and complex decision-making—qualities that machines currently struggle to replicate. While machines excel in repetitive and precise tasks, humans possess the ability to interpret nuanced situations, exercise judgment, and navigate complex, unstructured environments. By leveraging the strengths of both, Industry 5.0 aims to create a dynamic and adaptive manufacturing environment.

Moreover, the collaborative model enhances overall efficiency and productivity. Machines can handle routine and physically demanding tasks, freeing up human workers to focus on higher-level thinking, problem-solving, and innovation. This not only results in a more engaging and fulfilling work environment for human workers but also ensures that machines are used to their full potential, enhancing the overall capabilities of the manufacturing process.

In Industry 5.0, human-machine collaboration goes beyond mere coexistence—it fosters a synergy where the combination of human intuition, creativity, and adaptability with the precision and speed of machines leads to unprecedented levels of productivity and innovation. This collaborative approach not only addresses the challenges posed by automation but also paves the way for a more inclusive, dynamic, and sustainable future for industries.

Examples of How AI Augments Human Capabilities in the Workplace

AI serves as a powerful tool in augmenting human capabilities across various workplace scenarios, enhancing efficiency, and fostering innovation. One notable example is in healthcare, where AI assists medical professionals in diagnosing and treating diseases. Machine learning algorithms analyze vast datasets of medical records and imaging results, providing doctors with valuable insights for accurate and timely diagnoses. AI does not replace the expertise of healthcare professionals but acts as a supportive tool, improving diagnostic accuracy and enabling more personalized treatment plans. In the realm of customer service, chatbots and virtual assistants powered by AI enhance human interactions. These systems handle routine inquiries, freeing up human agents to focus on more complex and emotionally nuanced customer issues. AI-driven language processing allows these virtual assistants to understand and respond to user queries, creating a more efficient and responsive customer service experience (Aujla et.al, 2018).

Furthermore, in the field of creative work, AI aids professionals in content creation and design. For instance, AI algorithms can assist graphic designers by generating design suggestions based on input criteria. This accelerates the creative process, allowing designers to iterate and refine ideas more rapidly. The AI's ability to analyze vast datasets also contributes to market research and trend analysis, providing creatives with valuable insights to inform their work.

In these examples, AI acts as a supportive collaborator, augmenting human capabilities rather than replacing them. The synergy between human expertise and AI-driven efficiency leads to enhanced productivity, improved decision-making, and a more dynamic and adaptable work environment. As AI technology continues to advance, its role in augmenting human capabilities across diverse industries is likely to expand further, shaping a future where humans and intelligent systems work together harmoniously (Shi, W, 2019).

Case Studies Showcasing Successful Implementations of AI In Predictive Maintenance

Case Study 1: General Electric (GE) Aviation
   Background:
   General Electric (GE) Aviation, a leading aerospace manufacturer, faced challenges in minimizing downtime and optimizing the maintenance of its aircraft engines. Unplanned maintenance events not only disrupted flight schedules but also incurred significant costs. GE Aviation sought to leverage AI for predictive maintenance to enhance the reliability and efficiency of its engines.
   Implementation:
   GE Aviation implemented an AI-driven predictive maintenance solution called the Digital Twin. This involved creating a virtual replica of each aircraft engine using sensor data collected during its operation. Machine Learning algorithms were then applied to analyze this data in real-time, monitoring various parameters such as temperature, pressure, and vibration. The AI system learned the normal operating patterns of the engines and could detect deviations that might indicate potential issues.
   Results:
   The Digital Twin predictive maintenance system enabled GE Aviation to anticipate engine component failures before they occurred. By analyzing historical data and continuously learning from the performance of various engines, the AI system could provide early warnings for maintenance requirements. This proactive approach significantly reduced unplanned downtime, allowing airlines to plan maintenance activities during scheduled intervals. GE Aviation reported a substantial decrease in maintenance costs, improved engine reliability, and enhanced overall operational efficiency.
   Case Study 2: Siemens Gamesa Renewable Energy
   Background:

206

Siemens Gamesa, a global leader in renewable energy solutions, faced challenges in maintaining the optimal performance of its wind turbines. The company sought to enhance its predictive maintenance capabilities to minimize downtime, improve turbine reliability, and optimize maintenance costs.

Implementation:

Siemens Gamesa implemented an AI-driven predictive maintenance solution that incorporated advanced analytics and machine learning. The solution utilized sensors embedded in the turbines to collect data on various operational parameters, including rotor speed, temperature, and vibration. This data was then fed into AI algorithms that could analyze patterns and detect anomalies indicative of potential failures.

Results:

The AI-driven predictive maintenance system allowed Siemens Gamesa to move from a reactive to a proactive maintenance model. By predicting potential component failures, the company could schedule maintenance activities precisely when needed, minimizing turbine downtime. This led to increased availability of the wind turbines and improved overall energy output. Siemens Gamesa reported a substantial reduction in maintenance costs, optimized spare parts inventory, and enhanced customer satisfaction through improved turbine performance.

PREDICTIONS FOR THE FUTURE OF INDUSTRY 5.0 AND AI INTEGRATION

The future of Industry 5.0 promises a seamless integration of Artificial Intelligence (AI) into the fabric of manufacturing and industrial processes, fundamentally reshaping the way businesses operate. Several key predictions highlight the trajectory of this evolution and the potential impact on industries worldwide.

Advanced Human-Machine Collaboration:

Prediction: Industry 5.0 will witness even more advanced forms of human-machine collaboration, where AI systems become intuitive partners in decision-making processes.

Explanation: As AI technologies mature, they will become more adept at understanding human inputs, context, and preferences. The collaboration between humans and AI will evolve to a point where machines anticipate needs, offer real-time insights, and seamlessly integrate into the decision-making workflow. This advanced collaboration will result in more efficient and adaptive manufacturing environments.

AI-Driven Innovation Hubs:

Prediction: Industry 5.0 will foster the emergence of AI-driven innovation hubs, where research and development centers focus on pushing the boundaries of AI applications in manufacturing.

Explanation: Organizations will establish dedicated hubs to explore novel ways AI can enhance manufacturing processes. These hubs will serve as incubators for new AI algorithms, technologies, and applications, driving continuous innovation. The collaboration between industry players, academia, and AI specialists will fuel the development of cutting-edge solutions, positioning companies at the forefront of technological advancements.

Personalized and Sustainable Production:

Prediction: AI integration in Industry 5.0 will enable personalized and sustainable production, where manufacturing processes are tailored to individual preferences and environmental considerations.

Explanation: AI algorithms will analyze vast datasets to understand consumer preferences, allowing for the customization of products on a large scale. Simultaneously, AI will optimize production processes for sustainability, minimizing resource usage, waste, and energy consumption. This personalized and sustainable approach aligns with the growing demand for eco-friendly practices and consumer-centric production models.

Autonomous Decision-Making:

Prediction: Industry 5.0 will witness the rise of autonomous decision-making, where AI systems make complex decisions without human intervention.

Explanation: AI algorithms will continuously learn from vast datasets, historical patterns, and real-time data streams, allowing them to autonomously make decisions. This autonomy will extend to areas such as supply chain management, quality control, and process optimization. The result will be a more agile and responsive industrial ecosystem capable of adapting to dynamic market conditions.

Ethical AI Governance:

Prediction: The integration of AI in Industry 5.0 will prompt a heightened focus on ethical AI governance, ensuring responsible and transparent use of AI technologies.

Explanation: As AI systems become more embedded in critical industrial processes, there will be an increased emphasis on ethical considerations. Industries will develop robust frameworks for AI governance, addressing issues such as bias, privacy, and accountability. Ethical AI practices will become integral to Industry 5.0 strategies, fostering trust among stakeholders and mitigating potential risks.

The future of Industry 5.0 and AI integration holds great promise for transformative advancements in manufacturing and industry. The seamless collaboration between humans and machines, coupled with continuous innovation, personalized production, autonomous decision-making, and ethical governance, paints a picture of a dynamic, efficient, and sustainable industrial landscape. As these predictions unfold, industries embracing these technological trends are poised to lead the way in shaping the future of manufacturing and production.

CONCLUSION

In conclusion, the fusion of Research and Development (R&D) with the next phase of industrial evolution is not just a strategic choice; it is an imperative for sustained growth, competitiveness, and societal progress. R&D activities empower industries to push the boundaries of innovation, adopt emerging technologies, and optimize processes for efficiency and sustainability. As industries embark on the transformative journey of Industry 5.0, the commitment to robust R&D not only positions them as pioneers in technological advancements but also underscores a commitment to ethical practices, environmental responsibility, and the well-being of the workforce. The dynamic interplay between R&D initiatives and the evolving industrial landscape propels us towards a future where industries are not only efficient and resilient but also conscientious contributors to global progress.

Looking ahead, the symbiotic relationship between R&D and industrial evolution will continue to shape a landscape characterized by adaptability, technological prowess, and a commitment to creating positive societal impacts. As industries navigate the complexities of an ever-changing global environment, the strategic integration of R&D into business models will be the linchpin for staying ahead of the curve. The future unfolds with promise, driven by the collaborative efforts of researchers, innovators, and industry leaders who recognize the transformative potential of R&D in shaping a more sustainable, efficient, and ethically conscious industrial ecosystem.

REFERENCES

Aujla, G. S., Chaudhary, R., Kumar, N., Das, A. K., & Rodrigues, J. J. P. C. (2018). SecSVA: Secure storage, verification, and auditing of Big Data in the cloud environment. *IEEE Communications Magazine*, 56(1), 78–85. doi:10.1109/MCOM.2018.1700379

BizIntellia. (n.d.). *Benefits of Using Cloud Computing for Storing IoT Data*. BizIntellia. https://www.biz4intellia.com/blog/benefits-of-using-cloudcomputing-for-storing-IoT-data/

Díaz, M., Martín, C., & Rubio, B. (2016). State-of-the-art, challenges, and open issues in the integration of Internet of things and cloud computing. *Journal of Network and Computer Applications, 67*, 99–117. doi:10.1016/j.jnca.2016.01.010

Dinculeana, D., & Cheng, X. (2019). Devices, vulnerabilities and limitations of MQTT protocol used between IoT. *Applied Sciences (Basel, Switzerland), 9*(5), 848. doi:10.3390/app9050848

Gervais, J. (2019). *The future of IoT: 10 predictions about the Internet of Things*. Norton. https://us.norton.com/internetsecurity-iot-5-predictions-for-the-future-of-iot.html

Haripriya, A. P., & Kulothugan, K. (2016). ECC based self-certified key management scheme for mutual authentication in Internet of Things. In: *2016 International Conference on Emerging Technological Trends (ICETT),* (p. 1–6). Kollam, India: IEEE. 10.1109/ICETT.2016.7873657

Jiang, Q., Ma, J., Wei, F., Tian, Y., Shen, J., & Yang, Y. (2016). An untraceable temporal-credential-based two-factor authentication scheme using ECC for wireless sensor networks. *Journal of Network and Computer Applications, 76*, 37–48. doi:10.1016/j.jnca.2016.10.001

Kim, H., Kim, D. W., Yi, O., & Kim, J. (2019). Cryptanalysis of hash functions based on blockciphers suitable for IoT service platform security. *Multimedia Tools and Applications, 78*(3), 3107–3130. doi:10.1007/s11042-018-5630-4

Kumar, J. S., & Zaveri, M. A. (2018). Clustering approaches for pragmatic two-layer IoT architecture. *Wireless Communications and Mobile Computing, 2018*, 1–17. doi:10.1155/2018/8739203

Lee, C.-N., Huang, T.-H., Wu, C.-M., & Tsai, M.-C. (2017). The Internet of Things and its applications. In *Big Data Analytics for Sensor-Network Collected Intelligence* (pp. 256–279). Elsevier. doi:10.1016/B978-0-12-809393-1.00013-1

Lueth, K. L. (2020, November). *State of the IoT 2020: 12 billion IoT connections, surpassing non-IoT for the first time.* IoT Analytics. https://iot-analytics.com/state-of-the-iot-2020-12-billion-i ot-connections-surpassing-non-iot-forthe-first-time/

Sembroiz, D., Ricciardi, S., & Careglio, D. (2018). A novel cloud-based IoT architecture for smart building automation. In *Security and Resilience in Intelligent Data-Centric Systems and Communication Networks* (pp. 215–233). Elsevier. doi:10.1016/B978-0-12-811373-8.00010-0

Shi, W., Pallis, G., & Xu, Z. (2019). Edge computing. *Proceedings of the IEEE*, *107*(8), 1474–1481. doi:10.1109/JPROC.2019.2928287

Shivraj, V. L., Rajan, M. A., Singh, M., & Balamuralidhar, P. (2015). One time password authentication scheme based on elliptic curves for Internet of Things (IoT). In: *2015 5th National Symposium on Information Technology: Towards New Smart World (NSITNSW),* (pp. 1–6). Riyadh, Saudi Arabia: IEEE.

Zhao, G., Si, X., & Wang, J. (2011). A novel mutual authentication scheme for Internet of Things. In: *Proceedings of 2011 International Conference on Modelling, Identification and Control,* (pp. 563–566). Shanghai, China: IEEE. 10.1109/ICMIC.2011.5973767

Chapter 11
Precision Paradigm:
AI-Infused Evolution of Manufacturing in Industry 5.0

Dhinakaran Damodaran
https://orcid.org/0000-0002-3183-576X
Vel Tech Rangarajan Dr. Sagunthala R&D Institute of Science and Technology, India

A. Ramathilagam
P.S.R. Engineering Collge, India

Udhaya Sankar S. M.
https://orcid.org/0000-0003-0199-3502
R.M.K. College of Engineering and Technology, India

Edwin Raja S
https://orcid.org/0000-0002-2948-9669
Vel Tech Rangarajan Dr. Sagunthala R&D Institute of Science and Technology, India

ABSTRACT

In the era of Industry 5.0, the integration of artificial intelligence (AI) transforms manufacturing, reshaping operational efficiency, maintenance practices, and production dynamics. This chapter delves into the core of smart factories and intelligent manufacturing, exploring the synergy of AI, machine learning, and real-time data analytics that underpins Industry 5.0. From shop floors to supply chains, AI optimization drives data-driven decision-making, creating a precision manufacturing landscape for maximum efficiency. The chapter illuminates how AI revolutionizes equipment management through proactive predictive maintenance, reducing downtime and enhancing sustainability. AI's impact extends to manufacturing flexibility, where

DOI: 10.4018/979-8-3693-2615-2.ch011

smart factories adapt seamlessly to dynamic market demands, ensuring adaptability and agility. Through case studies and industry applications, this chapter unveils a future where AI and manufacturing converge, defining a transformative era of precision, adaptability, and efficiency in Industry 5.0.

INTRODUCTION

The rich tapestry of human history is woven with the threads of industrial revolutions, each leaving an indelible mark on the way we work, live, and interact with the world. At the heart of this historical continuum lies the concept of the Industrial Revolution, a transformative force that has shaped civilizations across centuries. This introduction serves as a gateway to explore the historical evolution of industry, with a particular focus on the epochal developments known as the Industrial Revolution, which unfolded in the late 18th and early 19th centuries. The Industrial Revolution stands as a testament to the ceaseless march of progress, marking a pivotal moment in human history where the foundations of modern industrialized societies were laid. At its core, this transformative period witnessed a seismic shift from agrarian and craft-based economies to industrial ones. Central to this metamorphosis were the groundbreaking advancements in steam power, machinery, and the mechanization of production processes (Bednar et al., 2020). To comprehend the profound impact of the Industrial Revolution, it is imperative to delve into the tangible manifestations of these epochal changes. Real-time examples abound in our contemporary world, illustrating the enduring influence of the Industrial Revolution on the fabric of societies and economies.

Consider the advent of steam power as a catalyst for monumental change. In the 18th century, James Watt's steam engine revolutionized industries such as textiles, transportation, and manufacturing. Fast forward to the present, and we witness a parallel transformation powered by the digital revolution. The evolution of computing and information technology mirrors the revolutionary impact of steam power, propelling industries into new realms of efficiency and interconnectedness. The mechanization of production processes during the Industrial Revolution laid the groundwork for the assembly line, a concept pioneered by Henry Ford in the early 20th century. Today, we witness a continuum of this innovation in smart factories, where artificial intelligence and automation redefine manufacturing processes in real-time. Factories equipped with interconnected sensors and AI algorithms optimize production, exemplifying the enduring legacy of mechanization. Moreover, the socioeconomic landscape underwent a profound reconfiguration during the Industrial Revolution. The shift from agrarian societies to urbanized industrial centers was not without challenges. Similarly, the present-day world grapples with the consequences of rapid urbanization, with smart cities emerging as a response to the need for sustainable and efficient urban living.

The Industrial Revolution's influence extends beyond economic realms to redefine societal structures. The advent of the factory system transformed labor dynamics, giving rise to new working classes and labor movements. Today, debates surrounding the impact of automation on employment echo the concerns raised during the Industrial Revolution, highlighting the enduring relevance of these historical shifts (Sankar et al., 2023). In essence, the Industrial Revolution serves as a touchstone for understanding the trajectory of human progress. Its echoes reverberate through time, resonating in the interconnected networks of the digital age, the efficiency of modern manufacturing, and the complexities of contemporary urban living. By examining real-time examples, we bridge the temporal gap, illuminating the enduring legacy of the Industrial Revolution and its ongoing influence on the course of history. This exploration invites us to navigate the continuum of progress, acknowledging the transformative power of industry and its perpetual role in shaping the world we inhabit.

BACKGROUND

Classic Manufacturing Environment

The Figure 1 shows a representation of a classic manufacturing environment from the work level/field level to the enterprise resource planning (ERP) management tool (ERP). The diagram is divided into four layers:

Work level/field level: This layer contains the sensors and actuators that interact with the physical manufacturing process.

Control level: This layer contains the programmable logic controllers (PLCs) and other devices that control the manufacturing process.

Process line level: This layer contains the manufacturing execution system (MES) and other devices that monitor and control the manufacturing process at the process line level.

ERP level: This layer contains the ERP system, which manages the overall manufacturing operation.

The layers are interconnected, and data flows between them in both directions. For example, the MES sends data to the ERP system about the production status, and the ERP system sends data to the MES about production orders.

The following is a more detailed explanation of each layer:

Figure 1. Classic manufacturing environment

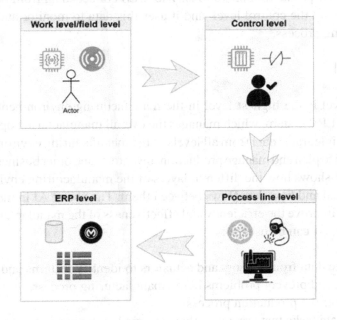

Work Level/Field Level

The work level/field level is the lowest level of the manufacturing environment. This layer contains the sensors and actuators that interact with the physical manufacturing process. Sensors collect data about the manufacturing process, such as temperature, pressure, and speed. Actuators control the manufacturing process, such as turning on and off motors and valves.

Control Level

The control level is the next layer in the manufacturing environment. This layer contains the PLCs and other devices that control the manufacturing process. PLCs are programmable devices that can be used to control a variety of manufacturing processes, such as assembly lines, robots, and machine tools.

Process Line Level

The process line level is the next layer in the manufacturing environment. This layer contains the MES and other devices that monitor and control the manufacturing process at the process line level. The MES is a software system that helps manufacturers to manage their production operations. The MES collects data from the PLCs and other devices on the control level, and it uses this data to monitor and control the manufacturing process.

ERP Level

The ERP level is the highest level in the manufacturing environment. This layer contains the ERP system, which manages the overall manufacturing operation. The ERP system integrates data from all levels of the manufacturing environment, and it uses this data to plan and manage production, inventory, and other business processes. The Figure 1 shows how the different layers of the manufacturing environment are interconnected and how data flows between them. The use of AI in manufacturing is helping to improve the efficiency and effectiveness of the manufacturing process. For example, AI can be used to:

- Analyze data from sensors and actuators to identify patterns and trends.
- Predict and prevent problems in the manufacturing process.
- Optimize the production process.
- Automate tasks that are currently performed by humans.
- AI is still in its early stages of development in manufacturing, but it has the potential to revolutionize the industry.

Industrial Revolutions

The Figure 2 depicts the progression of industrial revolutions, highlighting the key technological advancements and shifts in production processes that have transformed industries throughout time.

Figure 2. Industrial revolutions

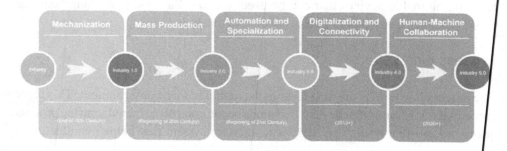

Industry 1.0: Mechanization (End of 18th Century)

The advent of steam engines marked the onset of Industry 1.0, ushering in an era of mechanized production. Factories emerged as hubs of manufacturing, relying on steam power to drive machinery and automate repetitive tasks.

Industry 2.0: Mass Production (Beginning of 20th Century)

Electricity revolutionized manufacturing in Industry 2.0, enabling the development of assembly lines and mass production techniques. Henry Ford's Model T epitomized this era, showcasing the efficiency and scale of mass-produced goods.

Industry 3.0: Automation and Specialization (Beginning of 21st Century)

Automation and computerization took center stage in Industry 3.0, leading to the digitization of production processes and the rise of specialized machines. Robotics and computer-aided design (CAD) became commonplace, transforming industrial practices.

Industry 4.0: Digitalization and Connectivity (2013+)

The Fourth Industrial Revolution, or Industry 4.0, saw the pervasive integration of digital technologies into manufacturing. The Internet of Things (IoT), cloud computing, and big data analytics revolutionized communication, collaboration, and decision-making in factories.

Industry 5.0: Human-Machine Collaboration (2020+)

Industry 5.0, also known as the "Symbiotic Revolution," emphasizes the harmonious integration of humans and machines. Artificial intelligence (AI), machine learning (ML), and augmented reality (AR) are employed to augment human capabilities and foster a collaborative environment. The Figure 2 effectively illustrates the continuous evolution of industrial technologies, highlighting the transformative impact of each revolution on manufacturing practices and the broader economy.

From Industry 4.0 to Industry 5.0

In the contemporary landscape, we find ourselves entrenched in the fourth industrial revolution, aptly termed Industry 4.0. This phase is distinguished by the amalgamation of digital technology, the Internet of Things (IoT), and artificial intelligence (AI) into the fabric of industrial and manufacturing processes (Srinivasan et al., 2023). It represents a transformative era where automation and connectivity redefine the operational paradigms of industries. Industry 4.0 introduces a realm of interconnected systems where machines, devices, and systems communicate seamlessly, creating an intricate network known as the Industrial Internet of Things (IIoT). The pillars of this revolution include advanced analytics, facilitating data-driven decision-making; automation and robotics, enhancing efficiency and reducing manual labor; and the evolution of traditional manufacturing plants into smart factories, where cyber-physical systems monitor and control physical processes in real-time.

As technology accelerates at an unprecedented pace, a new era, Industry 5.0, emerges. This evolution transcends mere automation, embodying a holistic approach to technology that revolves around the symbiotic relationship between humans and machines. Unlike its predecessors, Industry 5.0 recognizes and embraces the distinctive strengths of both human and machine intelligence. Artificial Intelligence assumes a pivotal role in this paradigm shift, acting as the catalyst propelling Industry 5.0 into uncharted territory. Here, the focus shifts from mere efficiency to fostering collaboration and coexistence between human ingenuity and machine capabilities, paving the way for a more interconnected and intelligent future.

Intelligent manufacturing, arising from the seamless integration of manufacturing and information technology, stands as the cornerstone of the latest industrial revolution. The advent of the new era of artificial intelligence introduces novel features, including deep learning, cross-disciplinary integration, human-machine collaboration, and group intelligence (Cullen-Knox et al., 2017). These attributes provide humanity with a fresh perspective for comprehending intricate systems and a transformative technology capable of reshaping nature and society. The profound amalgamation of the new generation of AI technology with advanced manufacturing forms a paradigmatic shift, giving rise to a new era in intelligent manufacturing technology. This phenomenon is not only pivotal to the ongoing industrial revolution but also serves as the driving force behind the fourth industrial revolution. Despite the overarching significance of intelligent manufacturing, the metallurgical industry finds itself in a critical phase, navigating the transition from the initial phase of intelligent manufacturing, known as digital manufacturing. The integration of the new generation of artificial intelligence technology into this sector remains at a nascent stage. Urgency underscores the need for a comprehensive examination and analysis of the key technologies underpinning the new generation of intelligent manufacturing, aligning these insights with the specific characteristics of the metallurgical industry. Yan et al. (2022) delves into the present developmental landscape of the metallurgical industry, scrutinizing the application and potential of key technologies such as intelligent machine tools, intelligent inspection technology, digital twin, and 3D printing technology within the metallurgical domain. The findings hold practical significance for steering the metamorphosis of the metallurgical industry toward intelligent transformation, serving as a valuable reference for stakeholders and decision-makers in related fields.

Wajid et al. (2022) represents the next transformative phase in the industrial landscape. While Industry 5.0, along with its precursor, share many concepts, Industry 5.0 is differentiated by three key components: adaptability, environmental responsibility, and human centricity. The author presents a cutting-edge digital manufacturing infrastructure framework that expands on Industry 4.0 principles. The objective is to enable AI-driven decision support while integrating fundamental components of reliability and human-centeredness, in line with the fundamental tenets of Industry 5.0. The suggested design acts as a link, guaranteeing a smooth transition and upholding a careful balance between using the apparent advantages of AI-centric modernization and retaining people's crucial role in important decision-making processes. By doing so, it addresses the evolving requirements of Industry 5.0, where the fusion of artificial intelligence with industrial processes necessitates a framework that not only optimizes efficiency but also prioritizes human involvement and upholds trust in decision-making processes.

Del Real Torres et al. (2022) addresses the contemporary challenges in intelligent manufacturing within the industry. Their work meticulously outlines the current status and potential of revolutionary technologies associated with I4.) and I5.0. The focus of the review is on artificial intelligence (AI), with particular attention given to Deep Reinforcement Learning (DRL) algorithms, recognized as an ideal solution to address the unpredictability and volatility prevalent in modern demand scenarios. The exploration of DRL algorithms involves a thorough examination of Reinforcement Learning (RL) concepts. The review traces their development, progressing through Artificial Neural Networks (ANNs) towards the advanced domain of DRL, emphasizing the expansive potential and versatility inherent in these algorithms. Significantly, given their data-based nature, the review delves into the modification of these algorithms to align with the specific operational requirements of the industry. Additionally, the review incorporates innovative concepts such as digital twins, responding to the absence of an environment model. The discussion highlights how the integration of digital twins further enhances the performance in addition to the application of DRL algorithms in intelligent manufacturing. The study highlights the proven suitability of DRL for several aspects of manufacturing industry operations and its advantages over traditional approaches. Notably, DRL is recognized for significantly enhancing the manufacturing process's robustness and flexibility. In order to ultimately realize the promise of these transformative tools, the analysis closes by highlighting the need for significant efforts in the higher education and industry spheres. One of the most critical steps in moving toward the anticipated I5.0 technological advancement is to start implementing them in the industrial sector.

In order to close an important expertise gap, Ghobakhloo et al. (2023) have conducted a crucial study that outlines the creation of a strategic plan that outlines the ways in which Industry 5.0 may advance sustainable manufacturing. The research begins with a content-focused examination of the literature, identifying 12 crucial roles that Industry 5.0 may play to fully improve sustainable manufacturing. Subsequently, a strategic roadmap is crafted, elucidating the intricate contextual relationships among these functions and providing insights into their synergistic utilization to maximize their impact on sustainability. A thorough examination of 85 works covering the whole automotive production process was carried out by Konstantinidis et al. (2023). Their research revealed that machine vision is essential to all of the scientific aspects of Industry 4.0, including the production of additives, virtual realities, independent robotics, and proactive upkeep. Artificial vision system computing methodologies and design elements were clustered as a result of the examination of 22 vision-based solutions spanning 47 automobile components. These methods included threshold-based approaches, as well as advanced reinforcement learning strategies, fit for the Industry 5.0 setting.

The authors suggest the Industry 5.0 Technology Readiness Evaluation Standard for machine vision technologies based on the knowledge they gained from their investigation. With this methodology, which systematically assesses nine practical elements spanning five escalating technological stages, the field of machine vision competence in the framework of Industry 5.0 may be evaluated in an organized manner. It is an effective instrument for pinpointing areas of weakness and potential for development, providing direction for the smooth incorporation of vision technology into a smart manufacturing environment. Becker et al. (2021) focuses on two critical research topics crucial for dispersed manufacturing: establishing a Chain of Trust (CoT) and using AI/ML to analyze and optimize production processes. In order to apply these ideas to a microelectronics manufacturing environment, they present the concepts of both CoT along with AI-based process analysis. The study focuses in particular on two standard methods: Solder Ball Application, which uses a high-mix/low-volume idea, along with Surface Mount Device (SMD) assembly, which uses extensively autonomous manufacturing equipment. Additionally, the digitalization process is designed to enhance overall product and process quality. The research underscores the importance of a Chain of Trust and the application of artificial intelligence in achieving a safer and more efficient distributed manufacturing environment.

In the framework of Industry 5.0, Salam et al. (2023) describe a novel deep-learning (DL) approach intended for the identification of web-based assaults. This study investigates the efficacy of several deep learning approaches in identifying suspicious patterns and categorizing attacks, such as parametric models, RNNs, and CNNs. When contrasted with both established DL techniques and conventional machine learning techniques, the suggested transformer-based system performs better. This dominance is observable in measurements like recall, accuracy, and precision. The results of the study highlight the effectiveness of DL, as seen by the noteworthy effectiveness of the suggested transformer-based system. A novel Industry 5.0 value-driven production automation ecological system, spearheaded by Javed et al. (2023), is defined by edge automation platforms constructed on localized clouds and employing an architecture centered around services. The innovation goes further by integrating cloud-based collaborative learning (CCL) in a number of domains, such as manufacturing line management, construction energy management, and logistic robot supervisors, along with assistance for human workers inside local clouds. This makes it easier to collaborate and share knowledge when developing manufacturing procedures.

Consequently, the process management system can produce the best possible process recipes that are driven by Industry 5.0. In addition to discussing energy oversight for environmental strategies that are sustainable and guaranteeing efficient, reliable, and economical manufacturing processes, the study highlights how critical it is to put the welfare of human workers first. This work has significant implications for future work because the suggested ecosystem can be implemented and evaluated in a variety of industrial use cases. This is a significant step forward in the development of flexible and effective Industry 5.0 manufacturing settings that incorporate automation, teamwork, and human welfare. The blockchain intelligent contract pyramid-driven multi-agent independent system for process control, created by Jiewu et al. (2023), is a ground-breaking method for improving the responsiveness and flexibility of oversight in resilient customized fabrication. This strategy is built on an agent-based architecture that uses blockchain technology and encapsulates factories as agents. In the multi-agent system, blockchain intelligent agreements are essential because they facilitate peer-to-peer negotiating and coordination of tasks.

To enable quick dispatching for each individual requirements and quickly dynamic schedule adjustments in reaction to internal interruptions, a quad-play blockchain intelligent contracts pyramid with decentralized oversight mechanisms is proposed. This innovative methodology represents a significant stride toward achieving resilient and individualized manufacturing. The integration of blockchain technology, smart contracts, and multi-agent systems, coupled with the designed pyramid structure and decentralized control patterns, underscores the adaptability of the manufacturing process. It empowers the system to promptly respond to individualized demands and unforeseen disruptions. The experiments conducted within the ManuChain system offer practical insights into the real-world implications and performance of the proposed approach.

In the context of Industry 5.0, Agote-Garrido et al. (2023) investigate the fundamental idea of social metabolism as a cornerstone for forming sociotechnical systems. The research carefully examines current sociotechnical design techniques and methodologies, illuminating works that use these structures to incorporate Industry 4.0 innovations into manufacturing workflows. The three key components of Industry 5.0 as well as the enabling infrastructure made possible by Industry 4.0 technologies are also examined. The researchers suggest an empirical framework for the design of industrial systems based on these discoveries. This model addresses the key elements of the Industry 5.0 framework while including social-technical structures to smoothly integrate the Industry 4.0 facilitating advancements. The suggested approach emphasizes how important it is to take sociotechnical systems into account early on in the design process, giving human-centricity, ecological responsibility, and robustness in manufacturing systems first priority.

The proposed model facilitates the implementation of an industrial setting that is in line with social needs by endorsing this comprehensive approach. It encourages an aware and flexible sector that prioritizes organizational resilience, conservation of the environment, along with well-being for humans in addition to embracing technology breakthroughs. Margherita et al. (2023) introduce a pioneering Smart Manufacturing Systems Design framework geared towards Industry 5.0. Augmented Digital Twin is the central idea of this architecture, capable of seamlessly integrating and digitizing all entities within the factory, including machines, robots, environments, interfaces, and people. This comprehensive digital representation facilitates the development of AI-driven applications aimed at supporting the user domain, fostering a co-evolution between humans and machines through systematic data sharing among physical in addition to the digital assets.

The framework enables the creation of a symbiotic relationship where machines and humans actively contribute to knowledge generation and learning from each other. This dynamic interaction promotes a virtuous co-evolution, optimizing factory productivity and enhancing workers' well-being. The framework's design specifically focuses on understanding the intricate interplay between humans and machines, fostering effective collaboration between individuals and Smart Manufacturing Systems (SMSs). To validate the effectiveness of the SMSD framework, the study involved collaboration with four industrial companies spanning diverse sectors. These companies shared a common interest in transcending the limitations of Industry 4.0 lines by incorporating human factors into the management of future SMSs. The framework's conception and validation represent a significant step toward the integration of human-centric considerations in the development and management of smart industrial systems for Industry 5.0.

Designed for the Industry 5.0 environment, Jain et al. (2022) presents an inventive framework called AI-based UAV-borne Protect Communicating with

Categorization. This model consists of two main stages: categorization based on DL and secure transmission depending on the steganography of images. Initially, the AIUAV-SCC model uses a new method of image steganography that combines encryption procedures, nano-bacterial colony optimisation for the best pixel selection, along with multilayer discrete wavelet processing. Then, in the second phase, the framework classifies securely obtained UAV images using the Bayesian optimization (BO)--b-based the SqueezeNet model. To assess the performance of the proposed model, extensive simulations are conducted using the UC Merced dataset (UCM) aerial dataset, and the results are scrutinized across various dimensions. The outcomes underscore the effectiveness of the AIUAV-SCC model, showcasing its superiority over compared methods when applied to the test UCM aerial dataset. This framework demonstrates promise in ensuring secure communication and efficient classification within the Industry 5.0 landscape, particularly in the context of UAV-borne applications.

Iqbal et al. (2022) put forth a comprehensive framework for the emerging technologies of I5.0. The study delves into the extension of I4.0 facilitated by I5.0 and examines the evolving role of humans in contributing to the manufacturing processes within this paradigm. The framework looks at important competencies that are important and desirable for workers in the manufacturing sector. The research examines the relationship between the Covid-19 pandemic and I5.0 as well as how this technological transformation fits together with the Sustainable Development Goals. The authors clarify how I5.0 might be conceived to support the goals of equitable growth. Lastly, the paper discusses issues that the industrial sector is facing and positions them as areas for additional investigation for I5.0. By presenting a comprehensive framework, Iqbal et al. contribute to the understanding of Industry 5.0's trajectory, emphasizing the human element, skill requirements, and the interconnectedness with global challenges and sustainable development goals.

Imran et al. (2022) identify significant fields of study that require more focus while doing an extensive evaluation of the Industry 5.0 supporting technologies. The transition of the manufacturing procedure from large-scale manufacturing to large-scale customization is discussed, as well as the expected use of digital twins and Cyber-Physical Systems (CPS). Finding the loopholes that prevent the current industrialization from reaching its full potential is the goal. The authors aim to bridge existing gaps within societal domains to facilitate the realization of Industry 5.0. Additionally, the research explores the potential of Explainable Artificial Intelligence (XAI) as a new domain. The study thus contributes by delving into various research challenges and opportunities associated with Industry 5.0, providing insights that can guide further developments in this transformative era. In a groundbreaking attempt to provide a thorough analysis of I5.0, Yadav et al. (2022) use insights learned from research to shed light on possible uses and accompanying technology.

The study presents new terms and concepts that are crucial to comprehending I5.0 while anchoring their investigation in historical contexts, such as instances from past industrial revolutions and viewpoints from capitalists. The study explores I5.0's specific uses, looking at areas including managing assets, cloud production, and smart healthcare, amid others as well. The study also looks at enabling technologies, which are essential to the support of I5.0 applications. Through this comprehensive approach, Yadav et al. contribute valuable insights and perspectives to the discourse around Industry 5.0, shedding light on its applications and the diverse array of technologies that can drive its successful implementation.

Challenges and Critiques During the Transition

The transition from Industry 4.0 to Industry 5.0, while promising unprecedented advancements in manufacturing, is accompanied by a set of challenges and critiques. The integration of diverse technologies, including Artificial Intelligence (AI), Internet of Things (IoT), and edge computing, introduces complexities in ensuring seamless functionality. Data security and privacy concerns arise as Industry 5.0's interconnected ecosystem generates vast amounts of data, necessitating robust cybersecurity measures. Workforce adaptability becomes crucial as employees need to navigate working alongside advanced technologies, necessitating skill set adaptations and comprehensive training programs. Ethical considerations, such as bias in AI algorithms and responsible deployment, gain prominence, demanding a balance between technological innovation and ethical responsibility. The absence of clear regulatory frameworks and interoperability challenges hinders the transition, posing risks to responsible and ethical practices. The upfront costs associated with implementing Industry 5.0 technologies, coupled with the need for a cultural shift towards collaboration and adaptability, present additional hurdles for organizations. Successfully addressing these challenges is imperative for unlocking the full potential of Industry 5.0 and ensuring a sustainable and responsible transition.

Industry 5.0: A Paradigm Shift

As the curtain rises on Industry 5.0, a new era unfolds — one that transcends the limitations of its predecessors and ushers in a paradigm shift with profound implications for the industrial landscape. At the heart of this transformation lies a triad of key principles, each shaping the essence of Industry 5.0 and redefining the relationship between technology and humanity. Industry 5.0 is distinguished by its unwavering focus on synergy — the harmonious collaboration between humans and machines. Unlike the automation-centric approaches of previous industrial revolutions, Industry 5.0 seeks to augment human capabilities through the seamless

Figure 3. Industry 5.0 principles

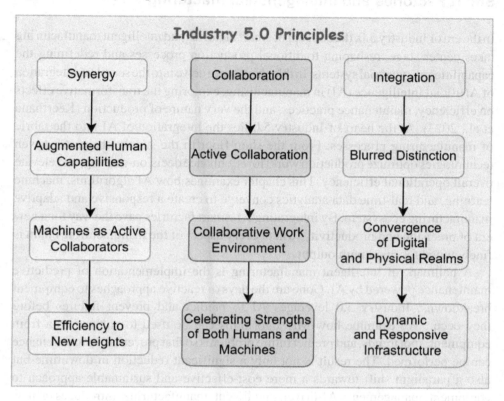

integration of artificial intelligence and robotics. Machines are not mere tools but active collaborators, working alongside human counterparts to enhance productivity, creativity, and problem-solving. This collaborative ethos not only propels efficiency to new heights but also fosters a work environment where the unique strengths of both humans and machines are celebrated.

The Figure 3 depicts the industry 5.0 Principles. In the tapestry of Industry 5.0, the distinction between the physical and digital realms becomes increasingly blurred. The integration of digital technologies with physical processes creates an interconnected ecosystem where data flows seamlessly between the virtual and tangible worlds. This convergence empowers real-time decision-making, predictive analytics, and a level of connectivity that was once the realm of science fiction. From smart factories to intelligent supply chains, Industry 5.0 paints a landscape where the boundaries between bits and atoms dissolve, giving rise to a dynamic and responsive industrial infrastructure.

Smart Factories and Intelligent Manufacturing

In the era of Industry 5.0, the concept of smart factories and intelligent manufacturing takes center stage, reshaping traditional production processes and redefining the capabilities of industrial systems. In this chapter, we delve into the seamless integration of Artificial Intelligence (AI) in manufacturing, exploring the transformative effects on efficiency, maintenance practices, and the very nature of production (Keerthana et al., 2023). At the heart of Industry 5.0 lies the integration of AI into the fabric of manufacturing processes. From the shop floor to the supply chain, AI-driven technologies optimize production workflows, enhance decision-making, and elevate overall operational efficiency. This chapter examines how AI algorithms, machine learning, and real-time data analytics converge to create a responsive and adaptive manufacturing ecosystem. By integrating AI, smart factories pave the way for a new era of precision and productivity, where every aspect of the manufacturing chain is finely tuned for maximum output.

A hallmark of intelligent manufacturing is the implementation of predictive maintenance powered by AI. Gone are the days of reactive approaches to equipment breakdowns; Industry 5.0 leverages AI to predict and prevent failures before they occur. We examine how ML algorithms can be used to analyze data from equipment, spot trends, and predict future problems so that preventative maintenance can be performed. The result is not only a significant reduction in downtime but also a paradigm shift towards a more cost-effective and sustainable approach to equipment management. AI-driven intelligent manufacturing introduces a new era of customization and flexibility (Harini et al., 2023). Smart factories adapt to dynamic market demands by leveraging AI to tailor production processes to specific needs. Whether it's producing small batches of highly customized products or swiftly adapting to changes in design specifications, AI enables manufacturing systems to be agile and responsive. This chapter explores how Industry 5.0 fosters a manufacturing environment where customization and flexibility are not just options but integral components of the production landscape.

TRANSFORMATIVE IMPACT ON INDUSTRIES AND SOCIETY

The transformative wave unleashed by Industry 5.0 extends its influence far beyond the confines of factory settings, leaving an indelible mark on entire industries and societal structures. Far more than a mere technological progression, Industry 5.0 represents a societal metamorphosis that intricately reshapes how businesses operate, supply chains function, and consumers experience life. In this exploration, our journey into the realm of Industry 5.0 will be enriched with real-time examples and applications, vividly illustrating its profound impact on both industries and society. This paradigm shift is evident in the redefinition of business models, where Industry 5.0 champions a collaborative approach between humans and machines. Picture collaborative robots, or cobots, seamlessly working alongside human counterparts on manufacturing floors, exemplifying optimized efficiency and precision in tasks that require the delicate touch of human expertise.

The transformative influence extends its tendrils into supply chain dynamics, where the integration of real-time data analytics and AI-driven logistics revolutionizes the movement of materials from production to consumption (Kumar et al., 2023). Imagine predictive analytics in supply chain management, enabling anticipatory adjustments to demand fluctuations and significantly reducing waste through precise inventory management. Consumer experiences undergo a profound transformation in the era of Industry 5.0, with products and services tailored through AI-driven insights. Think of online retailers utilizing AI algorithms to analyze consumer behavior, offering personalized recommendations, and thereby elevating the overall user experience. In the realm of education, Industry 5.0 prompts a reevaluation, demanding a skill set that harmonizes with intelligent technologies. Visualize virtual reality (VR) and augmented reality (AR) applications in education, providing students with immersive learning experiences and preparing them for a future driven by technology.

Healthcare undergoes a revolution with AI-driven diagnostics, telemedicine, and precision medicine. Envision AI algorithms analyzing medical images for early disease detection, resulting in quicker diagnoses and personalized treatment plans. Beyond technological advancement, Industry 5.0 emerges as a potent force in addressing global challenges. Consider collaborative efforts utilizing AI and data analytics for climate monitoring, resource optimization, and disaster response, contributing to a sustainable and resilient future. At the heart of Industry 5.0 lies human-machine collaboration, transforming manufacturing processes in real-time. Witness smart factories employing AI-driven systems for predictive maintenance, reducing downtime, and optimizing production schedules. The convergence of physical and digital realms materializes in smart cities that leverage data for efficient urban living. Picture smart city initiatives utilizing IoT sensors and AI algorithms to manage traffic flow, reduce energy consumption, and enhance overall urban sustainability.

THE RISE OF ARTIFICIAL INTELLIGENCE

As Industry 5.0 unfolds, it is propelled by the unprecedented rise of Artificial Intelligence (AI), a force that not only amplifies the capabilities of machines but fundamentally redefines the nature of industrial processes. In this chapter, we embark on a journey through the realms of AI, exploring the technologies that underpin

Industry 5.0, understanding their intricacies, and witnessing their transformative impact on the world of industry. At the core of Industry 5.0 lies a tapestry of AI technologies, each contributing to the intelligence that permeates the industrial landscape. From ML algorithms that enable systems to learn from data as well as improve over time, to computer vision that grants machines the ability to interpret and understand visual information, and natural language processing that facilitates communication between humans and machines—these technologies converge to create a dynamic and adaptive framework. Reinforcement learning, robotics, and sensor technologies further augment the AI arsenal, collectively forging a path towards a future where machines not only assist but actively collaborate with human counterparts.

AI's cornerstone, machine learning, enables systems to recognize patterns in data and come to well-informed conclusions without the need for explicit programming. Deep learning, a subclass of machine learning, is based on neural networks, which are inspired by the neural architecture of the human brain (Sai Aswin et al., 2023). With their multi-layered neural networks, DL algorithms are exceptional at digesting enormous volumes of data, identifying complex patterns, and providing insights that were previously unattainable. In Industry 5.0, these technologies become catalysts for innovation, enabling predictive maintenance, quality control, and adaptive manufacturing processes. The impact of AI on Industry 5.0 is not confined to theoretical frameworks but manifests in tangible and transformative applications across various sectors. Explore with us as we delve into real-world success stories where AI-driven solutions revolutionize manufacturing, logistics, healthcare, and more. Witness how predictive maintenance algorithms optimize machinery uptime, how neural networks enhance quality control in production lines, and how natural language processing facilitates intuitive human-machine interfaces. These applications not only showcase the potential of AI but also underscore its ability to drive efficiency, sustainability, and innovation across diverse industries.

AI-INFUSED EVOLUTION OF MANUFACTURING IN INDUSTRY 5.0

In the landscape of Industry 5.0, the pivotal integration of artificial intelligence (AI) stands as a cornerstone, orchestrating a profound transformation in manufacturing methodologies. The component diagram serves as a visual narrative, unraveling the intricate tapestry of key components and their interconnected relationships, offering a vivid portrayal of the dynamic and adaptive essence inherent in an AI-infused manufacturing ecosystem as shown in Figure 4. Within this evolving paradigm, Smart Factories emerge as the nerve centers, harnessing the power of sensors, IoT devices, and edge computing to usher in a new era of data-driven decision-making. The AI Integration Layer acts as the intelligence hub, where machine learning algorithms and data analytics converge to unlock unprecedented insights and efficiencies (Dhinakaran et al., 2023). Production Processes undergo a paradigm shift with the infusion of AI-driven Robotics, Adaptive Manufacturing, and Predictive Modeling, fostering agility and precision. Supply Chain Integration, fortified by AI-enhanced logistics, demand forecasting, and AI-driven procurement, achieves heightened responsiveness to market dynamics.

The Human-Machine Interface (HMI) takes center stage, adorned with augmented reality, natural language processing, and intuitive control, harmonizing the collaboration between human ingenuity and machine capabilities. Meanwhile, the Data Security and Governance realm, fortified by blockchain technology and robust cybersecurity measures, ensures the sanctity and resilience of critical information. As this component diagram unfolds, it encapsulates not just a technical blueprint but a narrative of innovation, efficiency, and adaptability that defines the essence of Industry 5.0—a narrative where artificial intelligence becomes the catalyst for a manufacturing revolution.

Smart Factories

Smart Factories, the vanguards of modern manufacturing, are meticulously equipped with an array of technological marvels that redefine industrial processes. Among these, Sensors and IoT Devices serve as vigilant sentinels, strategically deployed across the factory landscape. They diligently capture real-time data on a spectrum of parameters, encompassing machine performance, environmental conditions, and the nuanced quality of the end product. This granular data not only paints a vivid picture of the manufacturing environment but also becomes the bedrock for informed decision-making. At the forefront of efficiency enhancement is Edge Computing, a revolutionary paradigm that transcends traditional data processing. By enabling data to be processed closer to its source, Edge Computing minimizes latency, fostering real-time responsiveness. This not only accelerates decision-making but also optimizes resource utilization, marking a paradigm shift in the speed and precision of data-driven insights.

Underpinning the seamless orchestration of these technological marvels is Connectivity Infrastructure—a robust network that forms the backbone of the Smart Factory. This intricate web of connectivity ensures fluid communication between devices, creating a dynamic ecosystem where information flows seamlessly. It is this interconnectedness that propels the Smart Factory into an era of heightened productivity, where every component collaborates harmoniously, ushering in a new age of precision and efficiency in manufacturing processes.

Figure 4. AI-infused evolution of manufacturing in Industry 5.0

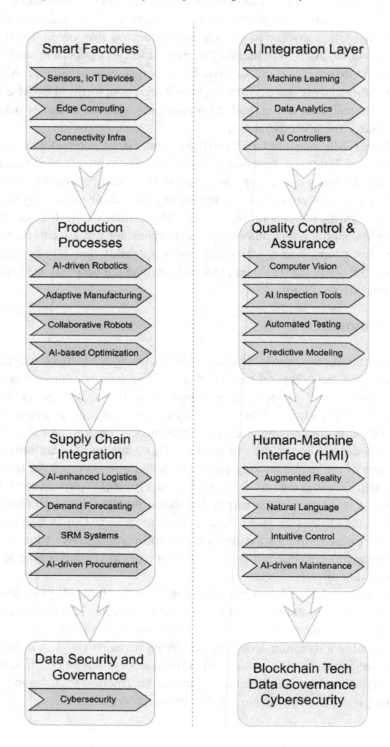

AI Integration Layer

Machine Learning, a cornerstone in the realm of artificial intelligence, empowers machines to evolve and adapt through the utilization of sophisticated algorithms and models. This transformative technology enables machines to glean invaluable insights from data, fostering an iterative learning process. As machines assimilate and analyze data patterns, their predictive and decision-making capabilities undergo continuous refinement, laying the groundwork for more informed and precise outcomes in various manufacturing scenarios. Data Analytics, a powerhouse of computational prowess, undertakes the formidable task of processing extensive datasets inherent in modern manufacturing environments. Through intricate analysis, this discipline sifts through the data deluge, distilling it into meaningful and actionable insights (Udhaya Sankar et al., 2023). These insights become the compass guiding optimization strategies for manufacturing processes. By uncovering patterns, trends, and correlations, Data Analytics paves the way for strategic decision-making, resource allocation, and efficiency enhancement within the manufacturing landscape.

At the heart of this orchestrated intelligence is AI Controllers, serving as the central command that harmonizes the symphony of components within the manufacturing ecosystem. These controllers act as the intelligence nexus, facilitating the seamless flow of information and instructions among diverse components. Through real-time coordination, AI Controllers optimize the collaboration between Machine Learning algorithms and Data Analytics insights, ensuring a dynamic and responsive manufacturing environment. In essence, AI Controllers emerge as the maestros, conducting the manufacturing orchestra toward greater efficiency, adaptability, and innovation.

Production Processes

In the realm of Industry 5.0, AI-driven Robotics emerges as a transformative force, infusing robots with advanced AI capabilities to elevate their precision and adaptability in diverse manufacturing tasks. By seamlessly integrating artificial intelligence into robotic systems, this technology empowers machines to learn, adapt, and execute tasks with a level of sophistication that significantly enhances the efficiency and accuracy of manufacturing processes. Adaptive Manufacturing represents a paradigm shift, offering real-time adjustments to manufacturing processes based on continuous data inputs. This dynamic approach enables the system to adapt swiftly to changing conditions, optimizing resource utilization and overall efficiency. The integration of Adaptive Manufacturing heralds a new era where production processes seamlessly align with the demands of the moment, fostering agility and responsiveness.

Collaborative Robots, or cobots, epitomize the symbiotic relationship between human operators and robotic precision. These robots work alongside their human counterparts, leveraging their strengths to create a harmonious and synergistic collaboration. Through this partnership, tasks are executed with a balance of human dexterity and robotic accuracy, enhancing overall productivity and workplace safety. AI-based Optimization takes center stage in the pursuit of operational excellence. Leveraging sophisticated AI algorithms, this component tirelessly analyzes and refines production processes to achieve peak efficiency. By identifying patterns, anomalies, and opportunities for improvement, AI-based Optimization contributes to the reduction of waste and the enhancement of overall operational effectiveness.

Predictive Modeling, powered by predictive analytics, propels manufacturing into a proactive realm of decision-making. This component anticipates future outcomes by analyzing historical data and patterns. By forecasting potential scenarios, Predictive Modeling equips decision-makers with insights to make informed choices, reducing uncertainties and fostering a more resilient and forward-thinking manufacturing environment. Together, these components weave a narrative of innovation and efficiency, where AI-driven Robotics, Adaptive Manufacturing, Collaborative Robots, AI-based Optimization, and Predictive Modeling converge to redefine the possibilities within the manufacturing landscape. This convergence not only optimizes processes but also lays the foundation for a manufacturing ecosystem that is agile, responsive, and poised for continual advancement in the era of Industry 5.0.

Supply Chain Integration

Supply Chain Integration in Industry 5.0 stands as a testament to the transformative impact of artificial intelligence, reshaping the traditional logistics and procurement landscape. Here, the integration of advanced technologies plays a pivotal role in optimizing and fortifying the efficiency of supply chain operations.

AI-enhanced Logistics

At the forefront, AI-enhanced Logistics revolutionizes the movement of goods and materials. Applying cutting-edge AI algorithms, logistics operations are streamlined and optimized. This not only accelerates the transit of goods but also ensures resource-efficient routes, minimizing costs and environmental impact. The result is a supply chain that operates with unprecedented efficiency and responsiveness to dynamic market conditions.

Demand Forecasting

The predictive prowess of Demand Forecasting is a linchpin in the Supply Chain Integration process. By leveraging AI to analyze market trends, historical data, and various influencing factors, this component accurately predicts market demands. This foresight allows for the adjustment of production schedules in real-time, mitigating the risks of overstock or shortages. The seamless synchronization between demand forecasting and production planning ensures a finely tuned and responsive supply chain.

SRM Systems (Supplier Relationship Management)

In the intricate web of global supply chains, Supplier Relationship Management (SRM) takes center stage. AI-driven SRM Systems manage relationships with suppliers, fostering transparency, reliability, and collaboration. Through real-time data analysis, these systems optimize supplier interactions, ensuring a continuous and resilient supply chain. This not only enhances operational efficiency but also minimizes disruptions, contributing to a robust and dependable supply chain ecosystem.

AI-driven Procurement

Automating and optimizing procurement processes, AI-driven Procurement injects efficiency into vendor selection and negotiation. By analyzing vast datasets, this component identifies optimal suppliers, negotiates favorable terms, and ensures compliance. The result is a streamlined and cost-effective procurement process that contributes to overall supply chain resilience and agility. In concert, these components create a harmonized supply chain ecosystem where AI-enhanced Logistics, Demand Forecasting, SRM Systems, and AI-driven Procurement converge. This convergence not only improves operational efficiency but also positions the supply chain as a strategic asset capable of adapting to the dynamic demands of Industry 5.0. In this era of intelligent manufacturing, Supply Chain Integration becomes a linchpin, ensuring that the flow of materials and goods is not just efficient but anticipatory and adaptive.

Human-Machine Interface (HMI)

In the intricate dance between human operators and advanced technologies, the Human-Machine Interface (HMI) in Industry 5.0 emerges as the nexus, reshaping how humans interact with and perceive the manufacturing environment. Here, a suite of sophisticated components within HMI not only enhances user experiences but also elevates the efficiency and responsiveness of manufacturing processes.

Augmented Reality (AR)

Augmented Reality revolutionizes the user's perception by seamlessly overlaying digital information onto the physical manufacturing environment. This immersive technology enhances the user's understanding of complex processes, providing real-time insights and visualizations. Whether guiding operators through intricate tasks or offering dynamic information displays, AR within HMI transforms the manufacturing landscape into an interactive and information-rich space.

Natural Language Processing (NLP)

Enabling seamless communication between humans and machines, Natural Language Processing breaks down communication barriers. By interpreting and responding to human language, machines become more intuitive to interact with, fostering a user-friendly environment. NLP within HMI transforms the way operators engage with manufacturing systems, making the communication process more natural and accessible.

Intuitive Control

At the heart of HMI, Intuitive Control simplifies the complexity of manufacturing processes. By providing user-friendly interfaces, operators gain the ability to monitor and control manufacturing operations seamlessly. This ease of control not only enhances operational efficiency but also empowers operators to adapt to dynamic situations swiftly. Intuitive Control ensures that even complex manufacturing tasks can be executed with precision and simplicity.

AI-driven Maintenance

Predictive Maintenance, empowered by AI algorithms, takes center stage in HMI, ensuring the optimal performance of manufacturing equipment. By analyzing data patterns and predicting potential issues, AI-driven Maintenance enables proactive measures, reducing downtime and enhancing overall equipment effectiveness. This predictive capability transforms maintenance practices from reactive to anticipatory, contributing to a more resilient and efficient manufacturing ecosystem.

As these components converge within the Human-Machine Interface, Industry 5.0 witnesses the harmonious collaboration between human ingenuity and machine capabilities. Augmented Reality, Natural Language Processing, Intuitive Control, and AI-driven Maintenance redefine the user experience, making it more intuitive, interactive, and adaptive. In this paradigm, HMI not only bridges the gap between humans and machines but becomes the catalyst for a manufacturing environment where collaboration and efficiency flourish.

Data Security and Governance

In the dynamic landscape of Industry 5.0, where data is a currency of paramount importance, the components within Data Security and Governance play a pivotal role in fortifying the integrity, confidentiality, and compliance of sensitive information.

Blockchain Technology

At the forefront of data security, Blockchain Technology introduces a decentralized and tamper-resistant ledger, ensuring the sanctity of data transactions. This distributed ledger not only enhances the security of data but also fosters transparency and trust. By decentralizing control and making data alterations virtually impossible, Blockchain emerges as a cornerstone in safeguarding the integrity and authenticity of critical information.

Data Governance

Guiding the responsible management of data, Data Governance establishes policies and procedures that govern the entire data lifecycle. From creation and storage to usage and disposal, this component ensures data quality, security, and compliance with regulatory standards. Data Governance becomes the custodian of best practices, creating a framework that not only protects data but also optimizes its utility for informed decision-making.

Cybersecurity

In the face of ever-evolving cyber threats, Cybersecurity stands as the shield protecting the manufacturing ecosystem. Implementing robust measures, this component safeguards sensitive information from unauthorized access, data breaches, and malicious activities. Cybersecurity becomes the vigilant guardian, continuously monitoring and mitigating risks to ensure the resilience and security of the entire manufacturing environment.

Interactions and Dynamics

The synergy among these components forms a holistic system where data seamlessly traverses through the manufacturing ecosystem. For example, data collected by sensors in Smart Factories undergoes processing by the AI Integration Layer, leading to optimized production processes. This refined data then flows into Supply Chain Integration, aligning demand forecasting with production capabilities. Meanwhile, the Human-Machine Interface (HMI) empowers human operators to interact with the system, providing input and receiving real-time feedback.

The Robust Data Security and Governance Layer

This layer acts as the guardian of confidentiality and integrity throughout the manufacturing process. Blockchain Technology, with its decentralized ledger, introduces an additional layer of transparency and trust in data transactions. As data becomes a strategic asset in Industry 5.0, the robust Data Security and Governance layer not only safeguards information but also fosters a resilient and compliant manufacturing ecosystem where data is a trusted and protected commodity.

CHALLENGES AND OPPORTUNITIES

As the unfolding narrative of Industry 5.0 paints a picture of technological marvels and seamless integration of artificial intelligence, it becomes imperative to traverse the intricate terrain of challenges and opportunities that accompany this transformative era. In this chapter, we embark on a journey delving into the ethical considerations surrounding the deployment of AI, confront the shifting landscape of employment and job displacement, and shed light on the vast opportunities for innovation and the emergence of novel business models. By grounding our exploration with real-time examples and applications, we aim to illuminate the nuanced facets of Industry 5.0.

Evolving Nature of Work and Job Displacement

As Industry 5.0 unfurls its transformative capabilities, the nature of work undergoes a metamorphosis, presenting both challenges and opportunities. Automation and AI-driven systems streamline processes, enhancing efficiency, yet the specter of job displacement looms large. Real-time examples highlight the urgency of addressing this paradigm shift. Consider the retail sector, where the integration of AI-powered checkout systems and inventory management technologies transforms the customer experience but concurrently displaces traditional cashier roles. The challenge lies in equipping the workforce with the skills necessary to navigate this evolving landscape, fostering a symbiotic relationship between human ingenuity and machine capabilities. However, Industry 5.0 also offers an array of opportunities to redefine work paradigms. Remote collaboration platforms powered by AI facilitate seamless communication and project management, transcending geographical constraints. The gig economy witnesses a surge, with platforms utilizing AI algorithms to match freelancers with diverse projects, providing newfound flexibility and opportunities for individuals to leverage their skills.

Ethical Considerations in AI Deployment

The ascent of AI in Industry 5.0 comes with a responsibility to address ethical concerns that accompany its deployment. From bias in algorithms to issues of privacy and data security, ethical considerations permeate every facet of AI integration (ElFar et al., 2021). This chapter scrutinizes the ethical dimensions of decision-making algorithms, the accountability of AI systems, and the implications for individuals and society at large. By dissecting these ethical quandaries, we lay the groundwork for a conscientious and responsible implementation of AI in Industry 5.0. The transformative power of AI in Industry 5.0 raises pertinent questions about the future of work. Automation, driven by intelligent machines, has the potential to reshape industries and redefine the nature of employment. We delve into the challenges posed by job displacement, explore the evolving skill sets required in the workforce, and examine strategies to navigate this shifting employment landscape. Through this exploration, we aim to foster a nuanced understanding of the intersection between automation, employment, and the societal implications of AI-driven advancements.

Another ethical dimension involves the societal impact of automation on employment. As Industry 5.0 introduces advanced technologies that may replace certain manual tasks, ethical guidelines must be established to manage potential job displacement. This includes strategies for upskilling the workforce, ensuring a just transition, and considering the broader socioeconomic implications of automation. Privacy emerges as a significant ethical concern in the interconnected ecosystem of Industry 5.0. The vast amounts of data generated by interconnected devices raise questions about data ownership, consent, and the potential misuse of personal information. Implementing robust data protection measures and privacy guidelines becomes imperative to safeguard individuals' rights. Moreover, ensuring the security of AI systems against malicious use or hacking is an ethical imperative. Guidelines for securing AI technologies and preventing them from being weaponized or manipulated for harmful purposes must be established to mitigate potential risks to individuals and society at large.

While AI presents challenges, it also serves as a catalyst for innovation, sparking the emergence of new business models and unprecedented opportunities. This chapter examines how businesses can harness the power of AI to drive innovation, enhance efficiency, and create value. From the advent of data-driven decision-making to the exploration of novel services and products, we uncover the myriad ways in which Industry 5.0 fosters an environment ripe for entrepreneurial ventures and strategic business evolution. By identifying these opportunities, organizations can position themselves at the forefront of the industrial revolution, driving positive change and sustainable growth.

Opportunities for Innovation and New Business Models

Amidst the challenges, Industry 5.0 unfolds a landscape fertile for innovation, birthing new business models that redefine industries. Real-time examples showcase the transformative power of innovation. In manufacturing, predictive maintenance powered by AI algorithms revolutionizes traditional practices. Smart factories leverage real-time data analytics to predict equipment failures, reducing downtime and maintenance costs. This not only enhances operational efficiency but also lays the groundwork for a sustainable and cost-effective approach to industrial maintenance. The healthcare sector undergoes a paradigm shift with the emergence of telemedicine, driven by AI-enabled diagnostics and personalized treatment plans. Virtual health platforms utilize AI algorithms to analyze patient data and provide remote consultations, offering accessibility and efficiency in healthcare delivery.

Furthermore, Industry 5.0 paves the way for sustainability-driven innovation. Smart cities, employing AI and IoT technologies, optimize resource utilization, manage energy consumption, and enhance overall urban resilience. Real-time data analytics enable cities to respond dynamically to environmental challenges, fostering a more sustainable and ecologically conscious future.

FUTURE RESEARCH DIRECTIONS

In charting the course for future research at the nexus of Industry 5.0 and artificial intelligence (AI) in manufacturing, several promising directions emerge, poised to redefine the landscape of smart and intelligent production. One pivotal avenue involves enhancing the explainability of AI algorithms, ensuring transparency and interpretability in decision-making processes. Concurrently, a focus on human-centric AI systems aims to harmonize technology with human skills, while ethical considerations demand robust frameworks addressing bias, fairness, and accountability in AI deployment. The resilience and security of smart manufacturing systems against evolving cyber threats represent a critical research frontier, demanding innovative strategies for safeguarding AI-infused technologies. Furthermore, research in AI for sustainable manufacturing seeks to optimize resource utilization and minimize environmental impact, aligning manufacturing practices with eco-friendly principles. Exploring the integration of AI at the edge of networks for real-time decision-making opens new horizons, with potential breakthroughs in overcoming challenges associated with edge computing.

Human-robot collaboration stands as a compelling research domain, investigating safe and intuitive interaction between humans and robots on the factory floor. The psychological aspects of this collaboration add a layer of complexity, prompting exploration into the nuanced dynamics of human-robot teamwork. Meanwhile, addressing the unique challenges faced by small and medium-sized enterprises (SMEs) in adopting AI technologies necessitates research into cost-effective solutions, scalability issues, and strategies for democratizing AI in SME-centric manufacturing ecosystems. AI-enhanced continuous improvement methodologies, such as Six Sigma or Lean Manufacturing, form another research trajectory, leveraging AI-driven insights to identify areas for improvement and optimize processes dynamically. Cross-industry collaboration facilitated by AI technologies offers a fertile ground for exploring knowledge sharing and technology transfer, fostering innovation and improved manufacturing practices across diverse sectors. Adaptive learning systems represent a visionary research focus, aiming to develop AI algorithms that can dynamically adapt to changing manufacturing environments, ensuring continuous evolution and learning. These multifaceted research directions collectively underscore the pivotal role of researchers in shaping a future where Industry 5.0 and AI converge to drive unprecedented efficiency, sustainability, and human-centric innovation in the manufacturing domain.

CONCLUSION

The chapter on Smart Factories and Intelligent Manufacturing serves as a comprehensive exploration of the transformative impact of Artificial Intelligence (AI) on the manufacturing landscape in the era of Industry 5.0. It unfolds the multifaceted aspects of this integration, reshaping traditional production processes and unlocking new potentials for efficiency and adaptability. The chapter begins by dissecting the seamless integration of AI into manufacturing workflows. It showcases how AI technologies, including machine learning and real-time data analytics, are harmoniously orchestrated to optimize production processes. The result is a manufacturing ecosystem finely tuned for precision and efficiency, marking a departure from traditional methods towards a more adaptive and responsive model. A significant focus is placed on the revolutionary impact of AI-driven predictive maintenance. The chapter outlines how machine learning algorithms analyze equipment data, anticipate patterns, and predict potential issues, leading to a substantial reduction in downtime. This shift from reactive to proactive maintenance not only enhances operational efficiency but also introduces a cost-effective and sustainable approach to managing industrial equipment.

The exploration extends to the realm of customization and flexibility in production. The chapter illustrates how AI empowers smart factories to adapt to dynamic market demands swiftly. Whether handling small-batch, highly customized products or responding to rapid changes in design specifications, AI enables manufacturing systems to be agile and responsive, ushering in a new era where customization is seamlessly woven into the fabric of production. Through real-world applications, case studies, and a forward-looking perspective, the chapter provides insights into the tangible impacts of AI in manufacturing. It underscores the potential for Industry 5.0 to redefine not only the efficiency of production but also the very nature of how goods are manufactured, setting the stage for a future where smart factories and intelligent manufacturing are integral components of a dynamic and interconnected industrial landscape.

REFERENCES

Agote-Garrido, A., Martín-Gómez, A. M., & Lama-Ruiz, J. R. (2023). Manufacturing System Design in Industry 5.0: Incorporating Sociotechnical Systems and Social Metabolism for Human-Centered, Sustainable, and Resilient Production. *Systems*, *11*(11), 537. doi:10.3390/systems11110537

Anish, T. P., Shanmuganathan, C., Dhinakaran, D., & Vinoth Kumar, V. (2023). Hybrid Feature Extraction for Analysis of Network System Security—IDS. In R. Jain, C. M. Travieso, & S. Kumar (Eds.), *Cybersecurity and Evolutionary Data Engineering. ICCEDE 2022. Lecture Notes in Electrical Engineering* (Vol. 1073). Springer. doi:10.1007/978-981-99-5080-5_3

Becker, K.-F., Voges, S., Fruehauf, P., Heimann, M., Nerreter, S., Blank, R., Erdmann, M., & (2021). Implementation of Trusted Manufacturing & AI-Based Process Optimization into Microelectronic Manufacturing Research Environments. *IMAPSource Proceedings 2021 (IMAPS Symposium):* (pp. 21–25). iMAPS. 10.4071/1085-8024-2021.1

Bednar, P. M., & Welch, C. (2020). Socio-technical perspectives on smart working: Creating meaningful and sustainable systems. *Information Systems Frontiers*, *22*(2), 281–298. doi:10.1007/s10796-019-09921-1

Cullen-Knox, C., Eccleston, R., Haward, M., Lester, E., & Vince, J. (2017). Contemporary challenges in environmental governance: Technology, governance and the social licence. *Environmental Policy and Governance*, *27*(1), 3–13. doi:10.1002/eet.1743

Dhinakaran, D., Selvaraj, D., Dharini, N., Raja, S. E., & Priya, C. S. L. (2023). Towards a Novel Privacy-Preserving Distributed Multiparty Data Outsourcing Scheme for Cloud Computing with Quantum Key Distribution. *International Journal of Intelligent Systems and Applications in Engineering*, *12*(2), 286–300.

Dhinakaran, D., Udhaya Sankar, S. M., Edwin Raja, S., & Jeno Jasmine, J. (2023). Optimizing Mobile Ad Hoc Network Routing using Biomimicry Buzz and a Hybrid Forest Boost Regression - ANNs [IJACSA]. *International Journal of Advanced Computer Science and Applications*, *14*(12). doi:10.14569/IJACSA.2023.0141209

ElFar, O. A., Chang, C.-K., Leong, H. Y., Peter, A. P., Chew, K. W., & Show, P. L. (2021). Prospects of Industry 5.0 in algae: Customization of production and new advance technology for clean bioenergy generation. *Energy Convers. Manag. X*, *10*, 100048. doi:10.1016/j.ecmx.2020.100048

Ghobakhloo, M., Iranmanesh, M., Foroughi, B., Tirkolaee, E. B., Asadi, S., & Amran, A. (2023). Industry 5.0 implications for inclusive sustainable manufacturing: An evidence-knowledge-based strategic roadmap. *Journal of Cleaner Production*, *417*, 138023. doi:10.1016/j.jclepro.2023.138023

Harini, M., Prabhu, D., Udhaya Sankar, S. M., Pooja, V., & Kokila Sruthi, P. (2023). *Levarging Blockchain for Transparency in Agriculture Supply Chain Management Using IoT and Machine Learning*. 2023 World Conference on Communication & Computing (WCONF), RAIPUR, India. 10.1109/WCONF58270.2023.10235156

Iqbal, M., Lee, C. K. M., & Ren, J. Z. (2022). Industry 5.0: From Manufacturing Industry to Sustainable Society. *2022 IEEE International Conference on Industrial Engineering and Engineering Management (IEEM)*, Kuala Lumpur, Malaysia. 10.1109/IEEM55944.2022.9989705

Jain, D. K., Li, Y., Er, M. J., Xin, Q., Gupta, D., & Shankar, K. (2022). Enabling Unmanned Aerial Vehicle Borne Secure Communication With Classification Framework for Industry 5.0. *IEEE Transactions on Industrial Informatics*, *18*(8), 5477–5484. doi:10.1109/TII.2021.3125732

Javed, S., Javed, S., Deventer, J. v., Mokayed, H., & Delsing, J. (2023). *A Smart Manufacturing Ecosystem for Industry 5.0 using Cloud-based Collaborative Learning at the Edge*. NOMS 2023-2023 IEEE/IFIP Network Operations and Management Symposium, Miami, FL, USA. 10.1109/NOMS56928.2023.10154323

Keerthana, M., Ananthi, M., Harish, R., Udhaya Sankar, S. M., & Sree, M. S. (2023). *IoT Based Automated Irrigation System for Agricultural Activities*. 2023 12th International Conference on Advanced Computing (ICoAC), Chennai, India. 10.1109/ICoAC59537.2023.10249426

Konstantinidis, F. K., Myrillas, N., Tsintotas, K. A., Mouroutsos, S. G., & Gasteratos, A. (2023). A technology maturity assessment framework for Industry 5.0 machine vision systems based on systematic literature review in automotive manufacturing. *International Journal of Production Research*, 1–37. doi:10.1080/00207543.2023 .2270588

Kumar, K. Y., Kumar, N. J., Udhaya Sankar, S. M., Kumar, U. J., & Yuvaraj, V. (2023). *Optimized Retrieval of Data from Cloud using Hybridization of Bellstra Algorithm*. 2023 World Conference on Communication & Computing (WCONF), RAIPUR, India. 10.1109/WCONF58270.2023.10234974

Leng, J., Sha, W., Lin, Z., Jing, J., Liu, Q., & Chen, X. (2023). Blockchained smart contract pyramid-driven multi-agent autonomous process control for resilient individualised manufacturing towards Industry 5.0. *International Journal of Production Research*, *61*(13), 4302–4321. doi:10.1080/00207543.2022.2089929

Olaizola, I. G., Quartulli, M., Garcia, A., & Barandiaran, I. (2022). *Artificial Intelligence from Industry 5.0 perspective: Is the Technology Ready to Meet the Challenge?* CUER. http://ceur-ws. org

Peruzzini, M., Prati, E., & Pellicciari, M. (2023). A framework to design smart manufacturing systems for Industry 5.0 based on the human-automation symbiosis. *International Journal of Computer Integrated Manufacturing*, 1–18. doi:10.1080/0951192X.2023.2257634

Sai Aswin, B. G., Vishnubala, S., Dhinakaran, D., Kumar, N. J., Udhaya Sankar, S. M., & Mohamed Al Faisal, A. M. (2023). *A Research on Metaverse and its Application*. 2023 World Conference on Communication & Computing (WCONF), RAIPUR, India. 10.1109/WCONF58270.2023.10235216

Salam, A., Ullah, F., Amin, F., & Abrar, M. (2023). Deep Learning Techniques for Web-Based Attack Detection in Industry 5.0: A Novel Approach. *Technologies*, *11*(4), 107. doi:10.3390/technologies11040107

Srinivasan, L., Selvaraj, D., & Udhaya Sankar, S. M. (2023). Leveraging Semi-Supervised Graph Learning for Enhanced Diabetic Retinopathy Detection. *SSRG International Journal of Electronics and Communication Engineering.*, *10*(8), 9–21. doi:10.14445/23488549/IJECE-V10I8P102

Taj, I., & Jhanjhi, N. Z. (2022). Towards Industrial Revolution 5.0 and Explainable Artificial Intelligence: Challenges and Opportunities. *International Journal of Computing and Digital Systems.*, *12*(1), 285–310. doi:10.12785/ijcds/120124

Torres, D. R., Alejandro, D. S. A., Roldán, Á. O., Bustos, A. H., & Luis, E. A. G. (2022). A Review of Deep Reinforcement Learning Approaches for Smart Manufacturing in Industry 4.0 and 5.0 Framework. *Applied Sciences (Basel, Switzerland)*, *12*(23), 12377. doi:10.3390/app122312377

Udhaya Sankar, S. M., Kumar, N. J., Dhinakaran, D., Kamalesh, S. S., & Abenesh, R. (2023). *Machine Learning System for Indolence Perception*. 2023 International Conference on Innovative Data Communication Technologies and Application (ICIDCA), Uttarakhand, India. 10.1109/ICIDCA56705.2023.10099959

Yadav, M., Vardhan, A., Chauhan, A. S., & Saini, S. (2023). *A Study on Creation of Industry 5.0: New Innovations using big data through artificial intelligence, Internet of Things and next-origination technology policy. Conference on Electrical, Electronics and Computer Science (SCEECS)*, Bhopal, India. 10.1109/SCEECS57921.2023.10063069

Yan, F., Liu, J., Yan, X., & Wang, G. (2022). *Application of Key Technologies of Intelligent Manufacturing in Metallurgical Industry Led by Artificial Intelligence*. 2022 International Conference on Cloud Computing, Big Data and Internet of Things (3CBIT), Wuhan, China. 10.1109/3CBIT57391.2022.00073

Chapter 12
Real–Time Applications of Artificial Intelligence Technology in Daily Operations

R. Renugadevi
R.M.K. Engineering College, India

J. Shobana
 https://orcid.org/0000-0001-9754-2604
SRM Institute of Science and Technology, India

K. Arthi
SRM Institute of Science and Teechnology, India

Kalpana A. V.
 https://orcid.org/0000-0003-2289-4968
SRM Institute of Science and Technology, India

D. Satishkumar
Nehru Institute of Technology, India

M. Sivaraja
Nehru Institute of Technology, India

ABSTRACT

Artificial intelligence (AI) is a system endowed with the capability to perceive its surroundings and execute actions aimed at maximizing the probability of accomplishing its objectives. It possesses the capacity to interpret and analyze data in a manner that facilitates learning and adaptation over time. Generative AI pertains to artificial intelligence models specifically designed for the creation of fresh content, spanning written text, audio, images, or videos. Its applications are diverse, ranging from generating stories mimicking a particular author's style to producing realistic images of non-existent individuals, composing music in the manner of renowned composers, or translating textual descriptions into video clips.

DOI: 10.4018/979-8-3693-2615-2.ch012

INTRODUCTION

Artificial intelligence (AI) is a system endowed with the capability to perceive its surroundings and execute actions aimed at maximizing the probability of accomplishing its objectives. It possesses the capacity to interpret and analyze data in a manner that facilitates learning and adaptation over time. Generative AI pertains to artificial intelligence models specifically designed for the creation of fresh content, spanning written text, audio, images, or videos. Its applications are diverse, ranging from generating stories mimicking a particular author's style to producing realistic images of non-existent individuals, composing music in the manner of renowned composers, or translating textual descriptions into video clips.

Traditional AI: Traditional AI relies on predetermined rules or algorithms to perform specific tasks. These rule-based systems lack the ability to learn from data or improve over time. In contrast, generative AI can learn from data and generate novel instances.

Machine Learning: Machine learning enables systems to learn from data rather than relying on explicit programming. Generative AI utilizes machine learning techniques, allowing it to learn from data and create new data instances.

Conversational AI: While generative AI and conversational AI may seem similar, especially when generative AI generates human-like text, their primary distinction lies in purpose. Conversational AI is tailored for creating interactive systems engaging in human-like dialogue, while generative AI encompasses the broader creation of various data types, not limited to text.

Artificial General Intelligence (AGI): AGI refers to highly autonomous systems, currently theoretical, that could surpass humans in most economically valuable tasks. While generative AI may be a component of AGI systems, it doesn't equate to AGI. Generative AI focuses on producing new data instances, while AGI implies a broader level of autonomy and capability.

The machine learning ideas include a wide range of computational methods that can be used to treat all types of application. The ML model's performance techniques such classification and regression trees (CART) and general additives models have been discussed, and the resulting data have then been compared. Additionally, a few potential CART expansions were suggested, including nearest neighbors, projection pursuit, bagging, random forest, boosting, and support vector machines. The primary reason for developing these extensions was the CART model's instability in response to variations in the training set. In training and learning procedures, the aggregated models produced by the extensions are more reliable. However, depending on the type of information need to be collected and the application of these extensions pertains to certain specific instances. ML is often a completely established statistical data analysis technique. It makes it possible to foresee and predict particular extremes

Figure 1. Applications of AI in daily operations

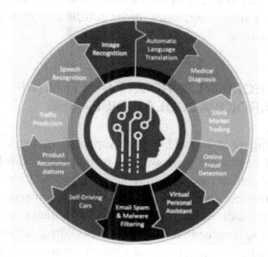

using any type of ecological data. Although machine learning has been applied to the building industry as well, model algorithms still require improvement. Applications of AI in day-to-day life is given in Figure 1.

AI IN SALES AND MARKETING

The lead-to-cash process encompasses all activities related to marketing and selling products and services. It includes the management and fulfillment of sales orders, provision of after-sales services, and the final stages of invoicing customers, managing accounts receivable, and collecting payments. Additionally, the process incorporates the foundational elements of managing customers and channels.

The specific execution of the lead-to-cash process varies primarily based on several factors:

Type of Customer: Distinctions between business-to-business (B2B) and business-to-consumer (B2C) transactions influence the nuances of the process.

Sales Channels:

The channels through which products and services are offered, such as direct sales, digital commerce platforms, and physical stores, introduce variations in the process.

Nature of Products and Services:

The type of products and services being sold contributes to the complexity of the process. This includes considerations such as simple tangible products, configurable products, engineered products, intangible products, one-time or recurring services, complex solutions, or projects. Understanding and adapting the lead-to-cash process according to these variables is essential for businesses to effectively navigate the diverse landscape of marketing, sales, order fulfillment, customer management, and financial transaction.

AI CLINICAL DECISION SUPPORT SYSTEM (AI-CDSS) FOR CARDIOVASCULAR DISEASES

The Artificial Intelligence Clinical Decision Support System (AI-CDSS) is a robust tool crafted to aid healthcare professionals in making well-informed, evidence-based decisions for patient care. By utilizing artificial intelligence algorithms and data analysis techniques, it delivers personalized recommendations and insights. The system encompasses various features and advantages, such as patient data analysis, diagnostics, treatment recommendations, drug interaction and adverse event detection, predictive analytics, real-time monitoring, alerts, and continuous learning and improvement.

The discussion in this model delves into the manifold applications of AI-driven decision-making systems in healthcare, with a specific focus on areas like cancer diagnosis and treatment, chronic disease management, medication optimization, surgical decision support, infectious disease outbreak management, radiology, medical imaging analysis, mental health support, and clinical trials and research. Furthermore, the paper underscores the relevance of established methodologies, including deep learning models such as Convolutional Neural Networks (CNNs) and Recurrent Neural Networks (RNNs), which have exhibited promise in cardiovascular disease prediction. However, it underscores the imperative for rigorous validation, evaluation, and careful consideration of ethical and regulatory aspects before deploying AI models in clinical practice. Ultimately, the integration of AI-driven decision-making systems holds the potential to transform chronic disease management, enhance patient outcomes, and elevate the overall quality of care for individuals with enduring health conditions.

AI IN FINANCE

AI plays a transformative role in the financial industry, providing benefits for both financial institutions and their customers. Here are key areas where AI is making a significant impact:

For Financial Institutions:

Market Research and Investment Analysis:

- AI enables rapid analysis of large datasets for market research.
- Identifies trends and forecasts future performance to assist investors in making informed decisions.
- Facilitates the evaluation of potential risks associated with investments.

Insurance Underwriting and Risk Assessment:

- Harvests and analyzes personal data to determine coverage and premiums.
- Enhances accuracy in underwriting processes by leveraging machine learning algorithms.
- Enables personalized risk assessments based on individual profiles.

Cybersecurity:

- Identifies and prevents fraudulent transactions by closely monitoring purchase behavior.
- Compares real-time transactions with historical data to flag anomalous activities.
- Automatically alerts institutions and customers to verify transactions and takes immediate action if needed.

For Banking Customers:
Improved Customer Experience:

- Automates basic banking activities such as payments, deposits, and transfers.
- Handles customer service requests efficiently, providing quick responses.
- Streamlines the overall banking experience, reducing the need for in-person interactions.

Contactless Banking and Endpoint Security:

- Supports the rise of online and contactless banking, offering convenience.
- Manages endpoint security to address vulnerabilities associated with online transactions.
- Ensures a secure and seamless experience for customers across various devices.

Credit Card and Loan Application Processes:

- Automates application processes for credit cards and loans.

- Provides near-instant responses for application acceptance or rejection.
- Enhances efficiency in the lending process, making it faster and more convenient for customers.

Personalized Financial Services:

- Utilizes AI to analyze customer behavior and preferences.
- Offers personalized financial advice and product recommendations.
- Enhances customer engagement and satisfaction through tailored services.

In summary, AI brings automation, efficiency, and enhanced security to financial institutions, enabling them to streamline processes, make data-driven decisions, and combat fraud. For banking customers, AI improves the overall experience by providing convenient, secure, and personalized financial services.

AI IN SUPPLY CHAIN

The Design to Operate (D2O) process encompasses the entire lifecycle of products within an end-to-end, interconnected, and interoperable supply chain. This comprehensive process spans from the initial design phase through planning, manufacturing, delivery, operational maintenance, and product operation.

Planning Stage:

- Defines supply chain, manufacturing, and service-fulfilment strategies.
- Plans demand, inventory, and supply.
- Aligns plans through sales and operations planning.
- Manages supply chain performance.
- Initiates operational procurement.

Production Stage:

- Involves production planning.
- Encompasses production operations.
- Includes quality management.
- Manages production performance for both tangible and intangible goods.

For Tangible Goods:

- Covers inbound and outbound deliveries.

- Encompasses order promising.
- Involves warehouse and inventory management.
- Addresses dock and yard logistics.
- Manages transportation.
- Includes logistics performance management.

For Services:

- Involves service planning and scheduling.
- Includes service execution and delivery.
- Encompasses service performance management.

Enabling and Foundational Activities:

- Involves data management.
- Emphasizes collaboration.
- Identifies and tracks materials.
- Addresses sustainable manufacturing operations.

By integrating these stages, the D2O process ensures a holistic approach to product lifecycle management within a supply chain context. It emphasizes seamless coordination and interoperability to enhance efficiency and effectiveness across all facets of the product journey, from conception to ongoing operation and maintenance.

AI IN HEALTH INFORMATION SYSTEMS

The integration of machine learning (ML) in healthcare holds the potential for significant improvements, yet it is crucial to acknowledge and address the drawbacks of artificial intelligence (AI) in this field. Ethical concerns must be carefully considered, particularly in light of potential biases inherent in computer algorithms (Ongen et al., 2017). Health-related predictions generated by AI may exhibit variations based on factors such as race, genetics, and gender, leading to the risk of overestimating or underestimating patient risk factors if not appropriately accounted for. In navigating these challenges, physicians play a vital role in ensuring the responsible development and application of AI algorithms (Lyall et al., 2018).

This systematic review predominantly relied on electronic medical records (EMRs) as the primary data collection method. ML techniques applied to EMRs enable the identification of patterns, supporting predictions and decisions related to diagnosis and treatment planning (Johnson et al., 2016). Combining EMR-based ML methods with other extensive medical data sources, such as genomics and medical imaging, has the potential to enhance clinical diagnosis and treatment systems through predictive algorithms (Zhang et al., 2017). EMR data typically include diverse information such as demographics, diagnoses, biochemical markers, vital signs, clinical notes, prescriptions, and procedures. This comprehensive dataset is easily obtainable and helps minimize errors associated with handling large amounts of information.

Several studies have previously explored medical diagnosis prediction tools utilizing EMRs (McCoy et al., 2015; Nguyen et al., 2017; Osborn et al., 2015; Rajkomar et al., 2018). In this systematic review, approximately 48% of the features in the diagnosis prediction model for perinatal complications were sourced from EMRs, with sociodemographic maternal characteristics being the most frequently used features. Consequently, this tool demonstrates the capacity to predict common perinatal complications within a specific population, contributing to the overall enhancement of perinatal public health. While ML in healthcare offers substantial benefits, it is imperative to remain vigilant about ethical considerations, biases, and the responsible oversight of these technologies in the hands of healthcare professionals.

AI IN AUTONOMOUS VEHICLES

A self-driving car, also known as an autonomous or driverless car, utilizes a combination of sensors, cameras, radar, and artificial intelligence (AI) to navigate from one point to another without human intervention. To be classified as fully autonomous, a self-driving car must be capable of navigating predetermined routes without human control, even on roads not specifically designed for autonomous vehicle use. Various companies, including Audi, BMW, Ford, Google, General Motors, Tesla, Volkswagen, and Volvo, are involved in the development and testing of autonomous cars. Google's Waymo project, for instance, has conducted extensive testing with a fleet of self-driving cars covering over 140,000 miles on California roads.

The functionality of self-driving cars relies on AI technologies, incorporating data from image recognition systems, machine learning, and neural networks. Neural networks analyze patterns in data, including images from car-mounted cameras, enabling the system to identify and interpret elements such as traffic lights, pedestrians, and street signs within its surroundings.

For example, Google's Waymo employs sensors, lidar, and cameras to create a 3D map of the car's environment. The AI software processes data from various sensors, Google Street View, and internal cameras to simulate human decision-making processes through deep learning. The car's software is designed to control actions like steering and braking, consulting Google Maps for additional information.

Self-driving cars exist on a spectrum of autonomy, ranging from partially autonomous systems with driver assistance features, such as adaptive cruise control and lane-centering steering, to fully autonomous vehicles. The U.S. National Highway Traffic Safety Administration (NHTSA) defines six levels of automation, starting from Level 0 (full human control) to Level 5 (complete automation without any human involvement), each representing different degrees of the car's ability to perform driving tasks.

AI IN EDUCATION

Artificial intelligence (AI) is a specialized field within computer science dedicated to developing software that can emulate intelligent behaviors and processes observed in humans. These behaviors encompass activities like reasoning, learning, problem-solving, and creative thinking. AI systems find application in diverse tasks, ranging from language translation, image recognition, and autonomous vehicle navigation to cancer detection and treatment. Generative AI, a subset of AI, goes beyond retrieving information and actively produces content and knowledge. Generative AI relies on "foundation models," which are systems trained on extensive datasets to acquire a broad base of knowledge adaptable to various specific purposes. The learning process for these models is self-supervised, where the model autonomously discovers patterns and relationships within the training data.

A notable type of foundation model is the Large Language Model (LLM), exemplified by OpenAI's GPT (Generative Pre-trained Transformer) series, Google's PaLM (Pathways Language Model), and Meta's LLaMA. OpenAI's GPT models, including GPT-3.5 and GPT-4, have undergone training on vast datasets comprising web content, books, Wikipedia articles, news articles, social media posts, and code snippets. For instance, GPT-3 was trained on a staggering 300 billion "tokens" or word pieces, utilizing over 175 billion parameters to shape its behavior—significantly surpassing the data used for the previous GPT-2 model.

LLMs analyze billions of sentences to develop a statistical understanding of language, encompassing the typical combinations of words and phrases, common associations between topics, and appropriate tones or styles for different contexts. This enables them to generate text resembling human language and perform a wide range of tasks, such as writing articles, answering questions, and analyzing unstructured data.

Examples of applications built on LLMs include ChatGPT, which is constructed upon GPT-3.5 and GPT-4, and Bard, utilizing Google's Pathways Language Model 2 (PaLM 2) as its foundational model. These models serve as the basis for various

AI applications, showcasing the versatility and capabilities of generative AI in diverse domains.

AI IN CYBERSECURITY

AI-based solutions in cybersecurity leverage machine learning algorithms to detect and respond to threats in real-time, encompassing both known and unknown security risks. These machine learning algorithms undergo training using extensive datasets, including historical threat data and information from network and endpoint activities. The goal is to identify patterns that might be challenging for humans to discern. This training enables AI-based solutions to autonomously recognize and address threats as they occur, eliminating the need for immediate human intervention. For instance, machine learning algorithms can analyze network traffic patterns, spotting anomalous behaviors that could indicate a cyberattack. Following detection, these algorithms can alert security personnel or even take automated actions to mitigate the identified threat. A key distinction between AI-based solutions and traditional cybersecurity approaches lies in their continuous learning and adaptive nature. Machine learning algorithms employed in these solutions can be regularly retrained on new data, allowing them to enhance their ability to detect and respond to emerging threats. This adaptability enables AI-based solutions to stay abreast of the evolving threat landscape, offering increasingly effective cybersecurity protection over time.

The integration of AI into cybersecurity signifies a significant shift in how organizations approach digital security. By utilizing machine learning algorithms to detect and respond to threats in real-time, AI-based solutions enhance protection against both known and unknown threats. This proactive approach helps organizations better secure their sensitive data and critical systems, contributing to a more robust cybersecurity posture. AI plays a crucial role in various aspects of cybersecurity, offering advanced capabilities for threat detection, response, and protection across different domains:

Malware Detection

Traditional antivirus software relies on signature-based detection, which may be limited to known malware variants. AI-based solutions, utilizing machine learning algorithms, analyze large datasets, including historical threat data, to identify patterns indicative of both known and unknown malware threats. These solutions can employ techniques like static and dynamic analysis for a more comprehensive approach to malware detection.

Phishing Detection

AI enhances phishing detection by analyzing email content and structure using machine learning algorithms. Unlike traditional approaches that rely on rules-based filtering or blacklisting, AI solutions can learn from vast amounts of data to detect patterns and anomalies indicative of phishing attacks. Additionally, behavioral analysis of user interactions with emails enables the identification of potential phishing threats, offering a more adaptive and effective approach.

Security Log Analysis

AI-based security log analysis utilizes machine learning algorithms to analyze large volumes of security log data in real-time. Unlike rule-based systems, AI algorithms can detect patterns and anomalies that may indicate a security breach, even without known threat signatures. This approach aids in quick identification and response to potential security incidents, reducing the risk of data breaches and insider threats.

Network Security

AI algorithms contribute to network security by monitoring for suspicious activity, identifying unusual traffic patterns, and detecting unauthorized devices on the network. Anomaly detection, based on historical traffic data, allows AI to learn normal network behavior and identify anomalous or suspicious activities, such as unusual port or protocol usage. Additionally, AI can monitor and alert security teams to potential threats posed by unauthorized devices on the network.

Endpoint Security

Endpoints, including laptops and smartphones, are protected by AI-based solutions that analyze endpoint behavior to detect potential threats. Unlike traditional antivirus software, which relies on signature-based detection, AI algorithms can identify unknown malware variants by analyzing their behavior. AI-based endpoint security solutions adapt and evolve over time, learning from new data to provide better protection against evolving and sophisticated cyber threats.

In summary, AI in cybersecurity offers real-time threat detection and response, adapts to evolving threats, and provides a more advanced and effective approach to safeguarding against various cyber risks. The continuous learning capabilities of AI algorithms contribute to a proactive and adaptive cybersecurity strategy.

CHALLENGES FACED WHEN USING ARTIFICIAL NEURAL NETWORK

Data Privacy

When employing Artificial Neural Network (ANN) models, access to extensive data, including sensitive information, is often necessary. Safeguarding data privacy and protection is critical to uphold the confidentiality and integrity of the data. Effective measures, such as robust data anonymization techniques, secure data storage, and compliance with relevant privacy regulations, are essential to tackle this challenge.

Expert Interpretation of Results

ANNs can be intricate and opaque, posing challenges for domain experts to interpret and comprehend the factors influencing the model's predictions. This lack of interpretability can impede trust and acceptance in practical applications. Ongoing efforts focus on developing techniques for interpreting and explaining ANN decisions, including feature importance analysis and model visualization.

Data Standardization

ANNs depend on high-quality and standardized data for training and validation. In environmental monitoring, data may originate from diverse sources, exhibiting variations in formats, units, and quality. Ensuring data standardization and normalization is vital to achieve reliable and accurate ANN models. Establishing data standards, employing data preprocessing techniques, and implementing quality control measures are necessary steps to address this challenge.

Limited Data Availability

ANNs require substantial labeled data for training, which may not always be readily available in environmental monitoring applications. Insufficient data can lead to overfitting or underfitting, resulting in suboptimal model performance. Techniques such as data augmentation, transfer learning, and active learning can help mitigate this challenge by optimizing the training process with the available data.

Computational Resources

Training and optimizing ANNs can be computationally demanding, especially for large-scale environmental monitoring applications. Adequate access to computational resources, such as high-performance computing clusters or cloud-based solutions, is essential to efficiently handle the complexity of ANN models.

Ethical Considerations

The ethical implications of utilizing ANNs in environmental monitoring demand careful consideration. Ensuring fairness, avoiding biases, and addressing potential discriminatory effects of ANN models are crucial. Regular audits, transparency in model development, and ongoing evaluation of the social and environmental impact of ANN applications are necessary to mitigate ethical concerns.

Addressing these challenges necessitates a multidisciplinary approach, involving experts from data science, environmental science, ethics, and policy-making. Collaborative efforts are essential to develop robust frameworks, guidelines, and best practices for modeling with ANNs in environmental monitoring, ensuring that data privacy, interpretability, standardization, and ethical considerations are appropriately addressed.

CONCLUSION

In conclusion, artificial intelligence (AI) applications encompass a wide range of fields, transforming our lifestyles, work environments, and interactions with technology. The far-reaching impact of AI is evident across diverse sectors, and its potential is continuously expanding with technological advancements. The following are key highlights summarizing the diverse applications of AI:

Healthcare: AI has made significant advancements in medical diagnostics, personalized treatment plans, drug discovery, and predictive analytics, enhancing patient care, accelerating research, and improving overall healthcare outcomes.

Finance: In the financial sector, AI is utilized for fraud detection, risk management, algorithmic trading, and customer service, contributing to increased efficiency, accuracy, and security in financial operations.

Education: AI plays a role in personalized learning experiences, intelligent tutoring systems, and educational analytics, adapting to individual learning styles and providing tailored support to both students and educators.

Manufacturing and Industry: AI-driven automation, predictive maintenance, and quality control optimize manufacturing processes, with robotics and AI technologies enhancing efficiency, reducing errors, and boosting overall productivity.

Autonomous Vehicles: The development of self-driving cars and autonomous vehicles relies heavily on AI technologies, including computer vision, machine learning, and sensor fusion. These advancements aim to enhance road safety and redefine transportation.

Natural Language Processing (NLP): AI applications in NLP enable virtual assistants, language translation, sentiment analysis, and chatbots, improving communication and accessibility across various industries.

Entertainment: AI contributes to content recommendation algorithms, virtual reality experiences, and the creation of realistic computer-generated imagery (CGI), enhancing user engagement and personalization in the entertainment industry.

Cybersecurity: AI is employed for threat detection, anomaly identification, and real-time response in cybersecurity, with machine learning algorithms adapting to evolving cyber threats to provide robust defense mechanisms.

Retail: AI applications in retail cover demand forecasting, personalized shopping experiences, inventory management, and recommendation systems, enhancing customer satisfaction and streamlining business operations.

Environmental Monitoring: AI is increasingly utilized in environmental science for data analysis, climate modeling, and resource management, contributing to the monitoring and mitigation of environmental challenges.

Despite the immense benefits offered by AI applications, it is crucial to consider ethical implications, ensure responsible deployment, and engage in ongoing research to address potential challenges. As AI technologies evolve, their responsible integration into various domains remains key to unlocking their full potential for the betterment of humanity.

REFERENCES

Johnson, W., Onuma, O., Owolabi, M., & Sachdev, S. (2016). Stroke: A Global Response Is Needed. *Bulletin of the World Health Organization*, *94*(9), 634–634A. doi:10.2471/BLT.16.181636 PMID:27708464

Lyall, L. M., Wyse, C. A., Morales, C. A. C., Lyall, D. M., Cullen, B., & Mackay, D. (2018). Seasonality of depressive symptoms in women but not in men: A cross-sectional study in the UK Biobank cohort. *Journal of Affective Disorders*, *229*, 296–305. doi:10.1016/j.jad.2017.12.106 PMID:29329063

McCoy, T. H., Castro, V. M., Rosenfield, H. R., Cagan, A., Kohane, I. S., & Perlis, R. H. (2015). A Clinical Perspective on the Relevance of Research Domain Criteria in Electronic Health Records. *The American Journal of Psychiatry*, *172*(4), 316–320. doi:10.1176/appi.ajp.2014.14091177 PMID:25827030

Nguyen, P., Tran, T., Wickramasinghe, N., & Venkatesh, S. (2017). A Convolutional Net for Medical Records. *IEEE Journal of Biomedical and Health Informatics*, *21*(1), 22–30. doi:10.1109/JBHI.2016.2633963 PMID:27913366

Ongen, H., Brown, A. A., Delaneau, O., Panousis, N. I., Nica, A. C., & Dermitzakis, E. T. (2017). Estimating the causal tissues for complex traits and diseases. *Nature Genetics*, *49*(12), 1676–1683. doi:10.1038/ng.3981 PMID:29058715

Osborn, D. P. J., Hardoon, S., Omar, R. Z., Holt, R. I. G., King, M., Larsen, J., Marston, L., Morris, R. W., Nazareth, I., Walters, K., & Petersen, I. (2015). Cardiovascular Risk Prediction Models for People with Severe Mental Illness. *JAMA Psychiatry*, *72*(2), 143–151. doi:10.1001/jamapsychiatry.2014.2133 PMID:25536289

Rajkomar, A., Oren, E., Chen, K., Dai, A. M., Hajaj, N., Hardt, M., Liu, P. J., Liu, X., Marcus, J., Sun, M., Sundberg, P., Yee, H., Zhang, K., Zhang, Y., Flores, G., Duggan, G. E., Irvine, J., Le, Q., Litsch, K., & Dean, J. (2018). Scalable and Accurate Deep Learning with Electronic Health Records. *NPJ Digital Medicine*, *1*(1), 18. doi:10.1038/s41746-018-0029-1 PMID:31304302

Zhang, Q., Zhou, D., & Zeng, X. (2017). HeartID: A Multiresolution Convolutional Neural Network for ECG-Based Biometric Human Identification in Smart Health Applications. *IEEE Access : Practical Innovations, Open Solutions*, *5*, 11805–11816. doi:10.1109/ACCESS.2017.2707460

Chapter 13
3D Localization in the Era of IIoT by Integrating Machine Learning With 3D Adaptive Stochastic Control Algorithm

Siva Kumar A.
SRM Institute of Science and Technology, India

G. Indra
R.M.K. College of Engineering and Technology, India

Umamageswaran Jambulingam
SRM Institute of Science and Technology, India

Ramyadevi K.
R.M.K. Engineering College, India

Praveen Kumar B.
GITAM School of Technology, India

Kalpana A. V.
iD https://orcid.org/0000-0003-2289-4968
SRM Institute of Science and Technology, India

ABSTRACT

In recent decades, technological advancements have reshaped industries, particularly in communication, consumer electronics, and medical electronics. The rise of the industrial internet of things (IIoT) has been transformative, revolutionizing industrial processes. Accurate node localization within IIoT-enabled systems is crucial for operational efficiency and informed decision-making. This chapter introduces the 3D adaptive stochastic control algorithm (ASCA), designed explicitly for IIoT environments with machine learning capabilities, including convolutional neural networks. The algorithm leverages received signal strength indicator data and sophisticated machine learning algorithms to accurately determine node positions in a 3D industrial space, offering a cost-effective and energy-efficient alternative. Comparative evaluations show the superiority of the 3D ASCA algorithm, with an average localization error ranging from 0.37 to 0.7, surpassing benchmarks set by other algorithms.

DOI: 10.4018/979-8-3693-2615-2.ch013

INTRODUCTION

The Industrial Internet of Things (IIoT) stands as a transformative force at the intersection of industrial processes and cutting-edge digital technologies. In essence, IIoT represents the integration of intelligent devices, sensors, and machinery within industrial environments, creating a networked ecosystem where data is seamlessly collected, exchanged, and analysed. Unlike traditional industrial systems, IIoT introduces a paradigm shift by interconnecting physical assets with digital intelligence, enabling unprecedented levels of automation, efficiency, and real-time decision-making. At its core, IIoT empowers industries to harness the power of data, generating valuable insights that drive operational optimization and innovation. Through interconnected devices and robust communication protocols, IIoT enables the monitoring and control of industrial processes with a level of granularity and precision previously unattainable. This not only enhances productivity but also opens doors to new possibilities in predictive maintenance, supply chain optimization, and overall resource management. Security and interoperability are paramount in the IIoT landscape, as critical infrastructure and sensitive data become integral components of the interconnected network. As industries increasingly adopt IIoT solutions, there is a growing emphasis on developing standardized protocols and robust cyber security measures to safeguard against potential threats.

In this era of digital transformation, IIoT represents a catalyst for the next industrial revolution, where industries evolve into agile, data-driven ecosystems. As businesses navigate this dynamic landscape, the promise of IIoT lies in its ability to drive efficiency, innovation, and competitiveness across a spectrum of industrial sectors. The journey into the IIoT era signifies not just a technological shift but a fundamental reimagining of how industries operate, ushering in an era where connectivity and intelligence converge to redefine the possibilities of industrial processes. The Industrial Internet of Things (IIoT) marks a revolutionary era in industrial landscapes, fusing traditional manufacturing and infrastructure with advanced digital technologies. At the heart of this transformation lies the seamless integration of intelligent devices, sensors, and machinery, creating a networked ecosystem where data flows ubiquitously. One of the pivotal applications within this paradigm is the integration of localization technologies with IIoT, redefining how industries manage and optimize their physical assets (Gillis, A.S., 2022).

In 2022, the industrial Internet of things (IIoT) industry was estimated to be worth over 544 billion US dollars worldwide. The market is anticipated to expand over the next several years, with a projected value of 3.3 trillion US dollars by 2030.

Due to growing industrial IoT usage in the manufacturing, energy, automotive, and agricultural sectors as well as rising investment and technical advancement, the industrial IoT marketplace is anticipated to expand at a consistent level of

Figure 1. IIoT market growth forecast from 2020 to 2028
[https://www.linkedin.com/pulse/industrial-iot-market-industr
y-size-share-growth-forecast-hi-manshu/]

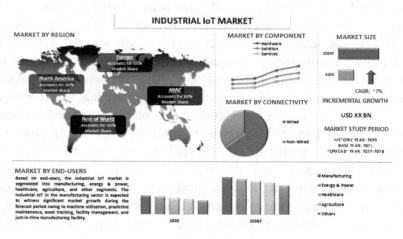

about 7%. Large players in the industry provide industrial IoT with cutting-edge features for difficult applications. For example, Cisco and the Newark firm teamed in February 2021 to deliver an industrial Internet of things network for rugged and non-carpeted environments throughout North America. The "Global Industrial IoT Market" research report by UnivDatos Market Insights (UMI) projects strong market growth throughout the course of the forecast period (2022–2028) as shown in Figure 1. This is mostly because manufacturing, oil and gas, and predictive maintenance are seeing increases in demand. Localization, the precise determination of the location of devices or assets, becomes a linchpin in the IIoT narrative. As IIoT connects disparate elements in industrial processes, the ability to pinpoint the exact location of machinery, tools, or resources becomes paramount. This integration not only enhances operational visibility but also empowers industries with granular insights into the movement and status of assets in real-time. In the context of IIoT, localization extends beyond traditional GPS-based solutions, encompassing sophisticated technologies such as Received Signal Strength Indicator (RSSI) as well as machine learning algorithms. These advancements not only refine location accuracy but also pave the way for novel applications, from predictive maintenance based on asset behaviour to dynamic supply chain optimizations driven by real-time spatial awareness. However, the journey toward a fully realized IIoT ecosystem with integrated localization is not without challenges. Security concerns, standardization efforts, and the need for seamless interoperability across diverse devices necessitate strategic considerations. As industries navigate this intricate landscape, the fusion of localization technologies with IIoT emerges as a transformative force, promising unparalleled efficiency, proactive decision-making, and a new era of industrial intelligence. In this evolving narrative, the convergence of IIoT and localization heralds a future where industries can unlock the full potential of their connected assets, ushering in an era of unprecedented control, agility, and innovation.

Current localization algorithms give top priority to the ability to choose and arrange nodes in the most advantageous locations in order to optimize coverage, data gathering, and connectivity. The position estimate capabilities of sensor nodes are critical to many Wireless Sensor Network (WSN) programs, including surveillance of targets, rescue operations, disaster assistance, and environmental monitoring. The accuracy of the localization system shows a decisive part in the localization approach's success (Alawi, A., R., 2011). A simple and practical localization technique that makes use of RSSI readings to calculate distances between the transmitter and sensor nodes is the Received Signal Strength Indicator (RSSI) ranging-based localisation system. GPS has revolutionized navigation and localization, but it has some drawbacks that restrict its usefulness in some environments. For instance, in metropolitan settings, large buildings, tunnels, as well as bridges can obstruct GPS signals, causing multipath errors, signal loss, and signal degradation. Hilly terrain, dense foliage, and wooded places can also interfere with signals, causing attenuation and mistakes. Meteorological phenomena that interfere with signals, including solar storms, can also lead to inaccurate location determination when using GPS. Security concerns about GPS spoofing as well as jamming have increased; the former modifies GPS signals to provide erroneous position data, while the latter disrupts transmissions (Al-Habashna et. al, 2022). An increasing number of systems are switching to RSSI-based localization from GPS-based systems in locations where the signal from GPS may be poor, obstructed, or non-existent. Unlike GPS, which relies on satellite signals, a wireless device's position is determined by measurements of signal strength. Consequently, it is particularly useful in confined areas where construction materials may obstruct or weaken GPS signals (Sadowski et. al, 2018) To provide accurate location estimations, a number of reference point, such as Bluetooth beacons or Wi-Fi access point, could be positioned tactically in these settings. One of the key benefits of RSSI-based techniques is that they are more cost-effective and require less power than GPS-based techniques. Furthermore, in complicated situations, RSSI-based approaches can provide more accurate location data even when other wireless devices are interfering with their signal (Yang et. al, 2021). Despite several limitations associated with RSSI-based localization, such as the possibility of mistakes resulting from external influences, advancements in machine learning as well as statistical modeling techniques are contributing to an increase in the accuracy of these systems. Because of this, the field of study surrounding RSSI-based localization is broad and promising and has the potential to have a substantial influence on a variety of applications, including asset monitoring, control of inventory, medical care, and crisis management (Potortì et. al).

There are currently some issues with the suggested statistical and stochastic localization techniques, mostly related to their poor accuracy (Filippoupolitis et. al, 2016). During the implementation phase, it is impractical to efficaciously implement these systems in real IoT devices due to their lack of precision, which presents a substantial challenge. But a different method has surfaced in the shape of computational learning-based algorithms, which present a viable way to deal with these problems (Tekler et. al, 2022). Many benefits can be obtained by using artificial intelligence for indoor localization. Primarily, the accuracy is significantly enhanced, surpassing the constraints of conventional statistical and probabilistic techniques. Machine learning algorithms have the capacity to learn large volumes of data efficiently. This enables them to recognise patterns and relationships that result in more reliable and precise localization. In addition to increased accuracy, machine learning approaches offer flexibility in use. Because these algorithms remain environment-adaptive, a wide range of indoor layouts, configurations, and architectural styles are possible. Another significant advantage of artificial intelligence for localization indoors is its affordability. The increasing prevalence of IoT devices and the burgeoning availability of cheaply priced computing resources can successfully meet the computational demands of machine learning algorithms (Zhuang et al., 2023). Machine learning-based localization methods are able to be extensively adopted and implemented due to their cost, which makes them accessible to a larger spectrum of users and applications. Another benefit that machine learning technologies offer is scalability. As the volume of data from sensors and Internet of things devices increases significantly, machine learning algorithms become more efficient at processing and interpreting the data. Furthermore, machine learning approaches might be advantageous for robust applications within buildings where accuracy and dependability are crucial. The needs of such complex environments are well handled by machine learning-based localization systems, which can be used to guide autonomous robots, make asset tracking easier, better allocate resources, or enable smooth navigation for people. Moreover, receiving signal strength (RSS) data integration with machine learning yields additional benefits. RSS allows for the utilisation of pre-existing infrastructure because it is dependent on the signals present in the surrounding environment. Because of the streamlined implementation process and reduced need for additional hardware deployment, this widespread utilisation reduces total costs.

Furthermore, a higher level of localization granularity is possible when machine learning and RSS are coupled. The system uses the detailed signal strength information from several access points or beacons to find IoT devices or individuals within an interior environment with great accuracy. This granularity enables more sophisticated applications, such as proximity-based interactions or real-time movement tracking. Compatibility is an additional advantage of combining machine learning and RSS. The present Internet of Things framework may readily integrate machine learning algorithms since a myriad of sensors as well as equipment can supply RSS information. The localization system's adaptability allows it to live in peace with different Internet of things applications and be easily integrated into a range of settings without requiring significant modifications.

For indoor localization, therefore, using machine learning-based algorithms has various advantages over traditional statistical and probabilistic methods. Because of its improved accuracy, flexibility, cost-effectiveness, scalability, suitability for dependable indoor applications, as well as additional benefits of RSS integration, machine learning is an appealing choice for building trustworthy and effective localization techniques for Internet of Things scenarios (Mohamed et al., 2022).

Self-Localization is termed as the inborn ability of humans or objects to locate themselves with respect to the surroundings. Aided Localization performs localization by incorporating electronics to locate the object. Aided Localization is further categorized into Outdoor Localization and Indoor Localization. In our day to day life, outdoor localization are familiar because of its everyday use on our smart phones or tables that helps to localize our current location by using GPS and Google maps. Indoor localization in real-time environments like residential monitoring, health monitoring, smart grids and smart metering, etc. is applied.

Extensive researches on Indoor Localization were conducted over the last 15 years, which offered diverse solutions that led to the development of several indoor localization systems using different technologies. For example, in Figure 2 shows the Radio Frequency (RF) based systems like RFID, Bluetooth, ZigBee, etc., are built based on different measurement methods like Received Signal Strength, Lateration and Angulation.

Researchers take much effort to solve the localization problem by implementing other positioning systems into the network. The following systems that are used as positioning systems are summed up as GPS, Cellular Network, Infrared device, Ultrasonic device and Micro inertial navigation.

BACKGROUND WORK ON LOCALIZATION TECHNIQUES

A huge amount of nodes are physically structured and interconnected with each other to communicate and carry out a significant task is called Wireless Sensor Network (WSN). These networks are said to be of cheap, self-configurable with no prior deployed infrastructure and easy to deploy. Hence, they are widely used in various application areas like environmental control, health monitoring, industry, earth

Figure 2. Various indoor localization systems

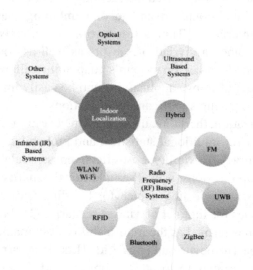

science monitoring, spatial and military surveillances. The majority of applications want the source of the sensed data. Sensor networks deploy a large number of sensory devices, with each having limited radio connectivity and processing power, to identify physical occurrences and phenomena of interest. These networks are then used by the sensors to communicate with one another. As seen in Figure 3, the sensor nodes must be aware of their locations in order to be able to pinpoint the precise location of an event. As a result, for a number of sensor network applications, sensor localization is crucial. The position or geographical coordinates of the nodes that collect data are referred to as the localization. It can be ascertained either GPS-free or using the Global Positioning System (GPS).

It is utterly unfeasible to outfit every node with GPS, as this would result in a larger and more expensive sensor node. When nodes have locations allocated to them, they are called beacon nodes; otherwise, they are called non-beacon nodes. It is the sending and receiving nodes' job to stay in sync with each other in order to calculate the distance. At least three to four beacon nodes are required in order to employ the trilateration technique or multilateration technique to locate a non-beacon node. Malicious nodes that infiltrate networks during the localization process supply inaccurate information, which reduces the accuracy of the localization and causes many WSNs to operate improperly. Because the majority of WSN algorithms in use today are susceptible to assaults in untethered environments, adversaries can readily interfere with location-dependent WSNs' regular operations by being aware

Figure 3. Need for localization

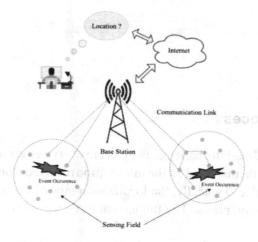

of the localization algorithms' weaknesses. In general following are the reasons to use localization in a WSN:

- **Location Stamps**: In order to localize the individual sensor measurements that is being collected.
- **Tracking**: To locate and track the events in the target environment.
- **Monitoring**: In order to monitor the diffusion of an event over a period of time, such as diffusion of forest fire over a given range.
- **Coverage Metrics**: To determine the coverage of a WSN that is determined from the range of active nodes in the WSN.
- **Load Balancing:** In densely packed networks, in order to conserve energy, some of the node in the network are shutdown. This can be done using localization of the nodes.
- **Clustering:** In application, where hierarchical routing and collaborative processing is used, only after localizing the nodes, clustering of the nodes can be done.
- **Routing:** Instead of node addresses, localization information is used in routing algorithm to provide efficient routing.
- **Spatial Querying**: Selective querying or information dissemination can be made without flooding the entire network. This in turn conserves the energy of the WSN and increases the lifetime of the WSN.

Figure 4. Basic process of localization

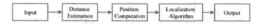

Localization Process

The goal of the localization challenge is to locate every sensor node in a particular WSN, or a subset of them. Based on the information received from the sensor nodes, this is put into practice. Typically, the localization procedure makes use of unique nodes known as beacon nodes. The fundamental steps of localization are depicted in Figure 4.

The process that occurs between selecting the localization algorithm and the nodes can be split up in a number of ways. Localization techniques can be roughly categorised into two groups based on computation: centralised and distributed localization systems, as seen in Figure 5. There are two more sorts of distributed localization techniques: range-based and range-free.

Centralized Localization Techniques

In the centralised localisation technique, all sensed data is collected and processed by the centred base station (BS), which then relays its findings downstream to the sensor nodes. Higher power and bandwidth usage during data transmission causes latency in the network. The centralised localisation approach's primary advantage is that it resolves the computation problem in practically all nodes. The fact that the data is not scalable and cannot be correctly retrieved is a drawback. These algorithms

Figure 5. Taxonomy of localization techniques

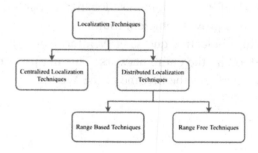

Figure 6. Centralized localization techniques

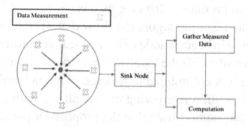

perform effectively for small-scale networks, however because they can access global information, they are more accurate than other algorithms.

The information is gathered from all of the nodes in the centralised localization strategy and sent to the sink node, or central unit, as depicted in Figure 6. The sink node determines each node's position and relays that information to the other nodes. The algorithms are able to execute sophisticated operations because it takes a significant amount of computational resources to do their operations on a central unit. By sending the results of computation to the central unit, the sensor nodes raise the cost of communication.

These include Semi-definite Programming (the SDP), Multidimensional Scaling (the MDS), The second-order Cone Programming (SOCP), Simulated Annealing based Localization (LBSA), and Multi-Dimensional Scaling-Mobile Assisted Programming (the MDS-MAP). The centralised localization technique is categorised based on the approach and they process the data at the central unit.

A novel centralised distance estimation approach was presented by Chaurasiya et al. (2014) to compute the dissimilarity matrix, which shows the distance among the network's nodes. The Multi-Dimensional Scaling oriented localization algorithm, which uses the node's coordinates in a local coordinate system, is the main contribution of the author's work. Helmert Transformation is utilised to transform a node's coordinates into a global system of coordinates. Since the nodes only communicate their local statistics to the centralised system that coordinates as localization server once for localization—thereby saving energy—the global coordinate system increases service time while reducing computational cost on individual nodes.

When a node joins the network, an algorithm creates a distance matrix that contains the node's location and the distances between it and any group of four other nodes. The technique yields the nodes' Cartesian coordinates in three-dimensional space. Many monitoring and automation systems depend heavily on localization accuracy, and the technique is only effective when there are a sufficient number of nodes in the network. Various node densities have been used to test the algorithm.

Additionally, it has been shown that if the node density drops below a specific level, the method is unable to produce a distance matrix.

The results of the localization algorithm show that: (i) the localization error has reduced with an increase in beacon nodes; (ii) the localization error has reduced with an increase in node density; (iii) the localization error has reduced with an increase in node locations within a communication range of the network; and (iv) the time needed to complete execution is growing exponentially with an increase in nodes in a network. The results demonstrate that the proposed algorithm outperforms the conventional and new centroid techniques in terms of computing overhead, energy efficiency, along with localization inaccuracy.

A unique and accurate distributed 3D localization method was used by Fan et al. (2017), and it functions well in situations when the network has an uneven shape. The method is divided into two stages: joint localization and segmentation. Applying the convex partitioning approach, the entire network is partitioned into many subnetworks during the first step. The spatial convex node classification algorithm is created that only uses the network connectivity statistics to aid in the network segmentation. In the subsequent stage, the Multi-Dimensional Scaling (MDS) technique precisely locates each subnetwork. By integrating coordinates between subnetworks and improving localization accuracy, the suggested work presents a novel 3D transform of coordinates algorithm and produces results that are error-free.

By using the maximum likelihood principle, Naddafzadeh et al. (2014) describe a resilient localization problem over an unbounded uncertain model that includes beacon placements. Since the resulting optimisation problem is non-convex, a convex relaxation results in It generates second-order cone programming, or SOCP. Through analysis, a set of nodes that are precisely placed using robust SOCP are identified, and a relationship is formed between the outcome of the proposed robust SOCP optimisation and the robust optimisation using Semi-Definite Programming (SDP). A robust SDP-SOCP localization technique has been suggested, which even in distributed large networks enjoys the advantages of increased SDP accuracy and decreased SOCP complexity.Shi et al. (2010) introduced the Sequential Greedy Optimisation (SGO) method as a greedy optimising algorithm that outperforms the traditional nonlinear Gauss-Seidel algorithm in dispersed networks. Both range-based and range-free node localization, which take into account faults in beacon placements, wireless range, along with range measurements, were taken into consideration. An edge-based SDP relaxing formulation is applied to the SGO algorithm in a proposed distributed node localization method that utilises Second-Order Cone Programming (SOCP). Furthermore, by using the SGO method on non-convex localization constructions, two distributed refinement strategies are demonstrated. Comprehensive simulation outcomes showcase the efficiency and precision of this methodology.

Figure 7. Distributed localization

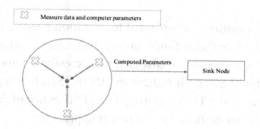

Distributed Localization Techniques

The location of each node can be computed by distributing the computation among the nodes as shown in Figure 7. It uses the computational energy of each node to run their operations but requires less energy when compared to centralized localization technique. But this technique requires huge inter-node communication to perform the computations.

The distributed localization technique is classified based on range measurements and are further be classified into range based and range free localization.

With the use of specialised hardware, this method uses the angle or distance among the nodes to determine the node's location. This method determines the precise distance between both transmitting and receiving nodes using distance estimation techniques. Therefore, this strategy uses a variety of distance estimation techniques to determine the location of each node in order to estimate the inter-node width or range. Following the computation of the distance, the position is determined using geometric concepts. Range-based localization's primary benefit is its accuracy, that is attained by adding more hardware, which raises the node's size and cost. The distributed localization technique uses sensor nodes that are placed throughout a network to find position information by communicating with one another and carrying out the necessary computing tasks. This kind of localisation is further split into range-based and range-free procedures, as was previously indicated.

Range-Free Localization Techniques

Schemes without a range are not dependent on the distance estimate metric. Many applications have been drawn to the usage of range-free localization technology due to its affordability and ease of implementation. However, in the field of range free localization, accuracy remains a research challenge.

Range Based Localization Techniques

To estimate a node's position, range-based localization needs a distance (or rather, angle) metric. To calculate the distance among the nodes or range, methods employ geometric concepts or some sort of distance metric estimate approach.

The Received Signal Strength Indicators (the RSSI), the angle of Arrival (the AOA), Times of Arrival (the TOA), and the Time Difference of Arrival (the TDOA) are a few examples of range-based localization approaches.

Received Signal Strength Indicator (RSSI)

The underlying idea of this technique is that a signal's energy diminishes as it travels farther away. The squares of a distance determine the attenuation (Bannour, A . et. al, 2021 & Kawecki et. al, 2022). The RSSI can be measured using Equation (1).

$$\text{RSSI}\left[\text{dBm}\right] = \text{RSSIsrc} - 10*n*\log10\left(\frac{\text{dist}}{\text{dist}_{\text{src}}}\right) \tag{1}$$

The unit of RSSI is dBm. The signal strength at the source is RSSI_{src} and n is a computational constant.

Angle of Arrival (AOA)

The AOA method bases the distance measurement on angle and demonstrated in the Figure 8 (Arbula et. al, 2020).

As depicted in Figure 8, the beacons A_1 and A_2 receives signal from M at the angles shown. The position of M based on the measure of the angles shown in Equations (2) and (3):

$$X_M = X_1 + \frac{d*sin\left(\pm\right) + sin\left(^2\right)}{sin\left(\pm +^2\right)} \tag{2}$$

$$Y_M = \frac{d*\cos\left(\pm\right)sin\left(^2\right)}{sin\left(\pm +^2\right)} \tag{3}$$

Figure 8. Angle-of-arrival

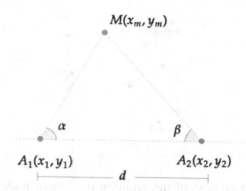

Time of Arrival (TOA)

As seen in Figure 9, the Time-of-Arrival (TOA) approach measures range by using the broadcast signal's time of arrival at the receiver. This method requires additional hardware support and requires synchronisation between the sender and the receiver in order to estimate arrival time accurately (Du et. al, 2018).

Time Difference of Arrival(TDOA)

Time Difference Of Arrival, otherwise known as multilateration, is a well-known approach for the geo-location of RF emitters. TDOA based algorithm requires three or more receivers and need to be synchronous with each other. It is used to identify the source of the signal from the receivers having different arrival times as shown in Figure 10.

Figure 9. Time-of-arrival

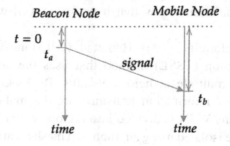

Figure 10. Time difference-of-arrival

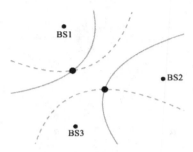

Localization is the process of determining the sensor nodes' location. Sensor nodes that have previously been assigned a location are referred to as beacon nodes; these locations can be determined using GPS, or the Global Positioning System, or through placing beacon nodes at predetermined locations. Unknown nodes, sometimes referred to as non-beacon nodes, are those that lack location data and typically get their information from beacon nodes.

Benslimane et al. (2014) introduced the Approximation Techniques (AT-Family) for wireless sensor networks, comprising three distributed localization methods: angle measurement (AT-Angle), where sensors calculate angles; signal measurements (AT-Dist), where sensors estimate distances; and free-measurement (AT-Free), where sensors lack measurement capability. AT-Free, AT-Dist, and AT-Angle determine each sensor's position and its accuracy. Notably, sensors can contribute to other nodes' location estimation and reject inaccurate positional input. This feature aids in managing measurement errors, a primary flaw in methods like AT-Dist and AT-Angle (Sarala et al., 2023). Simulations considering measurement errors and low beacon density demonstrate the effectiveness of these approaches in achieving accurate localization and energy-efficient operation.

Boustani et al. (2014) developed a novel and deterministic method using a CDMA (Code Division Multiple Access) driven jamming strategy to secure beacon-based position detection. Using this method, the fraudulent nodes are identified during localization and eliminated. Comparing this approach to other safe localization schemes, thorough simulations show that it performs well when it comes to of localization accuracy.

In order to locate unfamiliar nodes, Han et al. (2013) introduced a novel Two-Step Secured Localization (TSSL) method that uses the arrival time variance of localization against multiple malicious assaults. By looking at the positions, identities, and timing of information transmission, the malicious nodes can be located. Additionally, the WSN is divided into regions with different trust ratings and malicious nodes are isolated using an improved mesh creation mechanism. Test bed findings demonstrate that the TSSL locates malicious nodes effectively and with high accuracy in localization in simulation(Kumar. B.P et. al, 2023).

Yousefi et al. (2014) proposed a resilient two-stage distributed approach for localizing a cooperative sensor network without Non-Line-of-Sight (NLOS) identification. This method utilizes Time-of-Arrival (TOA) data and incorporates iterative optimization techniques. In the first stage, a convex relaxation of a Huber loss function is applied to mitigate the impact of outliers and estimate the true sensor locations. The gathered location data from the first stage is then used to further optimize the initial Huber cost function in the second stage, leading to more accurate location estimations. The optimization process in both stages employs a simple gradient descent method. The suggested convex relaxation demonstrates a lower Root Mean Squared Error (RMSE) compared to existing methods. In the second stage, the algorithm improves the RMSE value, particularly in Non-Line-of-Sight (NLOS) scenarios, almost reaching the level of other distributed techniques. The two-stage algorithm proves robust against outliers, particularly outperforming previous distributed techniques when the proportion of NLOS measurements is low.

In another approach by Wang et al. (2012), the focus is on the Received Signal Strength (RSS) based localization problem with unknown transmit power and unknown Path Loss Exponent (PLE). When the transmit power is unknown, a Weighed Least Squares (WLS) formulation, based on the Unscented Transformation (UT), is derived to evaluate the sensor node location and transmit power. In the case of an unknown Path Loss Exponent (PLE), an alternating estimation procedure is devised to estimate both the sensor node location and the PLE. This estimation technique is applicable when both the transmit power and PLE are unknown. Simulation results show the efficacy of this method compared to existing Semi-Definite Relaxation (SDR), Linear Least Squares (LLS), and Near Maximum Likelihood (NML) methods, particularly in terms of complexity (Kalpana et al., 2023). However, it's noted that range-based systems are impractical for Wireless Sensor Networks (WSNs) due to their high costs and unfulfilled promises.

Li et al. (2010) proposed a new range-free system dubbed the REndered Path (REP) protocol to replace the current system, which has low scalability and poor accuracy. By deriving the shortest paths connecting the nodes, REP exploits the geometric characteristics of the network to disseminate such information. In order to determine the distances between two nodes, REP constructs virtual shortest pathways using the virtual-hole concept. Based on simulation findings, REP has shown to be very scalable and effective, and it may be used in isotropic networks.

Yeredor et al. (2014) proposed a collaborative and recursive estimating scheme based on Time-of-Arrival (TOA) suitable for ad hoc networks characterized by partial and non-symmetric connections, without the necessity for precise sensor synchronization. The method initiates with a single transmission from each sensor, followed by the reception of Time-of-Arrival data from neighboring sensors (Dhanalakshmi, R, 2023). Subsequently, each sensor engages in an iterative process to evaluate its own timed offset and position, along with the timing offsets and positions of relevant neighboring sensors. The sensor then broadcasts its approximation results. Through this iterative process, the convergence of the technique facilitates the determination of the locations of adjacent sensors, especially when there is an average connection ratio of 1:3. This collaborative and recursive estimating scheme provides a practical approach for ad hoc networks with varying connectivity and asynchrony among sensors.

Lazos *et al.* (2005) has been discussed about a range-free, decentralized localization scheme that enables sensors to conclude their location in an un-trusted atmosphere with the help of small range of trusted entities. It requires only fewer reference points and has a lower communication cost (Kalpana, A.V., 2019). This method gives higher accurate results, than other methods, even in the presence of wormhole and Sybil attack, the algorithm performs in an assumption of trusted beacons.

Jadliwala et al. (2011) proposed an approach for beacon-based localization in a resilient environment, presenting three secure localization techniques, two based on heuristics, and one with polynomial time complexity. These methods demonstrate improved localization accuracy, even in the presence of dishonest beacon nodes. However, their applicability is limited to 2D environments.

In a similar vein, Liu et al. (2005) developed two methods for reliable localization in the presence of deceitful beacon nodes. The first method filters out malicious beacon nodes based on irregularities across multiple beacon signals, while the second employs an iteratively revised voting plan to accept signals from potentially malicious beacons. Both strategies contribute to enhancing localization reliability in the face of deceptive nodes.

Fiore et al. (2013) introduced the Neighbour Position Verification technique, allowing every node in a network to verify the position of its communicating neighbors without relying on pre-existing reliable nodes. This fully distributed and cooperative approach remains resilient against both independent and collaborating adversaries, even when the attacker possesses complete knowledge about the verifying node's vicinity. The protocol's effectiveness deteriorates only in the presence of a significant number of collaborating adversaries in the vicinity of the verifying node, as indicated by simulation findings, which demonstrate the protocol's ability to thwart a greater number of attacks even in adversarial scenarios.

The main objectives of the research are:

- To incarnate Euclidean distance and angular guaranteed configuration technique in 3D ASCA (Adaptive Stochastic Control Algorithm) algorithm for obtaining better localization of nodes

- To detect Sybil attack and securely transmit data through nodes by using Secure 3D algorithm
- To obtain robust 3D node localization utilizing barycentre intersection for node localization through 3D ROLOC(RObust LOCalization) algorithm
- To reduce multilateration error, achieve robustness and increase localization accuracy
- To develop a simple and energy efficient algorithm, to achieve higher localization accuracy in 3D space

PROPOSED METHODOLOGY

System Design

The technique of locating sensor nodes in a 3D region—such as hills, valleys, mountains, and other places—is known as 3D localization. In comparison with 2D localization algorithms, the following problems arise while implementing 3D localization algorithms:

(i)   In 3D localization, a minimum of four beacon nodes is required, whereas only three are needed for 2D localization.

(ii) Transmission signal degradation occurs due to obstructions, impacting localization accuracy by introducing slight variations in distances measured by the Received Signal Strength Indicator (RSSI).

(iii) Consider four beacon nodes denoted as A, B, C, and D at positions $(x_1, y_1, z_1), (x_2, y_2, z_2), (x_3, y_3, z_3), (x_4, y_4, z_4)$ respectively, as illustrated in Figure 11. The target position to be determined, denoted as K, has coordinates (x, y, z).

The distances between beacon nodes and unknown node are found as d_{KA}, d_{KB}, d_{KC} and d_{KD}, which is shown in Equation (4), (5), (6) and (7).

$$\sqrt{(x_1 - x)^2 + (y_1 - y)^2 + (z_1 - z)^2} - d_{KA} = 0 \qquad (4)$$

$$\sqrt{(x_2 - x)^2 + (y_2 - y)^2 + (z_2 - z)^2} - d_{KB} = 0 \qquad (5)$$

Figure 11. Distance based localization in the presence of cheating beacon which affects the localization process

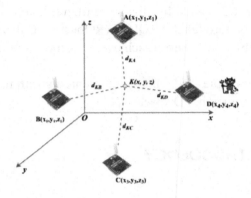

$$\sqrt{(x_3 - x)^2 + (y_3 - y)^2 + (z_3 - z)^2} - d_{KC} = 0 \qquad (6)$$

$$\sqrt{(x_4 - x)^2 + (y_4 - y)^2 + (z_4 - z)^2} - d_{KD} = 0 \qquad (7)$$

The coordinates of the unknown node K, represented as (x, y, z), are determined by solving the system of equations, as outlined in Equation (8). The process involves utilizing mathematical operations or algorithms to find values for x, y, and z that satisfy the given set of equations. Equation (8) encapsulates the relationships and constraints defining the spatial positioning of the unknown node K, and by solving these equations, one can ascertain the specific values for x, y, and z that characterize the location of node K in the given coordinate system as in Equation (4), (5), (6) and (7):

$$\begin{bmatrix} x \\ y \\ z \end{bmatrix} = \frac{1}{2} \begin{bmatrix} x_2 - x_1 & y_2 - y_1 & z_2 - z_1 \\ x_3 - x_1 & y_3 - y_1 & z_3 - z_1 \\ x_4 - x_1 & y_4 - y_1 & z_4 - z_1 \end{bmatrix} \cdot$$

Figure 12.

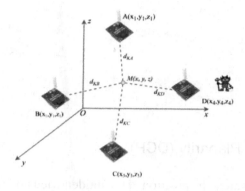

$$
\begin{bmatrix}
(x_2^2 - x_1^2) + (y_2^2 - y_1^2) + (z_2^2 - z_1^2) + \left(d_{KA}^2 - d_{KB}^2\right) \\
(x_3^2 - x_1^2) + (y_3^2 - y_1^2) + (z_3^2 - z_1^2) + \left(d_{KA}^2 - d_{KC}^2\right) \\
(x_4^2 - x_1^2) + (y_4^2 - y_1^2) + (z_4^2 - z_1^2) + \left(d_{KA}^2 - d_{KD}^2\right)
\end{bmatrix}
\tag{8}
$$

Quadrilateration is used to get the coordinates of the unidentified node K given (x, y, z). Figure 12 illustrates how the localization approach fails in location computation due to malicious beacons that fabricate location information. While beacon D forges the position coordinate and erroneously estimates the destination location, which is known as M rather than K, the nodes A, B, and C operate as expected.

Every beacon node has a controlling device to hold an authenticated Identity (ID) in addition to GPS. Every time a non-beacon node requests the destination location, the beacon node requests the authentication ID. Prior to transmitting any messages, this ID must be registered and obtain the public authentication key.

Terminologies and Notations

(a) Closeness Centrality (CC)

The closest path, or even the shortest path, between any two nodes is what is referred to as closeness centrality. Figure 13 shows this definition. From the nearest nodes, A, E, and G, node B receives location data.

Figure 13. Closeness centrality (CC)

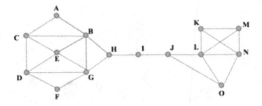

(b) Degree of Co-Planarity (DCP)

In a two-dimensional area, the position of the unidentified node can be found with at least three beacon nodes; in a three-dimensional area, quadrilateration—the use of at least four beacon nodes—is employed. It is always believed that the quadrilateration procedure is the intersection of three spheres, with the target node typically located in the centre of the intersection. When calculating the RSS values or the spherical region, errors may occur and there may not be a common intersecting region. As seen in Figure 14, the node E obtains numerous distance measurements using the available beacon nodes. The un-localized node has only two possible positions. The locations are identified as E and E', respectively. There is no distinction between nodes E and E' if they are not coplanar adjacent to the beacon node.

Coplanarity is the state in which all four beacon nodes are located in the same plane. Therefore, it must be confirmed that the confirmed beacon node is not situated in the same planes as the unknown node in order to determine its location. Figure 15 illustrates how DCP can be expressed as a ratio of four beacons nodes in 3D

Figure 14. The non-localized node cannot be localized due to DCP

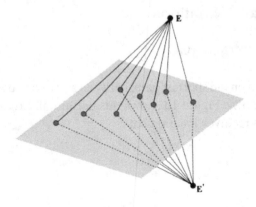

Figure 15. Degree of co-planarity (DCP)

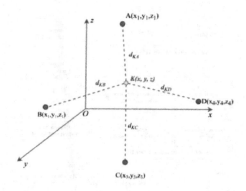

space. Equation illustrates how Mitrinovic et al. (1989) represented the tetrahedron's radius ratio in Equation (9).

$$= \frac{216v^2}{\sum_{i=0}^{3} s_i \sqrt{(a+b+c)(a+b-c)(a+c-b)(b+c-a)}} \tag{9}$$

Degree of Co-planarity (the DCP) can be embodied as presented in Equation (10).

$$\text{DCP} = \begin{cases} 0, coplanar \\ \rho, otherwise \end{cases} \tag{10}$$

where DCP $\in [1,0]$.

Thus, the greatest DCP value can be used to find the optimal localization unit.

PROPOSED 3D ASCA ALGORITHM

The algorithm comprises of three phases as shown in Figure 16 namely pre-processing, beacon node selection and 3D localization process. In the pre-processing phase, the nodes are deployed in the 3D region and initialized. The second phase called as beacon node selection phase, where the beacon nodes are selected based on few parameters to get the appropriate beacon node. In the third phase, the distance is estimated and subsequently location of the target node is obtained.

Figure 16. Proposed 3D adaptive stochastic control algorithm (3D - ASCA)

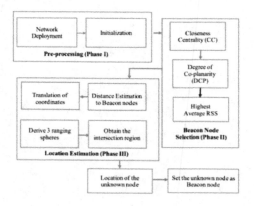

Pre-processing (Phase I)

Initially, the WSN is deployed in a 3D region as shown in

Figure 17. Then, the initialization parameters for the network should be set. The localization process is purely done based on RSSI. The path loss model should be determined based on the environment.

In wireless communication, RSSI is used as a function to specify the state of the power that is received in the sensor nodes. Equation (11) illustrates how Heurtefeux et al. (2012) demonstrated that, in an ideal setting, the RSSI value is inversely correlated to the distance between an unidentified node and beacon node.

Figure 17. Deployment of sensor nodes in a 3D region

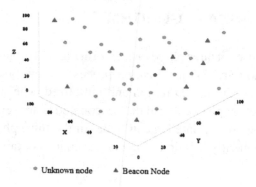

$$P_r(D) = P_t + G_t + G_r + 20log_{10}\left(\frac{\lambda}{4\pi D}\right) \tag{11}$$

where P_t = Transmission power of antenna in dBm

G_t = Antenna gain of the transmitting signal in dBm

P_r = Receiving power of antenna

G_r = Antenna gain of the receiving signal in dBm

λ = Wavelength of the signal in m

D = Distance between the antennas

The RSSI value is also based on the environmental conditions and therefore difficult to regulate antenna gains. Hence, a simplified formula used to relate the distance and RSSI as in Equation (12).

$P_r(D) = P_{r1} - \beta .log_{10}(D)$ (12)

where P_{r1} = receiving power for a specific distance (1m) in dBm

β = Path loss parameter

D = Distance between the transmitter and the receiver

It is always essential to consider P_{r1} and β as a vital parameter during experimental analysis, which are always determined empirically.

The RSSI method is used to find the path loss or signal attenuation which is derived from the distance between the transmission and receiving power. The path loss also gets varied for different environments and hence the characteristics of the real time environment can be determined as pass loss exponent (β) as in Equation (12).

The location of the experiment conducted in outdoor or indoor environment influences the path loss model. The RSSI characteristics are influenced by the multipath fading effect and signal propagation between the transmitter and receiver. The presence of multiple reflections causes electromagnetic waves to traverse different routes of varying lengths. These reflections contribute to the complexity of signal propagation, leading to fluctuations in received signal strength as the waves interfere constructively or destructively due to their varied paths. Understanding these phenomena is crucial for accurately interpreting RSSI measurements and optimizing communication systems in environments where multipath effects play a significant role.The multiple reflections are obtained due to multipath fading that affects the signal strength, which in turn obtained from distance between transmitter and receiver. Therefore, RSSI value purely depends upon the environment and the propagation model.

Figure 18. Two-ray received model

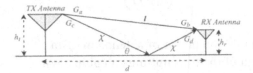

(a)The Free Space Propagation Model

The free space propagation exemplary should be used, as soon as transmitter and receiver continues in the Line-Of-Sight (the LOS) without any obstacles. Michael Tsai *et al.* (2011) specified the received power by the receiving antenna as in Equation (13), which is used to find the ratio of received and transmitted power.

$$P_r(D) = \frac{P_t G_t G_r \lambda^2}{(4\pi)^2 d^2} \tag{13}$$

Where P_t = Transmission power of antenna in dBm

G_t = Antenna gain of the transmitting signal in dBm

P_r = Receiving power of antenna

G_r = Antenna gain of the receiving signal in dBm

λ = Wavelength of the signal in m

d = Distance between the antennas

(b)The two-ray received power model

When there is attenuation in the ground reflecting path and a direct path from the transmitter as well as the receiver, as shown in Figure 18, the two-ray power received model ought to be applied.

The received power can be obtained using the Equation (14).

$$P_r(d) = P_t G_a G_b \frac{h_t^2 h_r^2}{d^2} \tag{14}$$

Where G_a, G_b = Free space model of the transmitter and the receiver antenna gain

h_t, h_r = Height of the transmitter and the receiver antenna

(c) Path-Loss Normal Shadowing Model

Path-loss normal shadowing model shows the relationship between distance and received power as in the Equation (15).

$$P_L(d) = P_L(d_0) + 10\beta log\left(\frac{d}{d_0}\right) + X_\sigma \tag{15}$$

where $P_L(d)$ = path loss in specific distance d in decibels and it derived from $log(Pt_{/}r)$

$P_L(d_0)$ = path loss for the specific distance d_0 (d_0 is one meter in indoor environment)

β = Path-loss exponent

X_σ = Random shadowing effect with zero mean and σ^2 is the variance

If the sensor node has transmitted power (P_t) and the received power (P_r) of the beacon node at the distance d can be obtained using the Equation (16).

$$P_r(d) = P_t - P_L(d) \tag{16}$$

Using Equation 15 and 16, $P_r(d)$ can be obtained as shown in Equation (17).

$$P_r(d) = P_t - P_L(d_0) - 10\beta log\left(\frac{d}{d_0}\right) + X_\sigma \tag{17}$$

Beacon Node Selection (Phase II)

In beacon node selection, the unknown node evaluates its position from three beacon nodes, so the selection of beacon node is an imperative criterion for the proposed algorithm. The beacon node is selected based upon Closeness Centrality and Degree of Co-planarity, which were already discussed in section 3.3.2. Apart from that, an additional factor called as Highest average RSS also taken into account to select the best beacon node.

Figure 19. Highest average RSS

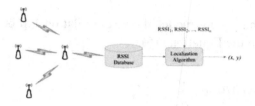

Highest Average RSS Value

Small scale signal variations affect the RSS value and proved a single RSS value cannot be effective in computing the distance. Therefore, an average or median value of RSS collected for more than 20 nodes at the same location will reduce the small scale signal variations as shown in Figure 19. It is very effective when the highest average RSS can be chosen as the input for the localization algorithms.

Location Estimation (Phase III)

The location of the unknown node can be estimated using the following methods:

(a) Trilateration for 2D
(b) Quadrilateration for 3D

(a) Trilateration for 2D

In a 2D space, the process of utilizing distance measurements obtained from three non-collinear beacon nodes, labeled as A, B, and C in Figure 20, to estimate the location of an unknown node is referred to as trilateration. This method relies on determining the point of intersection of three circles, each representing the calculated distance from one of the beacon nodes, to pinpoint the position of the unknown node. However, it's important to note that the RSSI values obtained through this trilateration method are dynamic and may contain errors attributed to signal propagation dynamics. These errors can arise due to factors such as multipath effects, interference, or signal attenuation, impacting the accuracy of the estimated node location.

(b)Quadrilateration for 3D

In a 3D space, the process of utilizing distance measurements obtained from more than three beacon nodes, specifically labeled as A, B, C, and D in Figure 21, to estimate

Figure 20. Trilateration

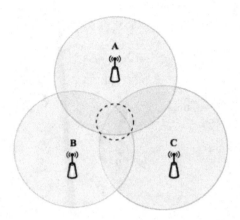

the location of the unknown node (denoted as E) is known as quadrilateration. This method involves determining the intersection point of spheres corresponding to the calculated distances from each of the beacon nodes. By considering the distances from these multiple beacons, quadrilateration enables the estimation of the three-dimensional coordinates of the unknown node E. This approach is particularly useful in scenarios where the three-dimensional positioning of a target node needs to be determined using distance information from multiple reference points. A node cannot locate its position unless it lies on the same plane as the others and no matter the number of distance measurements are available.

(c)Translation of Cartesian Coordinates to Spherical Coordinates

To identify the location of the unknown node in a 3D region, the Cartesian coordinates are converted into spherical coordinates as shown in Figure 22.

Using the Equation (18), (19) and (20), the spheres are constructed with the closure is also shown in Figure 22.

$$r = \sqrt{x^2 + y^2 + z^2} \tag{18}$$

$$\theta = \arccos\left(\frac{z}{r}\right) \tag{19}$$

Figure 21. Quadrilateration

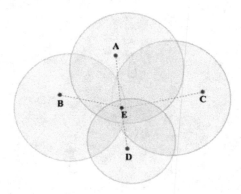

Figure 22. Translation of Cartesian coordinates to spherical coordinates

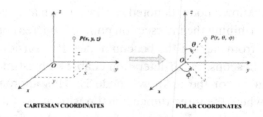

$$\varphi = \arctan\left(\frac{y}{x}\right) \tag{20}$$

Location Estimation

The spheres S_1, S_2 and S_3 as shown in Figure 23 overlap at $q_{1,2,3}^{(1)}$ and $q_{1,2,3}^{(2)}$ which is closer to the surface of the new sphere called S_4 and concluded as the target location of the unidentified node K which is shown in Equation (21).

Figure 23. Intersection of spheres with their closure

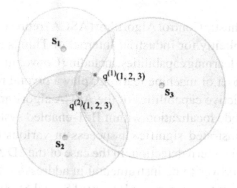

Proposed 3D ASCA Algorithm

Step 1:	Start the deployment of sensor nodes and initialise the network
Step 2:	Set the network with M unknown nodes and N sensor nodes
Step 3:	Select an unknown node, whose location to be found
Step 4:	Check whether there exists a minimum of 4 beacons for a 3D localization
Step 5:	Select a minimum of 4 beacons with highest average RSS, highest CC and highest DCP
Step 6:	Calculate the distance estimation to beacon nodes through Quadrilateration
Step 7:	Translate the Cartesian coordinates to polar coordinates
Step 8:	Derive 3 ranging spherical regions
Step 9:	Derive local positions for the reference beacon nodes
Step 10:	Estimate the distance to beacon using RSSI
Step 11:	Check whether any beacon exists without the estimated distance
Step 12:	Calculate the location of the unknown node
Step 13:	Set the un-localized node to localized node, which further helps in the localization process till all the nodes gets localized.

Integration of CNN With ASCA Algorithm

The 3D Adaptive Stochastic Control Algorithm (ASCA) represents a groundbreaking solution designed explicitly for Industrial Internet of Things (IIoT) environments, enriched with machine learning capabilities, including Convolutional Neural Networks (CNNs). CNNs, a subset of machine learning, play a pivotal role in enhancing the adaptability and predictive capabilities of the ASCA algorithm, particularly in the context of accurate node localization within IIoT-enabled systems.

CNNs have demonstrated significant success in various domains, notably in image recognition and pattern detection. In the case of the 3D ASCA algorithm, the integration of CNNs proves to be instrumental in addressing the challenges posed by industrial environments. Leveraging Received Signal Strength Indicator (RSSI) data, the ASCA algorithm utilizes sophisticated machine learning algorithms, with a specific emphasis on CNNs, to precisely determine node positions in a 3D industrial space.

The infusion of CNNs into the ASCA algorithm brings forth their inherent strengths in feature extraction and pattern recognition, making them well-suited for the intricate demands of industrial scenarios. This integration aligns with the evolving needs of the industrial sector, offering a sophisticated and intelligent solution for precise 3D localization within IIoT applications.

The Suggested CNN-Based Localization Framework's System Model

The system model comprises two phases, as illustrated in Figure 24: an online phase for determining the location of sensor nodes in the studied area using the trained model and an offline phase involving data gathering and pre-processing for input to the localization CNN framework, along with the training of the framework. In Figure 25, the investigated region is initially divided into distinct sections known as "classes." Each region is assigned a class label, represented by q, where $q \in 1$, 2, ..., Q, and Q denotes the maximum number of classes. The decision to define each class's area is influenced by available computing power and the desired level of localization accuracy for the application.

Preparing the RSSI Information

The term "preprocessing" of RSSI data describes the adjustments made to the input data prior to feeding it into the CNN model. The many procedures and methods (RSSI acquisition, RSSI normalization, and kurtosis computation) used to expedite training and produce strong classification performance are explained in depth.

Figure 24. Different phases of our CNN dependent localization technique

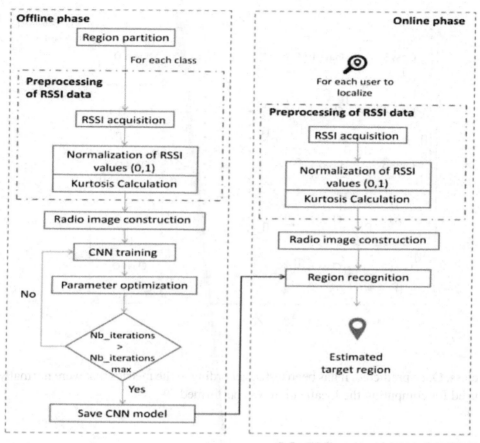

Target Localization and Model Training

Finding the ideal set of weights to optimize the accuracy of the model is referred to as training. It has to do with getting the highest possible categorization score. An optimization algorithm's backpropagation technique was employed for this. Following learning (in the offline training phase), our model could precisely locate a sensor node inside its designated area. After obtaining RSSI values and preprocessing the data, an RF image with the same dimensions and structure as those utilized in training was created in order to determine the position of a sensor. To predict the region that the sensor node corresponded to, this image was supplied to the trained model. Each class was given a probability for this, and the total of this probability was one. The class with the highest probability corresponding to it was the predicted

Figure 25. Region partition

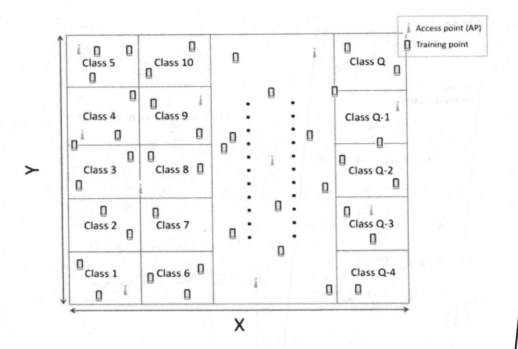

class. Once predicted, it has been tested according to the metrics that were normally used for computing the localization was performed.

PERFORMANCE METRICS

Average Localization Time

The average localization time is defined as the mean duration required by sensor nodes in a network to calculate their respective locations, expressed as a ratio to the total number of nodes in the network, as represented by Equation (22).

$$\text{Average localization time} = \frac{\sum Localization\_time}{Number\_of\_nodes} \tag{22}$$

Localization Error (LE)

The Localization Error (LE) is the disparity between the real coordinates and the estimated coordinates of nodes. The calculation of LE for each target node is demonstrated in Equation (23):

$$LE = \sqrt{(X_j - X_i)^2 + (Y_j - Y_i)^2 + (Z_j - Z_i)^2} \tag{23}$$

Average Location Error (ALE)

The average distance between the estimated location (X_j, Y_j, Z_j) and the actual location (X_i, Y_i, Z_i) of all the sensor nodes is estimated using Equation (24).

$$Average\ location\ error = \sum \frac{\sqrt{(X_j - X_i)^2 + (Y_j - Y_i)^2 + (Z_j - Z_i)^2}}{Number\_of\_nodes} \tag{24}$$

Detection Rate (DR)

Equation (25) illustrates the percentage, or "detection rate," of both the number of Sybil nodes and the proportion of detected Sybil nodes.

$$DR = \frac{Detected\_Sybil\_node}{Total\_Sybil\_nodes} \times 100\% \tag{25}$$

Localization Error Ratio (LER)

Equation (26) outlines the meaning of the Localization Error Ratio (LER), representing the ratio between the mean localization error and the communication range of nodes.

$$LER = \frac{\sum_{i=1}^{n} \sqrt{(x_n - x_n')^2 + (y_n - y_n')^2}}{n \times R} \tag{26}$$

Localized Node Proportion (LNP)

The fraction of nodes that are successfully localised (nSL) to the network's unknown nodes is known as LNP. Equation (27) displays the degree of positional coverage, which is referred to as LNP.

$$LNP = \frac{n_{SL}}{n} \qquad (27)$$

Bad Node Proportion (BNP)

Nodes classified as "bad" are those whose localization errors exceed their communication ranges. According to Equation (28), BNP is the ratio of the number of bad nodes ($n_{badnodes}$) to the total amount of nodes which have been successfully localised (nSL). The algorithm's stability metric is known as BNP.

$$BNP = \frac{n_{badnodes}}{n_{SL}} \qquad (28)$$

In addressing the challenge of localization, researchers have invested considerable efforts, primarily relying on received signal information for calculations involving Time of Arrival, Time Difference of Arrival, Angle of Arrival, and Received Signal Strength Indicator (RSSI). In the contemporary era of smart cities, a diverse range of applications in Wireless Sensor Networks (WSN) has emerged. The localization of indoor environments plays a crucial role in applications such as tracking and monitoring within smart buildings.

Achieving accurate indoor localization has become a significant success, demanding the development of low-cost and robust real-time systems. While Global Positioning System (GPS) has been widely adopted for outdoor 2D and 3D localization due to its simplicity, ease of integration, and high accuracy, its utility diminishes in confined indoor spaces. In such environments, GPS faces limitations and drawbacks, prompting the need for alternative localization systems tailored to the challenges posed by limited indoor space.:

(i)   More Energy Consumption
(ii)   Expensive when large number of nodes are used
(iii) Less accuracy in critical usage (upto 15m deviation)
(iv)   Takes long time to acquire a satellite fix

Figure 26. Effect of LNP with number of nodes

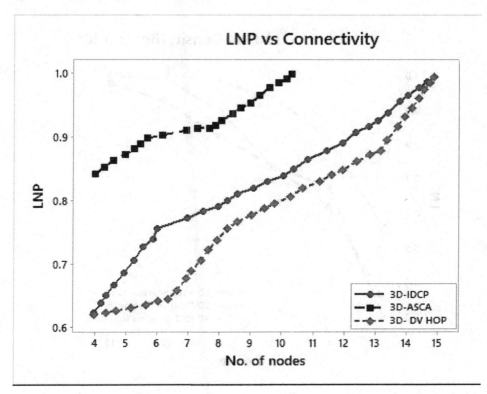

RSSI cannot definitely replace GPS, but wherever GPS not feasible, there RSSI is the right option to adhere. Hence, there is a greater attraction towards RSSI for indoor localization. The most common information available in every transceiver is RSSI. The proposed localization algorithm called 3D ASCA (Adaptive Stochastic Control Algorithm) locates the position of the node accurately and also gives a cost-effective solution in a 3D environment.

Effect of Localized Node Proportion

As the communication range expands, the Localization Node Placement (LNP) performance of the algorithm tends to show a gradual increase. This is attributed to the fact that the 3D ASCA method successfully localizes a larger number of nodes compared to both the 3D IDCP and 3D DV-Hop algorithms. The 3D-ASCA algorithm notably outperforms the other two approaches. Achieving a localization coverage of 1 when the network connectivity approaches 10 ensures that all unknown nodes can be accurately located. Figure 26 visually demonstrates the superior performance of

Figure 27. Effect of LNP with density (Beacon 10%)

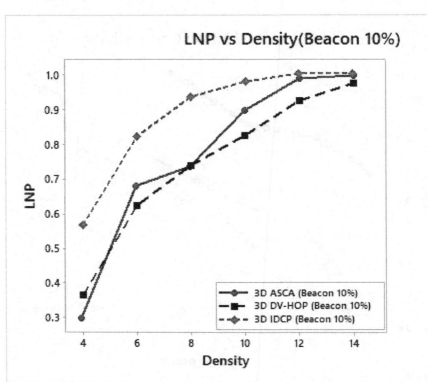

the 3D ASCA algorithm over the 3D IDCP and 3D DV-Hop algorithms in terms of localization coverage, showcasing improvements of up to 21% and 25%, respectively.

When the number of beacon nodes in the localization area gets increased, it increases the localized node proportion. Due to the introduction of Closeness Centrality and Degree of Co-planarity, the number of nodes that gets localized tends to be high in the proposed 3D ASCA algorithm, when compared to 3D DV-Hop and 3D IDCP. The effect of LNP for varying density of beacon nodes, say 10% and 20% is shown in Figure 27 and 28 respectively.

Effect of Bad Node Proportion (BNP)

The BNP of the suggested 3D ASCA is displayed in Figure 29 along with two alternative 3D localization algorithms, 3D-DV-Hop and 3D-IDCP. There are no malfunctioning nodes in the localization whenever the network connectivity reaches ten. Related to 3D DV-Hop and 3D-IDCP algorithms, 3D-ASCA has less problematic nodes. There are no malfunctioning nodes throughout the localization procedure

Figure 28. Effect of LNP with density (Beacon 20%)

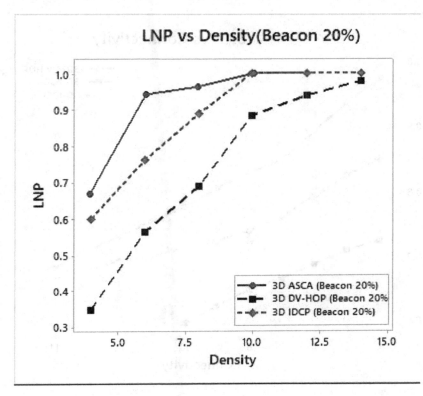

whenever the network connectivity reaches about 10. Locating a beacon node becomes more likely as the number of beacon nodes increases. As seen in Figure 30, the suggested 3D ASCA algorithms has a BNP that is less than 0.2, while the BNPs of the 3D IDCP as well as 3D DV-Hop are very high, ranging from 0.6 - 0.05 with a 10% beacon density.

In Figure 31, 3D IDCP algorithm has a BNP ranging from 0.32 to 0.03, whereas for 3D DV-Hop has a BNP ranging from 0.25 to 0.03. The proposed 3D ASCA algorithm has a lesser BNP ranging from 0.15 to 0.01, which is very minimal associated to 3D IDCP and 3D DV-Hop, due to the introduction of coordinate system. The polar coordinates make the localization process simpler and easier to calculate, when associated to 3D DV-Hop and 3D IDCP.

Effect of Localization Error Ratio (LER)

Figure 32 indicates the Localization Error Ratio (LER) for the proposed 3D ASCA algorithm with the other 3D IDCP and 3D DV-Hop. The sensor nodes are randomly

Figure 29. Effect of BNP vs connectivity

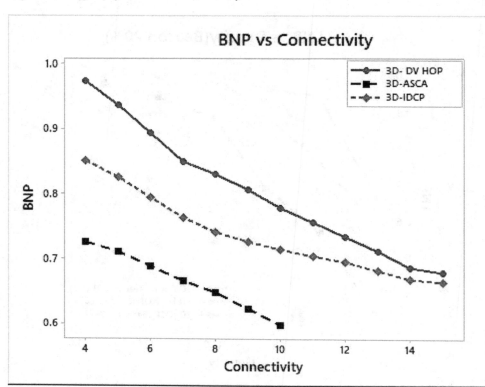

plus differently deployed each time. The LER is greatly reduced from 0.68 to 0.45 in the proposed 3D ASCA algorithm. The 3D IDCP has LER ranging from 0.75 to 0.58 and 3D DV-Hop lies in the range from 0.76 to 0.66, which is high compared to the proposed 3D ASCA algorithm.

The effect of beacon nodes with a density of 10% on Localization Error Ratio (LER) for the suggested 3D ASCA procedure is analysed and presented in Figure 33.

Initially, when there are less number of beacon nodes exists in the deployment region, the LER tends to be in the range from 0.52 to 0.42. When the node gets localized in an iterative manner, the proportion of beacon node gets increased, which in turn lowers the LER. The LER for the proposed 3D ASCA algorithm tends to be lower than 3D-DV Hop and 3D-IDCP algorithms.

The effect of beacon nodes with a density of 20% on Localization Error Ratio (LER) for the proposed 3D ASCA algorithm is analysed and shown in Figure 34. Initially, when there are a smaller number of beacon nodes exist in the deployment region, the LER tends to be in the range from 0.47 to 0.35. It is evident that by varying number of beacon nodes in the sensing region, lowers the LER.

Figure 30. Effect of BNP with density (Beacon 10%)

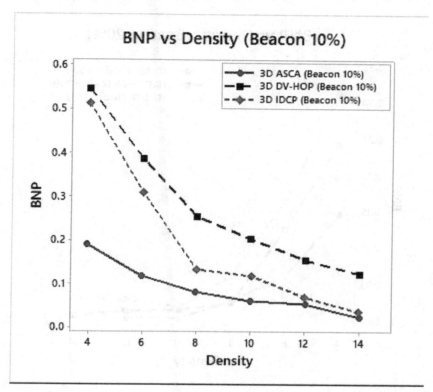

From the simulation outcomes, it stood proved that the proposed 3D ASCA system has a LER from 0.52 to 0.35 with 10% and 20% beacon nodes.

Test Bed

A PIC16F877A microcontroller which uses Tarang was used to create a sensing phenomenon, due to its availability and popularity. Tarang module works with a voltage of 3.3V, which has the processing unit consuming small voltage. Therefore, a second regulation circuit is not essential.

The power required is as low, so the frequency oscillator is set as 1MHz. Two USART interfaces are required to send data to Tarang module and for serial interfacing with the computer. The default baud rate of Tarang module is 9600 and used to configure the settings of the PIC for functioning on a baud rate of 1MHz frequency. The specifications of PIC16F877A are shown in Table 1.

Figure 31. Effect of BNP with density (Beacon 20%)

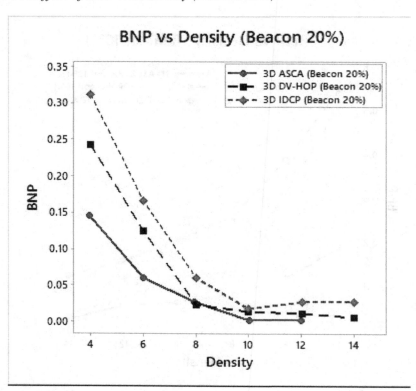

Addressable Universal Synchronous Asynchronous Receiver Transmitter (USART)

one of the two serial input/output modules in electronic systems. The USART offers versatile communication capabilities and can be configured to operate in different modes to accommodate various types of peripherals.

In its half-duplex synchronized mode, the USART can establish communication with peripherals such as Analog-to-Digital (A/D) or Digital-to-Analog (D/A) circuits, serial Electrically Erasable Programmable Read-Only Memory (EEPROMs), and similar devices. This mode is particularly useful when a system needs to exchange data bidirectionally, but not simultaneously, with these types of peripherals.

On the other hand, the USART can be configured as a full-duplex asynchronous system. In this mode, it is capable of simultaneous bidirectional communication, making it suitable for interfacing with peripherals like personal computers and Cathode Ray Tube (CRT) terminals. Full-duplex communication means that data

Figure 32. Effect of LER vs. beacon nodes

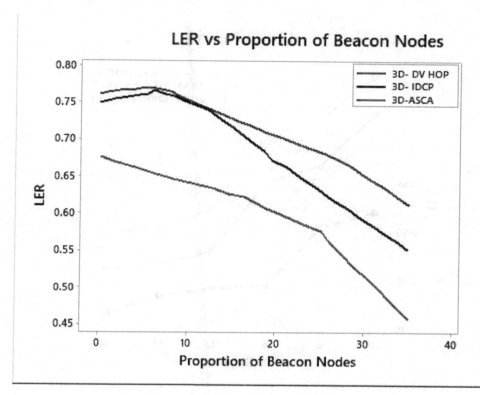

can be sent and received independently, allowing for more flexible and efficient interactions with devices that require constant data exchange.

By providing both half-duplex synchronized and full-duplex asynchronous modes, the USART proves to be a versatile and adaptable serial communication module, making it a valuable component in electronic systems where diverse peripherals with different communication requirements are employed.

The following modes used in USART:

- Asynchronous (full-duplex)
- Synchronous - Master (half-duplex)
- Synchronous - Slave (half-duplex)

To configure pins RC6/TX/CK and RC7/RX/DT as the Universal Synchronous Asynchronous Receiver Transmitter (USART), you need to set specific bits. Specifically, set bit SPEN (RCSTA<7>) to enable the USART module and configure

Figure 33. Effect of LER with density (Beacon 10%)

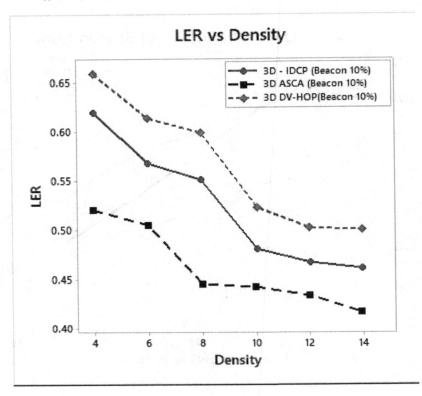

bits TRISC<7:6> to set the data direction for the respective pins. Additionally, the USART module supports multi-processor communication with 9-bit address detection.

USART Baud Rate Generator (BRG)

The USART Baud Rate Generator (BRG) accommodates both Asynchronous and Synchronous modes and is associated with an 8-bit baud rate generator. The period of a free-running timer is influenced by the value stored in the SPBRG register.

In Asynchronous mode, the baud rate is controlled by the BRGH (TXSTA<2>) bit, while in Synchronous mode, the BRGH bit is disregarded. The formulae for computing the baud rate in both Synchronous and Asynchronous modes, applicable specifically in Master mode, are provided in Table 2.

The SPBRG register's nearest integer value can be determined based on the provided baud rate and FOSC. This becomes particularly advantageous when high baud rates (BRGH = 1) are employed with slower baud clocks. In such cases, utilizing the formula FOSC/ (16 (X + 1)) can effectively reduce the baud rate.

Figure 34. Effect of LER with density (Beacon 20%)

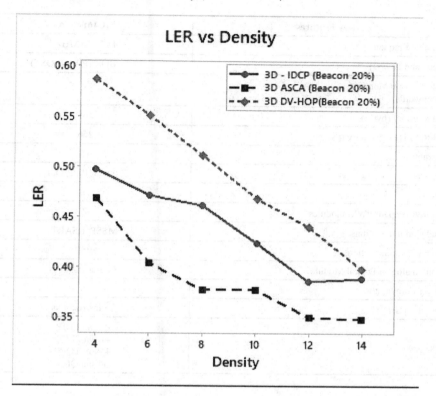

To reset the BRG timer and prevent it from waiting for a timer overflow, a new value can be written to the SPBRG register. Tarang P20, 2.4GHz, low power RF module of Melange Systems and appropriate for numerous IoT (Internet of Things) applications. Tarang P20 module is built with a 2.4GHz transceiver, RF-front end and a microcontroller. The specification for Tarang P20 is given in Table 3.

A host device can be communicated with the Tarang modules through a logic-level asynchronous serial port. A microcontroller or a PC can be interfaced with Tarang module using serial port as shown in Figure 35.

A successful serial communication can be established with the Tarang module by configuring the serial parameters properly in the module as well as the host side. 'HyperTerminal' is a terminal application to view and set the module and PC settings through AT command.

Table 1. Specifications of PIC16F877A

Key Features	PIC16F877A
Operating Frequency	DC - 20 MHz
Resets (and Delays)	POR, BOR (PWRT, OST)
Flash Program Memory (14-bit words)	8K
Data Memory (bytes)	368
EEPROM Data Memory (bytes)	256
Interrupts	15
I/O Ports	Ports A, B, C, D, E
Timers	3
Capture/Compare/PWM modules	2
Serial Communications	MSSP, USART
Parallel Communications	PSP
10-bit Analog-to-Digital Module	8 input channels
Analog Comparators	2
Instruction Set	35 Instructions
Packages	40-pin PDIP 44-pin PLCC 44-pin TQFP 44-pin QFN

Configuring the Baud Rate of Module

desired value, allowing the baud rate to be altered as 'NEW Baud rate' before committing it to memory in HyperTerminal. To save the configured settings to memory, the ATGWR command can be sent. Additionally, one of the modules can be configured with I/O pins designated as both input and output. To designate I/O pins as input or output, refer to the commands outlined in Table 4. After configuring the I/O pins as input, the default I/O pins are utilized as output pins. Verify the status

Table 2. Baud rate formula

SYNC	BRGH= 0 (Low Speed)	BRGH = 1 (High Speed)
0	$BRGH = FOSC/(64(X +2))$ (Asynchronous)	$Baud\ Rate = FOSC/(16(X +2))$
1	$BRGH = FOSC/(4(X +2))$ (Synchronous)	

Legend: X = value in SPBRG (0 to 25)

Table 3. Specification of Tarang P20

Parameters	Specifications
Supply Voltage	3.3 to 3.6 V
Serial UART Interface	CMOS Serial UART (3.3 V)
Serial Interface Data Rate	Configurable (1200 baud to 115200 baud)
RF Data Rate	250 kbps
Transmit Power Output	+19dBm Typical
Receiver Sensitivity	-101 dBm
Operating Frequency	ISM 2.4 GHz
No of Channels	16
MAC & PHY	IEEE 802.15.4/IEEE 802.15.4g
Channel Spacing	5 MHz
Supported Network Topologies	Star/Mesh
Addressing Options	PAN ID, Channel and Addresses
Antenna Options	Wire Antenna/External Antenna
Operating Temperature	-40 to 85 degree C

by sending the ATIOSx command, and upon receiving the 'OK' response from the module, save the configuration to memory using ATGWR and exit using ATGEX. Set up the modules according to the instructions in Figure 36, and activate the entire setup by pressing the switch. Verify if the LED on another module can be wirelessly controlled once the setup is powered on.

Figures 37 and 38 displays the I/O interface as input and output for 3D ASCA algorithm respectively, while Figure 39 displays the transfer of localization data using 3D ASCA algorithm.

Figure 35. Example of a serial interface

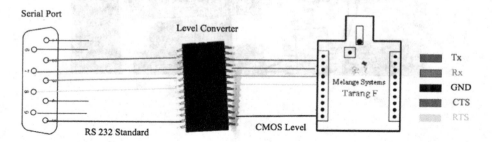

Table 4. AT commands

Command	Description
+++	Enter Command Mode
ATIDxx	I/O pin as input
ATGWR	Write to memory
ATGEX	Exit Command Mode

Figure 36. Example I/O interface in an application

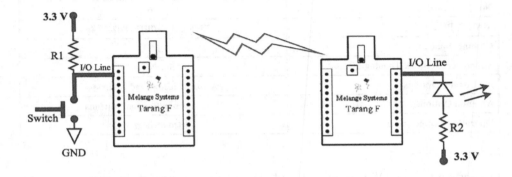

Figure 37. Input module of the 3D ASCA algorithm

Figure 38. Output module of the 3D ASCA algorithm

Figure 39. Transfer of 3D localization information using 3D ASCA

Effect of Localization Error Ratio on Node Number

For each node in the network, the localization error is analysed. Figure 40 show that the nodes 1, 4 and 8 represent the beacon nodes in the network, whereas the remaining nodes will have a slighter localization error rate higher than the beacon nodes.

Figure 40. Effect of LER for each node

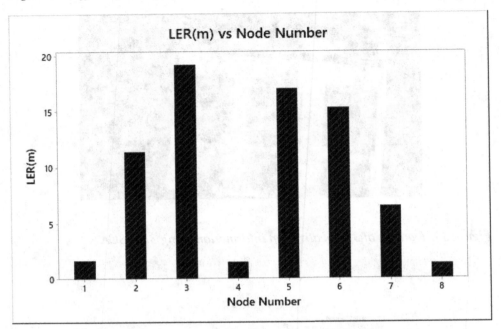

CONCLUSION

In conclusion, this chapter has presented the 3D Adaptive Stochastic Control Algorithm (ASCA), a pioneering solution for precise node localization within Industrial Internet of Things (IIoT) applications. The integration of small percentage of machine learning principles has proven to be a ground-breaking approach, surpassing benchmarks set by traditional algorithms. Moving forward, future work should focus on advancing the algorithm's capabilities. This includes exploring advanced machine learning techniques beyond CNNs, adapting the algorithm to dynamic industrial environments, incorporating multi-sensor fusion for robustness, optimizing energy efficiency, conducting real-world validations, enhancing security measures, ensuring scalability for large-scale deployments, and integrating human-in-the-loop considerations. By addressing these aspects, the 3D ASCA algorithm can evolve into an even more resilient and adaptable solution, contributing significantly to the on-going advancements in IIoT-enabled industrial processes.

REFERENCES

Al Alawi, R. (2011). *RSSI based location estimation in wireless sensors networks.* In: *Proceedings of the 17th IEEE International Conference on Networks*, Singapore.]10.1109/ICON.2011.6168517

Al-Habashna, A., Wainer, G., & Aloqaily, M. (2022). Machine learning-based indoor localization and occupancy estimation using 5G ultra-dense networks. *Simulation Modelling Practice and Theory, 118,* 102543. doi:10.1016/j.simpat.2022.102543

Arbula, D., & Ljubic, S. (2020). Indoor localization based on infrared angle of arrival sensor network. *Sensors (Basel), 20*(21), 6278. doi:10.3390/s20216278 PMID:33158151

Bannour, A., Harbaoui, A., & Alsolami, F. (2021). Connected objects geo-localization based on SS-RSRP of 5G networks. *Electronics (Basel), 10*(22), 2750. doi:10.3390/electronics10222750

Benslimane, A., Clement, S., Konig, J. C., & Mohammed, B. (2014). Cooperative localization techniques for Wireless Sensor Networks: Free, signal and angle based techniques. *Wireless Communications and Mobile Computing, 14*(17), 1627–1646. doi:10.1002/wcm.2303

Boustani, A., Alamatsaz, N., Jadliwala, M., & Namboodiri, V. (2014) LocJam: A novel jamming-based approach to secure localization in wireless networks. *11th Consumer Communications and Networking Conference (CCNC).* IEEE Publications [DOI: 10.1109/CCNC.2014.6866592].

Chaurasiya, K. V., Neeraj, J., & Nandi, G. C. (2014). A novel distance estimation approach for 3D localization in wireless sensor network using multidimensional scaling. *Information Fusion, 15,* 5–18. doi:10.1016/j.inffus.2013.06.003

Du, H., Zhang, C., Ye, Q., Xu, W., Kibenge, P.L. & Yao, K. (2018). A hybrid outdoor localization scheme with high-position accuracy and low-power consumption. *EURASIP Journal on Wireless Communications and Networking, 4.*]. doi:10.1186/s13638-017-1010-4

Fan, J., Hu, Y., Luan, T. H., & Dong, M. (2017). Disloc: A convex partitioning based approach for distributed 3-D localization in wireless sensor networks. *IEEE Sensors Journal, 17*(24), 8412–8423. doi:10.1109/JSEN.2017.2763155

Filippoupolitis, A., Oliff, W., Loukas, G., & Low Energy, B. (2016). *Based occupancy detection for emergency management.* In: *Proceedings of the 2016 15th International Conference on Ubiquitous Computing and Communications and 2016 International Symposium on Cyberspace and Security (IUCC-CSS)*, Granada, Spain.

Gillis, A. S. (2022) What is IOT (Internet of things) and how does it work?— Definition. *IoT agenda*. Techtarget.com. https://www.techtarget.com/iotagenda/definition/Internet-of-Things-IoT

Han, G., Jiang, J., Shu, L., Guizani, M., & Nishio, S. (2013). A two-step secure localization for wireless sensor networks. [Journal]. *The Computer Journal, 56*(10), 1154–1166. doi:10.1093/comjnl/bxr138

Jadliwala, M., Zhong, S., Upadhyaya, S. J., Qiao, U. C., & Hubaux, J. (2011). Secure distance-based localization in the presence of cheating beacon nodes. *IEEE Transactions on Mobile Computing, 9*(6), 810–823. doi:10.1109/TMC.2010.20

Kalpana, A. V. (2019). A unique approach to 3D localization in wireless sensor network by using adaptive stochastic control algorithm *Applied Mathematics & Information Sciences, 13*, 621–628. doi:10.18576/amis/130414

Kawecki, R., Hausman, S., & Korbel, P. (2022). Performance of fingerprinting-based indoor positioning with measured and simulated RSSI reference maps. *Remote Sensing (Basel), 14*(9), 1992. doi:10.3390/rs14091992

Kumar, B. P., Kalpana, A. V., & Nalini, S. (2023). Gated attention based deep learning model for analysing the influence of social media on education. *Journal of Experimental & Theoretical Artificial Intelligence.*

Kumar, B. P., Umamageswaran, J., Kalpana, A. V., & Dhanalakshmi, R. (2023). *Mobility and Behavior based Trustable Routing in Mobile Wireless Sensor Network.* 7th International Conference on Computing Methodologies and Communication (ICCMC), Erode, India.

Lazos, L., & Poovendran, R. (2005). SeRLoc: Robust localization for wireless sensor networks. *ACM Transactions on Sensor Networks, 1*(1), 73–100. doi:10.1145/1077391.1077395

Li, M., & Liu, Y. (2010). Rendered path: Range-free localization in anisotropic sensor networks with holes. *IEEE/ACM Transactions on Networking, 18*(1), 320–332. doi:10.1109/TNET.2009.2024940

Liu, D., Ning, P., Liu, A., Wang, W., & Du, W. (2005). Attack-resistant location estimation in sensor networks. *ACM Transactions on Information and System Security, 11*(4), 1–39. doi:10.1145/1380564.1380570

Mohamed, A., Tharwat, M., Magdy, M., Abubakr, T., Nasr, O., & Youssef, M. (2022). DeepFeat: Robust large-scale multi-features outdoor localization in LTE networks using deep learning. *IEEE Access : Practical Innovations, Open Solutions, 10*, 3400–3414. doi:10.1109/ACCESS.2022.3140292

Naddafzadeh-Shirazi, G., Shenouda, M. B., & Lampe, L. (2014). Second order cone programming for sensor network localization with Anchor position uncertainty. *IEEE Transactions on Wireless Communications, 13*(2), 749–763. doi:10.1109/TWC.2013.120613.130170

Potortì, F., Park, S., Jiménez Ruiz, A. R. J., Barsocchi, P., Girolami, M., Crivello, A., Lee, S. Y., Lim, J. H., Torres-Sospedra, J., Seco, F., Montoliu, R., Mendoza-Silva, G. M., Pérez Rubio, M. D. C., Losada-Gutiérrez, C., Espinosa, F., & Macias-Guarasa, J. (2017). Comparing the performance of indoor localization systems through the Evaal framework. *Sensors (Basel), 17*(10), 2327. doi:10.3390/s17102327 PMID:29027948

Sadowski, S., & Spachos, P. (2018). RSSI-based indoor localization with the Internet of things. *IEEE Access : Practical Innovations, Open Solutions, 6*, 30149–30161. doi:10.1109/ACCESS.2018.2843325

Sarala, B., Sumathy, G., Kalpana, A. V., & Jasmine Hephzipah, J. J. (2023) Glioma brain tumor detection using dual convolutional neural networks and histogram density segmentation algorithm. *Biomedical Signal Processing and Control, 85.*] doi:10.1016/j.bspc.2023.104859

Shi, Q., He, C., Chen, H., & Jiang, L. (2010). Distributed wireless sensor network localization via sequential greedy optimization algorithm. *IEEE Transactions on Signal Processing, 58*(6), 3328–3340. doi:10.1109/TSP.2010.2045416

Tekler, Z. D., & Chong, A. (2022). Occupancy prediction using deep learning approaches across multiple space types: A minimum sensing strategy. *Building and Environment, 226*, 109689. doi:10.1016/j.buildenv.2022.109689

Wang, G., Chen, H., Li, Y., & Jin, M. (2012). On received-signal-strength based localization with unknown transmit power and path loss exponent. *IEEE Wireless Communications Letters, 1*(5), 536–539. doi:10.1109/WCL.2012.072012.120428

Yang, T., Cabani, A., & Chafouk, H. (2021). A survey of recent indoor localization scenarios and methodologies. *Sensors (Basel), 21*(23), 8086. doi:10.3390/s21238086 PMID:34884090

Yeredor, A. (2014) Decentralized TOA-based localization in non-synchronized wireless networks with partial. *Asymmetric Connectivity 15th international workshop on signal processing advances in wireless communications*. IEEE.

Zhuang, D., Gan, V. J. L., Duygu Tekler, Z. D., Chong, A., Tian, S., & Shi, X. (2023). Data-driven predictive control for smart HVAC system in IoT-integrated buildings with time-series forecasting and reinforcement learning. *Applied Energy, 338*, 120936. doi:10.1016/j.apenergy.2023.120936

Compilation of References

Abdulwahid, A. H., Pattnaik, M., Palav, M. R., Babu, S. T., Manoharan, G., & Selvi, G. P. (2023, April). Library Management System Using Artificial Intelligence. In *2023 Eighth International Conference on Science Technology Engineering and Mathematics (ICONSTEM)* (pp. 1-7). IEEE.

Agote-Garrido, A., Martín-Gómez, A. M., & Lama-Ruiz, J. R. (2023). Manufacturing System Design in Industry 5.0: Incorporating Sociotechnical Systems and Social Metabolism for Human-Centered, Sustainable, and Resilient Production. *Systems*, *11*(11), 537. doi:10.3390/systems11110537

Ahmadi-Assalemi, G., Al-Khateeb, H., Maple, C., Epiphaniou, G., Alhaboby, Z. A., & Alkaabi, S. (2020). Digital twins for precision healthcare. Cyber Defence in the Age of AI Smart Societies and Augmented Humanity. Springer.

Akinsolu, M. O. (2023). Applied Artificial Intelligence in Manufacturing and Industrial Production Systems: PEST Considerations for Engineering Managers. *IEEE Engineering Management Review*, *51*(1), 52–62. doi:10.1109/EMR.2022.3209891

Al Alawi, R. (2011). *RSSI based location estimation in wireless sensors networks*. In: *Proceedings of the 17th IEEE International Conference on Networks*, Singapore.]10.1109/ICON.2011.6168517

Alenizi, F. A., Abbasi, S., Mohammed, A. H., & Rahmani, A. M. (2023). The Artificial Intelligence Technologies in Industry 4.0: A Taxonomy, Approaches, and Future Directions. *Computers & Industrial Engineering*, *185*, 109662. doi:10.1016/j.cie.2023.109662

Alexopoulos, K., Hribrenik, K., Surico, M., Nikolakis, N., Al-Najjar, B., Keraron, Y., & Makris, S. (2021). *Predictive maintenance technologies for production systems: A roadmap to development and implementation.*

Al-Habashna, A., Wainer, G., & Aloqaily, M. (2022). Machine learning-based indoor localization and occupancy estimation using 5G ultra-dense networks. *Simulation Modelling Practice and Theory*, *118*, 102543. doi:10.1016/j.simpat.2022.102543

Alqahtani, H., & Kumar, G. (2024). Machine Learning for Enhancing Transportation Security: A Comprehensive Analysis of Electric and Flying Vehicle Systems. *Engineering Applications of Artificial Intelligence*, *129*, 107667. https://www.sciencedirect.com/science/article/pii/S095219762 3018511. doi:10.1016/j.engappai.2023.107667

Al-Safi, J. K. S., Bansal, A., Aarif, M., Almahairah, M. S. Z., Manoharan, G., & Alotoum, F. J. (2023, January). Assessment Based On IoT For Efficient Information Surveillance Regarding Harmful Strikes Upon Financial Collection. In *2023 International Conference on Computer Communication and Informatics (ICCCI)* (pp. 1-5). IEEE. 10.1109/ICCCI56745.2023.10128500

Ameen, N., Tarhini, A., Reppel, A., & Anand, A. (2021). Customer experiences in the age of artificial intelligence. *Computers in Human Behavior, 114,* 106548. doi:10.1016/j.chb.2020.106548 PMID:32905175

Andras, E., Mazzone, E., van Leeuwen, F. W. B., De Naeyer, G., van Oosterom, M. N., Beato, S., Buckle, T., O'Sullivan, S., van Leeuwen, P. J., Beulens, A., Crisan, N., D'Hondt, F., Schatteman, P., van Der Poel, H., Dell'Oglio, P., & Mottrie, A. (2020). Artificial intelligence and robotics: A combination that is changing the operating room. *World Journal of Urology, 38*(10), 2359–2366. doi:10.1007/s00345-019-03037-6 PMID:31776737

Angelopoulos, A., Michailidis, E. T., Nomikos, N., Trakadas, P., Hatziefremidis, A., Voliotis, S., & Zahariadis, T. (2020). Tackling Faults in the Industry 4.0 Era-A Survey of Machine-Learning Solutions and Key Aspects. *Sensors (Basel), 20*(1), 109. doi:10.3390/s20010109 PMID:31878065

Anish, T. P., Shanmuganathan, C., Dhinakaran, D., & Vinoth Kumar, V. (2023). Hybrid Feature Extraction for Analysis of Network System Security—IDS. In R. Jain, C. M. Travieso, & S. Kumar (Eds.), *Cybersecurity and Evolutionary Data Engineering. ICCEDE 2022. Lecture Notes in Electrical Engineering* (Vol. 1073). Springer. doi:10.1007/978-981-99-5080-5_3

Anonymous. (2019). The Revolution of Artificial Intelligence and the Significance of Adopting AI in Different Industries. Int*ernational Journal of Recent Technology and Engineering*.

Arbula, D., & Ljubic, S. (2020). Indoor localization based on infrared angle of arrival sensor network. *Sensors (Basel), 20*(21), 6278. doi:10.3390/s20216278 PMID:33158151

Arinez, J. F., Chang, Q., Gao, R. X., Xu, C., & Zhang, J. (2020). Artificial intelligence in advanced manufacturing: Current status and future outlook. *Journal of Manufacturing Science and Engineering, 142*(11), 142. doi:10.1115/1.4047855

Arunmozhi Manimuthu, V. G. (2022). Design and development of automobile assembly model using federated artificial intelligence with smart contract. *International Journal of Production Research, 60*(1), 111–135. doi:10.1080/00207543.2021.1988750

Attarha, S., Narayan, A., Hage Hassan, B., Krüger, C., Castro, F., Babazadeh, D., & Lehnhoff, S. (2020). Virtualization management concept for flexible and fault-tolerant smart grid service provision. *Energies, 13*(9), 2196. doi:10.3390/en13092196

Aujla, G. S., Chaudhary, R., Kumar, N., Das, A. K., & Rodrigues, J. J. P. C. (2018). SecSVA: Secure storage, verification, and auditing of Big Data in the cloud environment. *IEEE Communications Magazine, 56*(1), 78–85. doi:10.1109/MCOM.2018.1700379

Azari, M. S., Flammini, F., Santini, S., & Caporuscio, M. (2023). A Systematic Literature Review on Transfer Learning for Predictive Maintenance in Industry 4.0. *IEEE Access : Practical Innovations, Open Solutions*, *11*, 12887–12910. doi:10.1109/ACCESS.2023.3239784

Azhari, M. E., Toumanari, A., Latif, R., & El Moussaid, N. (2018). Round Estimation Period for Cluster-Based Routing in Mobile Wireless Sensor Networks. *International Journal of Advanced Intelligence Paradigms*, *10*(4), 374–390. https://www.scopus.com/inward/record.uri?eid=2-s2.0-85048034 340&doi=10.1504%2FIJAIP.2018.092034&partnerID=40&md5=c9e1d8c 8ea8aed7e82c2f82de9960806. doi:10.1504/IJAIP.2018.092034

Bai, Q., Li, S., Yang, J., Song, Q., Li, Z., & Zhang, X. (2020). Object detection recognition and robot grasping based on machine learning: A survey. *IEEE Access : Practical Innovations, Open Solutions*, *8*, 181855–181879. doi:10.1109/ACCESS.2020.3028740

Bannour, A., Harbaoui, A., & Alsolami, F. (2021). Connected objects geo-localization based on SS-RSRP of 5G networks. *Electronics (Basel)*, *10*(22), 2750. doi:10.3390/electronics10222750

Bao, Y., Tang, Z., Li, H., & Zhang, Y. (2019). Computer vision and deep learning–based data anomaly detection method for structural health monitoring. *Structural Health Monitoring*, *18*(2), 401–4215. doi:10.1177/1475921718757405

Baviskar, D., Ahirrao, S., Potdar, V., & Kotecha, K. (2021). Efficient automated processing of the unstructured documents using artificial intelligence: A systematic literature review and future directions. *IEEE Access : Practical Innovations, Open Solutions*, *9*, 72894–72936. doi:10.1109/ACCESS.2021.3072900

Baygin, M., Karakose, M., Sarimaden, A., & Erhan, A. K. I. N. (2017, September). Machine vision based defect detection approach using image processing. In 2017 international artificial intelligence and data processing symposium (IDAP) (pp. 1-5). IEEE. doi:10.1109/IDAP.2017.8090292

Becker, K.-F., Voges, S., Fruehauf, P., Heimann, M., Nerreter, S., Blank, R., & Erdmann, M. (2021). Implementation of Trusted Manufacturing & AI-Based Process Optimization into Microelectronic Manufacturing Research Environments. *IMAP Source Proceedings 2021 (IMAPS Symposium)*. iMaps. 10.4071/1085-8024-2021.1

Bednar, P. M., & Welch, C. (2020). Socio-technical perspectives on smart working: Creating meaningful and sustainable systems. *Information Systems Frontiers*, *22*(2), 281–298. doi:10.1007/s10796-019-09921-1

Behgounia, F., & Zohuri, B. (2020). Artificial intelligence integration with nanotechnology. *Journal of Nanosciences Research & Reports, 117*.

Belk, R. (2021). Ethical issues in service robotics and artificial intelligence. *Service Industries Journal*, *41*(13-14), 860–876. doi:10.1080/02642069.2020.1727892

Benslimane, A., Clement, S., Konig, J. C., & Mohammed, B. (2014). Cooperative localization techniques for Wireless Sensor Networks: Free, signal and angle based techniques. *Wireless Communications and Mobile Computing*, *14*(17), 1627–1646. doi:10.1002/wcm.2303 bin Abdullah, M. R., & Iqbal, K. (2022). A Review of Intelligent Document Processing Applications Across Diverse Industries. *Journal of Artificial Intelligence and Machine Learning in Management, 6*(2), 29-42.

BizIntellia. (n.d.). *Benefits of Using Cloud Computing for Storing IoT Data.* BizIntellia. https://www.biz4intellia.com/blog/benefits-of-using-cloudcom puting-for-storing-IoT-data/

Bordeleau, F.-È., Mosconi, E., & Santa-Eulalia, L. A. (2018). Business intelligence in industry 4.0: State of the art and research opportunities. *Proc. 51st Hawaii Int. Conf. Syst. Sci.,* (pp. 1-10). IEEE. 10.24251/HICSS.2018.495

Boustani, A., Alamatsaz, N., Jadliwala, M., & Namboodiri, V. (2014) LocJam: A novel jamming-based approach to secure localization in wireless networks. *11th Consumer Communications and Networking Conference (CCNC).* IEEE Publications [DOI: 10.1109/CCNC.2014.6866592].

Brenes, R. F., Johannssen, A., & Chukhrova, N. (2022). An Intelligent Bankruptcy Prediction Model Using a Multilayer Perceptron. *Intelligent Systems with Applications, 16.* https://www.sciencedirect.com/science/article/pii/S266730532 2000734

Cai, Y., Starly, B., Cohen, P., & Lee, Y.-S. (2017, January). Sensor data and information fusion to construct digital-twins virtual machine tools for cyber-physical manufacturing. *Procedia Manufacturing, 10,* 1031–1042. doi:10.1016/j.promfg.2017.07.094

Campagna, G., & Rehm, M. (2023). *Analysis of Proximity and Risk for Trust Evaluation in Human-Robot Collaboration.* 2023 32nd IEEE International Conference on Robot and Human Interactive Communication (RO-MAN), Busan, Korea. 10.1109/RO-MAN57019.2023.10309470

Carayannis, E. G., Christodoulou, K., Christodoulou, P., Chatzichristofis, S. A., & Zinonos, Z. (2021). Known unknowns in an era of technological and viral disruptions—Implications for theory, policy, and practice. *Journal of the Knowledge Economy,* 1–24.

Chaurasiya, K. V., Neeraj, J., & Nandi, G. C. (2014). A novel distance estimation approach for 3D localization in wireless sensor network using multidimensional scaling. *Information Fusion,* · *15,* 5–18. doi:10.1016/j.inffus.2013.06.003

Çınar, Z. M., Abdussalam Nuhu, A., Zeeshan, Q., Korhan, O., Asmael, M., & Safaei, B. (2020). Machine learning in predictive maintenance towards sustainable smart manufacturing in industry 4.0. *Sustainability (Basel), 12*(19), 8211. doi:10.3390/su12198211

Colombo, A. W., Karnouskos, S., Yu, X., Kaynak, O., Luo, R. C., Shi, Y., Leitao, P., Ribeiro, L., & Haase, J. (2021). A 70-year industrial electronics society evolution through industrial revolutions: The rise and flourishing of information and communication technologies. *IEEE Industrial Electronics Magazine, 15*(1), 115–126. doi:10.1109/MIE.2020.3028058

Cooke, P. (2020). Gigafactory logistics in space and time: Tesla's fourth gigafactory and its rivals. *Sustainability (Basel), 12*(5), 2044. doi:10.3390/su12052044

Corallo, A., Lazoi, M., & Lezzi, M. (2020). Cybersecurity in the context of industry 4.0: A structured classification of critical assets and business impacts. *Computers in Industry, 114*, 103165. doi:10.1016/j.compind.2019.103165

Cullen-Knox, C., Eccleston, R., Haward, M., Lester, E., & Vince, J. (2017). Contemporary challenges in environmental governance: Technology, governance and the social licence. *Environmental Policy and Governance, 27*(1), 3–13. doi:10.1002/eet.1743

Dai, W., Nishi, H., Vyatkin, V., Huang, V., Shi, Y., & Guan, X. (2019, December). Industrial edge computing: Enabling embedded intelligence. *IEEE Industrial Electronics Magazine, 13*(4), 48–56. doi:10.1109/MIE.2019.2943283

Das, K., Wang, Y., & Green, K. E. (2021). Are robots perceived as good decision makers? A study investigating trust and preference of robotic and human linesman-referees in football. *Paladyn : Journal of Behavioral Robotics, 12*(1), 287–296. doi:10.1515/pjbr-2021-0020

De Azambuja, A. J. G., Plesker, C., Schützer, K., Anderl, R., Schleich, B., & Almeida, V. R. (2023). Artificial Intelligence-Based Cyber Security in the context of Industry 4.0—A survey. *Electronics (Basel), 12*(8), 1920. doi:10.3390/electronics12081920

de Oliveira, J., Bastos-Filho, C., & Oliveira, S. (2022). Non-Invasive Embedded System Hardware/Firmware Anomaly Detection Based on the Electric Current Signature. *Advanced Engineering Informatics, 51*. https://www.sciencedirect.com/science/article/pii/S147403462 1002676

Deviprasad, S., Madhumithaa, N., Vikas, I. W., Yadav, A., & Manoharan, G. (2023). The Machine Learning-Based Task Automation Framework for Human Resource Management in MNC Companies. *Engineering Proceedings, 59*(1), 63.

Dhamija, P., & Bag, S. (2020). Role of artificial intelligence in operations environment: A review and bibliometric analysis. *The TQM Journal, 32*(4), 869–896. doi:10.1108/TQM-10-2019-0243

Dhanalakshmi, R., Kalpana, A. V., Umamageswaran, J., & Kumar, B. P. (2023). Health Information Broadcast Distributed Pattern Association based on Estimated Volume. *2023 Third International Conference on Artificial Intelligence and Smart Energy (ICAIS)*. 10.1109/ICAIS56108.2023.10073672

Dhinakaran, D., & Joe Prathap, P. M. (2022). Protection of data privacy from vulnerability using two-fish technique with Apriori algorithm in data mining. *The Journal of Supercomputing, 78*(16), 17559–17593. doi:10.1007/s11227-022-04517-0

Dhinakaran, D., Selvaraj, D., Dharini, N., Raja, N. S. E., & Priya, C. S. L. (2023). Towards a Novel Privacy-Preserving Distributed Multiparty Data Outsourcing Scheme for Cloud Computing with Quantum Key Distribution. *International Journal of Intelligent Systems and Applications in Engineering, 12*(2), 286–300.

Dhinakaran, D., Udhaya Sankar, S. M., Edwin Raja, S., & Jeno Jasmine, J. (2023). Optimizing Mobile Ad Hoc Network Routing using Biomimicry Buzz and a Hybrid Forest Boost Regression - ANNs [IJACSA]. *International Journal of Advanced Computer Science and Applications, 14*(12). doi:10.14569/IJACSA.2023.0141209

Díaz, M., Martín, C., & Rubio, B. (2016). State-of-the-art, challenges, and open issues in the integration of Internet of things and cloud computing. *Journal of Network and Computer Applications, 67*, 99–117. doi:10.1016/j.jnca.2016.01.010

Dinculeana, D., & Cheng, X. (2019). Devices, vulnerabilities and limitations of MQTT protocol used between IoT. *Applied Sciences (Basel, Switzerland), 9*(5), 848. doi:10.3390/app9050848

Ding, H., Gao, R. X., Isaksson, A. J., Landers, R. G., Parisini, T., & Yuan, Y. (2020). State of AI-based monitoring in smart manufacturing and introduction to focused section. *IEEE/ASME Transactions on Mechatronics, 25*(5), 2143–2154. doi:10.1109/TMECH.2020.3022983

Dondos, A., & Papanagopoulos, D. (1996). Three Models for Chain Conformation of Block Copolymers in Solution and in Solid State. *Journal of Polymer Science. Part B, Polymer Physics, 34*(7), 1281–1288. doi:10.1002/(SICI)1099-0488(199605)34:7<1281::AID-POLB10>3.0.CO;2-6

Du, H., Zhang, C., Ye, Q., Xu, W., Kibenge, P.L. & Yao, K. (2018). A hybrid outdoor localization scheme with high-position accuracy and low-power consumption. *EURASIP Journal on Wireless Communications and Networking, 4*.]. doi:10.1186/s13638-017-1010-4

Durai, S., Krishnaveni, K., & Manoharan, G. (2022, May). Designing entrepreneurial performance metric (EPM) framework for entrepreneurs owning small and medium manufacturing units (SME) in Coimbatore. In AIP Conference Proceedings (Vol. 2418, No. 1). AIP Publishing.

El Namaki, M. S. S. (2016). *How companies are applying AI to the business strategy formulation.* The Conversation.

Elallid, B. B., Benamar, N., Hafid, A. S., Rachidi, T., & Mrani, N. (2022). A comprehensive survey on the application of deep and reinforcement learning approaches in autonomous driving. *Journal of King Saud University. Computer and Information Sciences, 34*(9), 7366–7390. doi:10.1016/j.jksuci.2022.03.013

ElFar, O. A., Chang, C.-K., Leong, H. Y., Peter, A. P., Chew, K. W., & Show, P. L. (2021). Prospects of Industry 5.0 in algae: Customization of production and new advance technology for clean bioenergy generation. *Energy Convers. Manag. X, 10*, 100048. doi:10.1016/j.ecmx.2020.100048

Ennaji, O., Vergütz, L., & El Allali, A. (2023). Machine Learning in Nutrient Management: A Review. *Artificial Intelligence in Agriculture, 9*, 1–11. https://www.sciencedirect.com/science/article/pii/S258972172300017X

Etengu, R., Tan, S. C., Chuah, T. C., Lee, Y. L., & Galán-Jiménez, J. (2023). AI-Assisted Traffic Matrix Prediction Using GA-Enabled Deep Ensemble Learning for Hybrid SDN. *Computer Communications*, *203*, 298–311. https://www.sciencedirect.com/science/article/pii/S014036642 3000920. doi:10.1016/j.comcom.2023.03.014

Fahle, S., Prinz, C., & Kuhlenkötter, B. (2020). Systematic review on machine learning (ML) methods for manufacturing processes–Identifying artificial intelligence (AI) methods for field application. *Procedia CIRP*, *93*, 413–418. doi:10.1016/j.procir.2020.04.109

Fan, J., Hu, Y., Luan, T. H., & Dong, M. (2017). Disloc: A convex partitioning based approach for distributed 3-D localization in wireless sensor networks. *IEEE Sensors Journal*, *17*(24), 8412–8423. doi:10.1109/JSEN.2017.2763155

Felzmann, H., Fosch-Villaronga, E., Lutz, C., & Tamò-Larrieux, A. (2020). Towards transparency by design for artificial intelligence. *Science and Engineering Ethics*, *26*(6), 3333–3361. doi:10.1007/s11948-020-00276-4 PMID:33196975

Filippoupolitis, A., Oliff, W., Loukas, G., & Low Energy, B. (2016). *Based occupancy detection for emergency management*. In: *Proceedings of the 2016 15th International Conference on Ubiquitous Computing and Communications and 2016 International Symposium on Cyberspace and Security (IUCC-CSS)*, Granada, Spain.

Ganz, C., & Isaksson, A. J. (2023). Trends in Automation. In *Springer Handbook of Automation* (pp. 103–117). Springer International Publishing. doi:10.1007/978-3-030-96729-1_5

Garg, S., Mahajan, N., & Ghosh, J. (2022). Artificial Intelligence as an emerging technology in Global Trade: the challenges and Possibilities. In *Handbook of Research on Innovative Management Using AI in Industry 5.0* (pp. 98–117). IGI Global. doi:10.4018/978-1-7998-8497-2.ch007

Geetha, M. (2022). *Covid-19: a special review of MSMES for sustaining entrepreneurship. Msmes and covid - 19 the opportunities, challenges and recovery measures, 2022*. Trueline Academic And Research Centre.

Geetha, M., Murthi, P., & Poongodi, K. (2023). Inventory Management of Construction Project Through ABC Analysis: A Case Study. *National conference on Advances in Construction Materials and Management, Springer Nature Singapore*, (pp. 95-105). IEEE.

Gervais, J. (2019). *The future of IoT: 10 predictions about the Internet of Things*. Norton. https://us.norton.com/internetsecurity-iot-5-predictions-for-the-future-of-iot.html

Ge, W., Lueck, C., Suominen, H., & Apthorp, D. (2023). Has Machine Learning Over-Promised in Healthcare?: A Critical Analysis and a Proposal for Improved Evaluation, with Evidence from Parkinson's Disease. *Artificial Intelligence in Medicine*, *139*, 102524. https://www.sciencedirect.com/science/article/pii/S093336572 3000386. doi:10.1016/j.artmed.2023.102524 PMID:37100503

Ghobakhloo, M., Iranmanesh, M., Foroughi, B., Tirkolaee, E. B., Asadi, S., & Amran, A. (2023). Industry 5.0 implications for inclusive sustainable manufacturing: An evidence-knowledge-based strategic roadmap. *Journal of Cleaner Production, 417*, 138023. doi:10.1016/j.jclepro.2023.138023

Gillis, A. S. (2022) What is IOT (Internet of things) and how does it work?—Definition. *IoT agenda.* Techtarget.com. https://www.techtarget.com/iotagenda/definition/Internet-of-Things-IoT

Gocev, I., Grimm, S., & Runkler, T. (2020). Supporting Skill-based Flexible Manufacturing with Symbolic AI Methods. *IECON 2020 The 46th Annual Conference of the IEEE Industrial Electronics Society, Singapore,* (pp. 769-774). IEEE. 10.1109/IECON43393.2020.9254797

Goldman, C. V., Baltaxe, M., Chakraborty, D., Arinez, J., & Diaz, C. E. (2023). Interpreting learning models in manufacturing processes: Towards explainable AI methods to improve trust in classifier predictions. *Journal of Industrial Information Integration, 33*, 100439. doi:10.1016/j.jii.2023.100439

Hadian, H., Chahardoli, S., Golmohammadi, A. M., & Mostafaeipour, A. (2020). A Practical Framework for Supplier Selection Decisions with an Application to the Automotive Sector. *International Journal of Production Research, 58*(10), 2997–3014. doi:10.1080/00207543.2019.1624854

Han, G., Jiang, J., Shu, L., Guizani, M., & Nishio, S. (2013). A two-step secure localization for wireless sensor networks. [Journal]. *The Computer Journal, 56*(10), 1154–1166. doi:10.1093/comjnl/bxr138

Harini, M., Prabhu, D., Udhaya Sankar, S. M., Pooja, V., & Kokila Sruthi, P. (2023). Levarging Blockchain for Transparency in Agriculture Supply Chain Management Using IoT and Machine Learning. 2023 World Conference on Communication & Computing (WCONF), RAIPUR, India. pp. 1-6. 10.1109/WCONF58270.2023.10235156

Haripriya, A. P., & Kulothugan, K. (2016). ECC based self-certified key management scheme for mutual authentication in Internet of Things. In: *2016 International Conference on Emerging Technological Trends (ICETT),* (p. 1–6). Kollam, India: IEEE. 10.1109/ICETT.2016.7873657

Heilala, J., & Singh, K. (2023). *Evaluation Planning for Artificial Intelligence-based Industry 6.0 Metaverse Integration.* 6th International Conference on Intelligent Human Systems Integration (IHSI 2023) Integrating People and Intelligent Systems, Venice, Italy.

Heo, S. (2016). Climate Change and Concerted Actions by Mankind. *J. Korean Soc. Trends Perspectibes, 96*, 214–220.

He, S., Leanse, L. G., & Feng, Y. (2021). Artificial intelligence and machine learning assisted drug delivery for effective treatment of infectious diseases. *Advanced Drug Delivery Reviews, 178*, 113922. doi:10.1016/j.addr.2021.113922 PMID:34461198

Howard, J. (2019). Artificial intelligence: Implications for the future of work. *American Journal of Industrial Medicine, 62*(11), 917–926. doi:10.1002/ajim.23037 PMID:31436850

Iqbal, M., Lee, C. K. M., & Ren, J. Z. (2022). Industry 5.0: From Manufacturing Industry to Sustainable Society. *2022 IEEE International Conference on Industrial Engineering and Engineering Management (IEEM)*, Kuala Lumpur, Malaysia. 10.1109/IEEM55944.2022.9989705

Jacob, T. Cassady, Chris Robinson, and Dan O. Popa. (2020). Increasing user trust in a fetching robot using explainable AI in a traded control paradigm. In *Proceedings of the 13th ACM International Conference on PErvasive Technologies Related to Assistive Environments (PETRA '20).* Association for Computing Machinery, New York, NY, USA. 10.1145/3389189.3393740

Jadliwala, M., Zhong, S., Upadhyaya, S. J., Qiao, U. C., & Hubaux, J. (2011). Secure distance-based localization in the presence of cheating beacon nodes. *IEEE Transactions on Mobile Computing, 9*(6), 810–823. doi:10.1109/TMC.2010.20

Jagatheesaperumal, S. K., Rahouti, M., Ahmad, K., Al-Fuqaha, A., & Guizani, M. (2021). The duo of artificial intelligence and big data for industry 4.0: Applications, techniques, challenges, and future research directions. *IEEE Internet of Things Journal, 9*(15).

Jaichandran, R., Krishna, S. H., Madhavi, G. M., Mohammed, S., Raj, K. B., & Manoharan, G. (2023, January). Fuzzy Evaluation Method on the Financing Efficiency of Small and Medium-Sized Enterprises. In *2023 International Conference on Artificial Intelligence and Knowledge Discovery in Concurrent Engineering (ICECONF)* (pp. 1-7). IEEE.

Jain, D. K., Li, Y., Er, M. J., Xin, Q., Gupta, D., & Shankar, K. (2022). Enabling Unmanned Aerial Vehicle Borne Secure Communication With Classification Framework for Industry 5.0. *IEEE Transactions on Industrial Informatics, 18*(8), 5477–5484. doi:10.1109/TII.2021.3125732

Javaid, M., Haleem, A., Singh, R. P., & Suman, R. (2022). Artificial intelligence applications for industry 4.0: A literature-based study. *Journal of Industrial Integration and Management, 7*(01), 83–111. doi:10.1142/S2424862221300040

Javed, S., Javed, S., Deventer, J. v., Mokayed, H., & Delsing, J. (2023). *A Smart Manufacturing Ecosystem for Industry 5.0 using Cloud-based Collaborative Learning at the Edge.* NOMS 2023-2023 IEEE/IFIP Network Operations and Management Symposium, Miami, FL, USA. 10.1109/NOMS56928.2023.10154323

Jiang, Q., Ma, J., Wei, F., Tian, Y., Shen, J., & Yang, Y. (2016). An untraceable temporal-credential-based two-factor authentication scheme using ECC for wireless sensor networks. *Journal of Network and Computer Applications, 76*, 37–48. doi:10.1016/j.jnca.2016.10.001

Jiang, X., Lin, G.-H., Huang, J.-C., Hu, I.-H., & Chiu, Y.-C. (2021). Performance of Sustainable Development and Technological Innovation Based on Green Manufacturing Technology of Artificial Intelligence and Block Chain. *Mathematical Problems in Engineering, 2021*, 1–11. https://www.scopus.com/inward/record.uri?eid=2-s2.0-85104505059&doi=10.1155%2F2021%2F5527489&partnerID=40&md5=545252da8e6b7413a04ff7dd925c3c8d. doi:10.1155/2021/5527489

Jiang, X., Satapathy, S. C., Yang, L., Wang, S.-H., & Zhang, Y.-D. (2020). S.-.H. Wang, Y.-.D. Zhang, "A survey on artificial intelligence in Chinese sign language recognition,". *Arabian Journal for Science and Engineering*, *45*(12), 9859–9894. doi:10.1007/s13369-020-04758-2

Joe Prathap, P. M. (2022). Preserving data confidentiality in association rule mining using data share allocator algorithm. *Intelligent Automation & Soft Computing*, *33*(3), 1877–1892. doi:10.32604/iasc.2022.024509

Johnson, M., & Anderson, R. (2021). AI in Healthcare: A U.S. Hospital's Journey. *Journal of Healthcare Innovation*, *5*(3), 123–140.

Johnson, W., Onuma, O., Owolabi, M., & Sachdev, S. (2016). Stroke: A Global Response Is Needed. *Bulletin of the World Health Organization*, *94*(9), 634–634A. doi:10.2471/BLT.16.181636 PMID:27708464

Joung, T. H., Kang, S. G., Lee, J. K., & Ahn, J. (2020). The IMO initial strategy for reducing Greenhouse Gas (GHG) emissions, and its follow-up actions towards 2050. *J. Int. Marit. Saf. Environ. Aff. Shipp.*, *4*(1), 1–7. doi:10.1080/25725084.2019.1707938

Kalpana, A. V. (2019). A unique approach to 3D localization in wireless sensor network by using adaptive stochastic control algorithm *Applied Mathematics & Information Sciences*, *13*, 621–628. doi:10.18576/amis/130414

Kalpana, A. V., Venkataramanan, V., Charulatha, G., & Geetha, G. (2023). An Intelligent Voice-Recognition Wheelchair System for Disabled Persons. *2023 International Conference on Sustainable Computing and Smart Systems (ICSCSS)*. 10.1109/ICSCSS57650.2023.10169364

Kamble, S. S., Gunasekaran, A., & Gawankar, S. A. (2018, July). Sustainable industry 4.0 framework: A systematic literature review identifying the current trends and future perspectives. *Process Safety and Environmental Protection*, *117*, 408–425. doi:10.1016/j.psep.2018.05.009

Kawecki, R., Hausman, S., & Korbel, P. (2022). Performance of fingerprinting-based indoor positioning with measured and simulated RSSI reference maps. *Remote Sensing (Basel)*, *14*(9), 1992. doi:10.3390/rs14091992

Keerthana, M., Ananthi, M., Harish, R., Udhaya Sankar, S. M., & Sree, M. S. (2023). IoT Based Automated Irrigation System for Agricultural Activities. *2023 12th International Conference on Advanced Computing (ICoAC)*, Chennai, India. 10.1109/ICoAC59537.2023.10249426

Keserwani, H. P. T., Rekha; P. R., Jyothi; Manoharan, Geetha; Mane, Pallavi; Gupta, Shashi Kant (2021). Effect Of Employee Empowerment On Job Satisfaction In Manufacturing Industry. Turkish Online *Journal of Qualitative Inquiry, 12*(3).

Kim, H., Kim, D. W., Yi, O., & Kim, J. (2019). Cryptanalysis of hash functions based on blockciphers suitable for IoT service platform security. *Multimedia Tools and Applications*, *78*(3), 3107–3130. doi:10.1007/s11042-018-5630-4

Kim, S. W., Kong, J. H., Lee, S. W., & Lee, S. (2022). Recent advances of artificial intelligence in manufacturing industrial sectors: A review. *International Journal of Precision Engineering and Manufacturing, 23*(1), 1–19. doi:10.1007/s12541-021-00600-3

Konstantinidis, F. K., Myrillas, N., Tsintotas, K. A., Mouroutsos, S. G., & Gasteratos, A. (2023). A technology maturity assessment framework for Industry 5.0 machine vision systems based on systematic literature review in automotive manufacturing. *International Journal of Production Research*, 1–37. doi:10.1080/00207543.2023.2270588

Kumar, B. P., Umamageswaran, J., Kalpana, A. V., & Dhanalakshmi, R. (2023). *Mobility and Behavior based Trustable Routing in Mobile Wireless Sensor Network*. 7th International Conference on Computing Methodologies and Communication (ICCMC), Erode, India.

Kumar, K. Y., Kumar, N. J., Udhaya Sankar, S. M., Kumar, U. J., & Yuvaraj, V. (2023). *Optimized Retrieval of Data from Cloud using Hybridization of Bellstra Algorithm*. 2023 World Conference on Communication & Computing (WCONF), RAIPUR, India. 10.1109/WCONF58270.2023.10234974

Kumaravel, D., Tomar, P. S., Tulanovna, K. D., Sharma, S., Pawar, K. P., & Patil, P. P. (2023). Advanced Manufacturing Process Development for Automation in the Industry using Artificial Intelligence-Based System. *2023 3rd International Conference on Advance Computing and Innovative Technologies in Engineering (ICACITE)*, (pp. 955-958). IEEE.

Kumar, B. P., Kalpana, A. V., & Nalini, S. (2023). Gated attention based deep learning model for analysing the influence of social media on education. *Journal of Experimental & Theoretical Artificial Intelligence*.

Kumar, J. S., & Zaveri, M. A. (2018). Clustering approaches for pragmatic two-layer IoT architecture. *Wireless Communications and Mobile Computing, 2018*, 1–17. doi:10.1155/2018/8739203

Lakshmi, G. A., Gummadi, A., & Changala, R. (2023). Block Chain and Machine Learning Models to Evaluate Faults in the Smart Manufacturing System. *International Journal of Scientific Research in Science and Technology, 10*(5), 247–255. doi:10.32628/IJSRST2321438

Lammie, C., & Azghadi, M. R. (2020). Memtorch: a simulation framework for deep memristive cross-bar architectures. 2020 IEEE international symposium on circuits and systems (ISCAS). IEEE.

Lazos, L., & Poovendran, R. (2005). SeRLoc: Robust localization for wireless sensor networks. *ACM Transactions on Sensor Networks, 1*(1), 73–100. doi:10.1145/1077391.1077395

Le Nguyen, T., & Do, T. T. H. (2019). Artificial intelligence in healthcare: a new technology benefit for both patients and doctors. *2019 Portland International Conference on Management of Engineering and Technology (PICMET)*, IEEE. 10.23919/PICMET.2019.8893884

Lee, E. S., Bae, H. C., Kim, H. J., Han, H. N., Lee, Y. K., & Son, J. Y. (2020). Trends in AI technology for smart manufacturing in the future. *Electronics and telecommunications trends, 35*(1), 60-70.

Lee, B. R., & Kim, I. S. (2018). The role and collaboration model of human and artificial intelligence considering human factor in financial security. *Journal of the Korea Institute of Information Security & Cryptology, 28*(6), 1563–1583.

Lee, C.-N., Huang, T.-H., Wu, C.-M., & Tsai, M.-C. (2017). The Internet of Things and its applications. In *Big Data Analytics for Sensor-Network Collected Intelligence* (pp. 256–279). Elsevier. doi:10.1016/B978-0-12-809393-1.00013-1

Lee, M. S., Grabowski, M. M., Habboub, G., & Mroz, T. E. (2020). The impact of artificial intelligence on quality and safety. *Global Spine Journal, 10*(1, suppl), 99S–103S. doi:10.1177/2192568219878133 PMID:31934528

Lee, W. J., Wu, H., Yun, H., Kim, H., Jun, M. B., & Sutherland, J. W. (2019). Predictive maintenance of machine tool systems using artificial intelligence techniques applied to machine condition data. *Procedia CIRP, 80*, 506–511. doi:10.1016/j.procir.2018.12.019

Lee, Y., & Roh, Y. H. (2023). An Expandable Yield Prediction Framework Using Explainable Artificial Intelligence for Semiconductor Manufacturing. *Applied Sciences (Basel, Switzerland), 13*(4), 2660. doi:10.3390/app13042660

Leng, J., Sha, W., Lin, Z., Jing, J., Liu, Q., & Chen, X. (2023). Blockchained smart contract pyramid-driven multi-agent autonomous process control for resilient individualised manufacturing towards Industry 5.0. *International Journal of Production Research, 61*(13), 4302–4321. doi:10.1080/00207543.2022.2089929

Li, B. H., Hou, B. C., Yu, W. T., Lu, X. B., & Yang, C. W. (2017). Applications of artificial intelligence in intelligent manufacturing: A review. *Frontiers of Information Technology & Electronic Engineering, 18*(1), 86–96. doi:10.1631/FITEE.1601885

Li, J. (2018). hua: Cyber security meets artificial intelligence: A survey. *Front. Inf. Technol. Electron. Eng., 19*(12), 1462–1474. doi:10.1631/FITEE.1800573

Li, J., Zhou, Y., Yao, J., & Liu, X. (2021). An empirical investigation of trust in AI in a Chinese petrochemical enterprise based on institutional theory. *Scientific Reports, 11*(1), 13564. doi:10.1038/s41598-021-92904-7 PMID:34193907

Li, M., & Liu, Y. (2010). Rendered path: Range-free localization in anisotropic sensor networks with holes. *IEEE/ACM Transactions on Networking, 18*(1), 320–332. doi:10.1109/TNET.2009.2024940

Lim, J. Y., Lim, J. Y., Baskaran, V. M., & Wang, X. (2023). A Deep Context Learning Based PCB Defect Detection Model with Anomalous Trend Alarming System. *Results in Engineering, 17*, 100968. https://www.sciencedirect.com/science/article/pii/S2590123023000956. doi:10.1016/j.rineng.2023.100968

Linaza, M. T., Posada, J., Bund, J., Eisert, P., Quartulli, M., Döllner, J., Pagani, A., Olaizola, I. G., Barriguinha, A., & Moysiadis, T. (2021). Data-driven artificial intelligence applications for sustainable precision agriculture. *Agronomy (Basel), 11*, 1227. doi:10.3390/agronomy11061227

Liu, K., Zhang, J. J., Tan, B., & Feng, D. (2021). *Can We Trust Machine Learning for Electronic Design Automation?* 2021 IEEE 34th International System-on-Chip Conference (SOCC), Las Vegas, NV, USA. 10.1109/SOCC52499.2021.9739485

Liu, D., Ning, P., Liu, A., Wang, W., & Du, W. (2005). Attack-resistant location estimation in sensor networks. *ACM Transactions on Information and System Security, 11*(4), 1–39. doi:10.1145/1380564.1380570

Lourens, M., Raman, R., Vanitha, P., Singh, R., Manoharan, G., & Tiwari, M. (2022, December). Agile Technology and Artificial Intelligent Systems in Business Development. In *2022 5th International Conference on Contemporary Computing and Informatics (IC3I)* (pp. 1602-1607). IEEE. 10.1109/IC3I56241.2022.10073410

Lourens, M., Sharma, S., Pulugu, R., Gehlot, A., Manoharan, G., & Kapila, D. (2023, May). Machine learning-based predictive analytics and big data in the automotive sector. In *2023 3rd International Conference on Advance Computing and Innovative Technologies in Engineering (ICACITE)* (pp. 1043-1048). IEEE. 10.1109/ICACITE57410.2023.10182665

Lueth, K. L. (2020, November). *State of the IoT 2020: 12 billion IoT connections, surpassing non-IoT for the first time.* IoT Analytics. https://iot-analytics.com/state-of-the-iot-2020-12-billion-iot-connections-surpassing-non-iot-forthe-first-time/

Lyall, L. M., Wyse, C. A., Morales, C. A. C., Lyall, D. M., Cullen, B., & Mackay, D. (2018). Seasonality of depressive symptoms in women but not in men: A cross-sectional study in the UK Biobank cohort. *Journal of Affective Disorders, 229*, 296–305. doi:10.1016/j.jad.2017.12.106 PMID:29329063

Madrid, J. A. (2023). The Role of Artificial Intelligence in Automotive Manufacturing and Design. *International Journal of Advanced Research in Science. Tongxin Jishu.*

Mahbub, M., & Shubair, R. M. (2023). Contemporary Advances in Multi-Access Edge Computing: A Survey of Fundamentals, Architecture, Technologies, Deployment Cases, Security, Challenges, and Directions. *Journal of Network and Computer Applications, 219*, 103726. https://www.sciencedirect.com/science/article/pii/S1084804523001455. doi:10.1016/j.jnca.2023.103726

Manoharan, G., Durai, S., Ashtikar, S. P., & Kumari, N. (2024). Artificial Intelligence in Marketing Applications. In Artificial Intelligence for Business (pp. 40-70). Productivity Press.

Manoharan, G., Durai, S., Rajesh, G. A., Razak, A., Rao, C. B., & Ashtikar, S. P. (2023). A study of postgraduate students' perceptions of key components in ICCC to be used in artificial intelligence-based smart cities. In *Artificial Intelligence and Machine Learning in Smart City Planning* (pp. 117–133). Elsevier. doi:10.1016/B978-0-323-99503-0.00003-X

Manoharan, G., Durai, S., Rajesh, G. A., Razak, A., Rao, C. B., & Ashtikar, S. P. (2023). A study on the perceptions of officials on their duties and responsibilities at various levels of the organizational structure in order to accomplish artificial intelligence-based smart city implementation. In *Artificial Intelligence and Machine Learning in Smart City Planning* (pp. 1–10). Elsevier. doi:10.1016/B978-0-323-99503-0.00007-7

Mardiani, E., Judijanto, L., & Rukmana, A. Y. (2023). Improving Trust and Accountability in AI Systems through Technological Era Advancement for Decision Support in Indonesian Manufacturing Companies. *West Science Interdisciplinary Studies.*, *1*(10), 1019–1027. doi:10.58812/wsis.v1i10.301

Mattera, G., Nele, L., & Paolella, D. A. (2023). Monitoring and control the Wire Arc Additive Manufacturing process using artificial intelligence techniques: A review. *Journal of Intelligent Manufacturing*, 1–31.

Maurya, P., Gaikawad, C., & Salvi, S. (2022). Visual Inspection for Industries. International Journal of Advanced Research in Science. *Tongxin Jishu.*

Ma, X., & Zhang, Y. (2022). Digital Innovation Risk Management Model of Discrete Manufacturing Enterprise Based on Big Data Analysis. *Journal of Global Information Management*, *30*(7), 1–14. https://www.scopus.com/inward/record.uri?eid=2-s2.0-85114305 136&doi=10.4018%2FJGIM.286761&partnerID=40&md5=bed6df90a7355 98eedf350ce3a3b8268. doi:10.4018/JGIM.286761

May, M. C., Neidhöfer, J., Körner, T., Schäfer, L., & Lanza, G. (2022). Applying Natural Language Processing in Manufacturing. *Procedia CIRP*, *115*, 184–189. doi:10.1016/j.procir.2022.10.071

McCormick, M. E. (2007). *Ocean Wave Energy Conversion*. Dover Publications Inc.

McCoy, T. H., Castro, V. M., Rosenfield, H. R., Cagan, A., Kohane, I. S., & Perlis, R. H. (2015). A Clinical Perspective on the Relevance of Research Domain Criteria in Electronic Health Records. *The American Journal of Psychiatry*, *172*(4), 316–320. doi:10.1176/appi.ajp.2014.14091177 PMID:25827030

Medida, L. H., & Renugadevi, R. (2023). Machine Learning Techniques for Predicting Pregnancy Complications. In D. Satishkumar & P. Maniarasan (Eds.), *Predicting Pregnancy Complications Through Artificial Intelligence and Machine Learning* (pp. 116–125). IGI Global. doi:10.4018/978-1-6684-8974-1.ch008

Meenaakumari, M., Jayasuriya, P., Dhanraj, N., Sharma, S., Manoharan, G., & Tiwari, M. (2022, December). Loan Eligibility Prediction using Machine Learning based on Personal Information. *2022 5th International Conference on Contemporary Computing and Informatics (IC3I)* (pp. 1383-1387). IEEE. 10.1109/IC3I56241.2022.10073318

Melnyk, L. H., Dehtyarova, I. B., Dehtiarova, I. B., Kubatko, O. V., & Kharchenko, M. O. (2019). *Economic and social challenges of disruptive technologies in conditions of industries 4.0 and 5.0: the EU Experience*. Research Gate.

Meyer, W., & Isenberg, R. (1990). Knowledge-based factory supervision: EP 932 results. *International Journal of Computer Integrated Manufacturing, 3*(3-4), 206–233. doi:10.1080/09511929008944450

Miltiadou, D., Perakis, K., Sesana, M., Calabresi, M., Lampathaki, F., & Biliri, E. (2023). A novel Explainable Artificial Intelligence and secure Artificial Intelligence asset sharing platform for the manufacturing industry. *2023 IEEE International Conference on Engineering, Technology and Innovation (ICE/ITMC)*, (pp. 1-8). IEEE. 10.1109/ICE/ITMC58018.2023.10332346

Mohamed, A., Tharwat, M., Magdy, M., Abubakr, T., Nasr, O., & Youssef, M. (2022). DeepFeat: Robust large-scale multi-features outdoor localization in LTE networks using deep learning. *IEEE Access : Practical Innovations, Open Solutions, 10*, 3400–3414. doi:10.1109/ACCESS.2022.3140292

Moyne, J., & Iskandar, J. (2017). Big data analytics for smart manufacturing: Case studies in semiconductor manufacturing. *Processes, 5*(3), 39., J., & Iskandar, J. (2017). Big data analytics for smart manufacturing: Case studies in semiconductor manufacturing. *Processes (Basel, Switzerland), 5*(3), 39. doi:10.3390/pr5030039

Mypati, O., Mukherjee, A., Mishra, D., Pal, S. K., Chakrabarti, P. P., & Pal, A. (2023). A critical review on applications of artificial intelligence in manufacturing. *Artificial Intelligence Review, 56*(S1), 661–768. doi:10.1007/s10462-023-10535-y

Naddafzadeh-Shirazi, G., Shenouda, M. B., & Lampe, L. (2014). Second order cone programming for sensor network localization with Anchor position uncertainty. *IEEE Transactions on Wireless Communications, 13*(2), 749–763. doi:10.1109/TWC.2013.120613.130170

Nahavandi, S. (2019). Industry 5.0—A human-centric solution. *Sustainability (Basel), 11*(16), 4371. doi:10.3390/su11164371

Nguyen, P., Tran, T., Wickramasinghe, N., & Venkatesh, S. (2017). A Convolutional Net for Medical Records. *IEEE Journal of Biomedical and Health Informatics, 21*(1), 22–30. doi:10.1109/JBHI.2016.2633963 PMID:27913366

Nikitas, A., Michalakopoulou, K., Njoya, E. T., & Karampatzakis, D. (2020). Artificial intelligence, transport and the smart city: Definitions and dimensions of a new mobility era. *Sustainability (Basel), 12*(7), 2789. doi:10.3390/su12072789

Ning, H., Yin, R., Ullah, A., & Shi, F. (2021). A survey on hybrid human-artificial intelligence for autonomous driving. *IEEE Transactions on Intelligent Transportation Systems, 23*(7), 6011–6026. doi:10.1109/TITS.2021.3074695

Olaizola, I. G., Quartulli, M., Garcia, A., & Barandiaran, I. (2022). *Artificial Intelligence from Industry 5.0 perspective: Is the Technology Ready to Meet the Challenge?* CUER. http://ceur-ws.org

Ongen, H., Brown, A. A., Delaneau, O., Panousis, N. I., Nica, A. C., & Dermitzakis, E. T. (2017). Estimating the causal tissues for complex traits and diseases. *Nature Genetics, 49*(12), 1676–1683. doi:10.1038/ng.3981 PMID:29058715

Compilation of References

Osborn, D. P. J., Hardoon, S., Omar, R. Z., Holt, R. I. G., King, M., Larsen, J., Marston, L., Morris, R. W., Nazareth, I., Walters, K., & Petersen, I. (2015). Cardiovascular Risk Prediction Models for People with Severe Mental Illness. *JAMA Psychiatry*, *72*(2), 143–151. doi:10.1001/jamapsychiatry.2014.2133 PMID:25536289

Panesar, S., Cagle, Y., Chander, D., Morey, J., Fernandez-Miranda, J., & Kliot, M. (2019). Artificial intelligence and the future of surgical robotics. *Annals of Surgery*, *270*(2), 223–226. doi:10.1097/SLA.0000000000003262 PMID:30907754

Patel, A. I., Khunti, P. K., Vyas, A. J., & Patel, A. B. (2022). Explicating artificial intelligence: Applications in medicine and pharmacy. *Asian Journal of Pharmacy and Technology*, *12*(4), 401–406. doi:10.52711/2231-5713.2022.00061

Penumuru, D. P., Muthuswamy, S., & Karumbu, P. (2020). Identification and classification of materials using machine vision and machine learning in the context of industry 4.0. *Journal of Intelligent Manufacturing*, *31*(5), 1229–1241. doi:10.1007/s10845-019-01508-6

Peruzzini, M., Prati, E., & Pellicciari, M. (2023). A framework to design smart manufacturing systems for Industry 5.0 based on the human-automation symbiosis. *International Journal of Computer Integrated Manufacturing*, 1–18. doi:10.1080/0951192X.2023.2257634

Plathottam, S. J., Rzonca, A., Lakhnori, R., & Iloeje, C. O. (2023). A review of artificial intelligence applications in manufacturing operations. *Journal of Advanced Manufacturing and Processing*, *5*(3), 10159. doi:10.1002/amp2.10159

Ponduri, S. B., Ahmad, S. S., Ravisankar, P., Thakur, D. J., Chawla, K., Chary, D. T., & Sharma, S. (2024). A Study on Recent Trends of Technology and its Impact on Business and Hotel Industry. *Migration Letters : An International Journal of Migration Studies*, *21*(S1), 801–806.

Potortì, F., Park, S., Jiménez Ruiz, A. R. J., Barsocchi, P., Girolami, M., Crivello, A., Lee, S. Y., Lim, J. H., Torres-Sospedra, J., Seco, F., Montoliu, R., Mendoza-Silva, G. M., Pérez Rubio, M. D. C., Losada-Gutiérrez, C., Espinosa, F., & Macias-Guarasa, J. (2017). Comparing the performance of indoor localization systems through the Evaal framework. *Sensors (Basel)*, *17*(10), 2327. doi:10.3390/s17102327 PMID:29027948

Praveen Kumar, B., kalpana, A. V., & Nalini, S. (2023, March 16). Gated Attention Based Deep Learning Model for Analyzing the Influence of Social Media on Education. *Journal of Experimental & Theoretical Artificial Intelligence*, 1–15. doi:10.1080/0952813X.2023.2188262

Prhashanna, A., & Dormidontova, E. E. (2020). Micelle Self-Assembly and Chain Exchange Kinetics of Tadpole Block Copolymers with a Cyclic Corona Block. *Macromolecules*, *53*(3), 982–991. https://www.scopus.com/inward/record.uri?eid=2-s2.0-85079191834&doi=10.1021%2Facs.macromol.9b02398&partnerID=40&md5=363787355c3e21af9e1a1e2187b3e234. doi:10.1021/acs.macromol.9b02398

Pughazendi, N., Rajaraman, P., & Mohammed, M. (2023). Graph Sample and Aggregate Attention Network Optimized with Barnacles Mating Algorithm Based Sentiment Analysis for Online Product Recommendation. *Applied Soft Computing, 145*. https://www.sciencedirect.com/science/article/pii/S156849462 3005501

Rajkomar, A., Oren, E., Chen, K., Dai, A. M., Hajaj, N., Hardt, M., Liu, P. J., Liu, X., Marcus, J., Sun, M., Sundberg, P., Yee, H., Zhang, K., Zhang, Y., Flores, G., Duggan, G. E., Irvine, J., Le, Q., Litsch, K., & Dean, J. (2018). Scalable and Accurate Deep Learning with Electronic Health Records. *NPJ Digital Medicine, 1*(1), 18. doi:10.1038/s41746-018-0029-1 PMID:31304302

Ranasinghe, K., Sabatini, R., Gardi, A., Bijjahalli, S., Kapoor, R., Fahey, T., & Thangavel, K. (2022). Advances in Integrated System Health Management for mission-essential and safety-critical aerospace applications. *Progress in Aerospace Sciences, 128*, 100758. doi:10.1016/j.paerosci.2021.100758

Razak, A., Nayak, M. P., Manoharan, G., Durai, S., Rajesh, G. A., Rao, C. B., & Ashtikar, S. P. (2023). Reigniting the power of artificial intelligence in education sector for the educators and students competence. In *Artificial Intelligence and Machine Learning in Smart City Planning* (pp. 103–116). Elsevier. doi:10.1016/B978-0-323-99503-0.00009-0

Renugadevi, R., & Sethukarasi, T. (2023). A Novel and Efficient Multi-Band Wireless Communication System for Healthcare Management System. *2023 International Conference on Intelligent and Innovative Technologies in Computing, Electrical and Electronics (IITCEE)*. 10.1109/IITCEE57236.2023.10090991

Sadowski, S., & Spachos, P. (2018). RSSI-based indoor localization with the Internet of things. *IEEE Access : Practical Innovations, Open Solutions, 6*, 30149–30161. doi:10.1109/ACCESS.2018.2843325

Sai Aswin, B. G., Vishnubala, S., Dhinakaran, D., Kumar, N. J., Udhaya Sankar, S. M., & Mohamed Al Faisal, A. M. (2023). *A Research on Metaverse and its Application*. 2023 World Conference on Communication & Computing (WCONF), Raipur, India. 10.1109/WCONF58270.2023.10235216

Saikia, P. G. & Rakshit, D. (2020). Designing a Clean and Efficient Air Conditioner with AI Intervention to Optimize Energy-Exergy Interplay. *Energy and AI, 2*. https://www.sciencedirect.com/science/article/pii/S266654682 030029X

Salam, A., Ullah, F., Amin, F., & Abrar, M. (2023). Deep Learning Techniques for Web-Based Attack Detection in Industry 5.0: A Novel Approach. *Technologies, 11*(4), 107. doi:10.3390/technologies11040107

Sarala, B., Sumathy, G., Kalpana, A. V., & Jasmine Hephzipah, J. J. (2023) Glioma brain tumor detection using dual convolutional neural networks and histogram density segmentation algorithm. *Biomedical Signal Processing and Control, 85.*] doi:10.1016/j.bspc.2023.104859

Sarker, S., Jamal, L., Ahmed, S. F., & Irtisam, N. (2021). Robotics and artificial intelligence in healthcare during COVID-19 pandemic: A systematic review. *Robotics and Autonomous Systems*, *146*, 103902. doi:10.1016/j.robot.2021.103902 PMID:34629751

Satpathy, A., Samal, A., Gupta, S., Kumar, S., Sharma, S., Manoharan, G., Karthikeyan, M., & Sharma, S. (2024). To Study the Sustainable Development Practices in Business and Food Industry. *Migration Letters : An International Journal of Migration Studies*, *21*(S1), 743–747. doi:10.59670/ml.v21iS1.6400

Sayedahmed, H. A. M., Fahmy, I. M. A., & Hefny, H. A. 2021. "Impact of Fuzzy Stability Model on Ad Hoc Reactive Routing Protocols to Improve Routing Decisions." In *Advances in Intelligent Systems and Computing*, Springer, Cham, 441–54. https://link.springer.com/chapter/10.1007/978-3-030-58669-0_40 (October 14, 2020).

Selvaraj, D., Udhaya Sankar, S. M., Pavithra, S., & Boomika, R. (2023). Assistive System for the Blind with Voice Output Based on Optical Character Recognition. In: Gupta, D., Khanna, A., Hassanien, A.E., Anand, S., Jaiswal, A. (eds) *International Conference on Innovative Computing and Communications. Lecture Notes in Networks and Systems*. Springer, Singapore. 10.1007/978-981-19-3679-1_1

Sembroiz, D., Ricciardi, S., & Careglio, D. (2018). A novel cloud-based IoT architecture for smart building automation. In *Security and Resilience in Intelligent Data-Centric Systems and Communication Networks* (pp. 215–233). Elsevier. doi:10.1016/B978-0-12-811373-8.00010-0

Semeraro, F., Griffiths, A., & Cangelosi, A. (2023). Human–robot collaboration and machine learning: A systematic review of recent research. *Robotics and Computer-integrated Manufacturing*, *79*, 102432. doi:10.1016/j.rcim.2022.102432

Senapati, B., & Rawal, B. S. (2023). Quantum Communication with RLP Quantum Resistant Cryptography in Industrial Manufacturing. *Cyber Security and Applications*. Science Direct. https://www.sciencedirect.com/science/article/pii/S2772918423000073

Shahbazi, Z., & Byun, Y.-C. (2021). Smart Manufacturing Real-Time Analysis Based on Blockchain and Machine Learning Approaches. *Applied Sciences (Basel, Switzerland)*, *11*(8), 3535. doi:10.3390/app11083535

Shah, Y., Verma, Y., Sharma, U., Sampat, A., & Kulkarni, V. (2023). *Supply Chain for Safe & Timely Distribution of Medicines using Blockchain & Machine Learning. 5th International Conference on Smart Systems and Inventive Technology (ICSSIT)*, Tirunelveli, India. 10.1109/ICSSIT55814.2023.10061049

Shameem, A., Ramachandran, K. K., Sharma, A., Singh, R., Selvaraj, F. J., & Manoharan, G. (2023, May). The rising importance of AI in boosting the efficiency of online advertising in developing countries. In *2023 3rd International Conference on Advance Computing and Innovative Technologies in Engineering (ICACITE)* (pp. 1762-1766). IEEE. 10.1109/ICACITE57410.2023.10182754

Sharifi, Z., & Shokouhyar, S. (2021). Promoting Consumer's Attitude toward Refurbished Mobile Phones: A Social Media Analytics Approach. *Resources, Conservation and Recycling, 167*, 105398. https://www.sciencedirect.com/science/article/pii/S092134492 1000057. doi:10.1016/j.resconrec.2021.105398

Shi, Q., He, C., Chen, H., & Jiang, L. (2010). Distributed wireless sensor network localization via sequential greedy optimization algorithm. *IEEE Transactions on Signal Processing, 58*(6), 3328–3340. doi:10.1109/TSP.2010.2045416

Shivraj, V. L., Rajan, M. A., Singh, M., & Balamuralidhar, P. (2015). One time password authentication scheme based on elliptic curves for Internet of Things (IoT). In: *2015 5th National Symposium on Information Technology: Towards New Smart World (NSITNSW),* (pp. 1–6). Riyadh, Saudi Arabia: IEEE.

Shi, W., Pallis, G., & Xu, Z. (2019). Edge computing. *Proceedings of the IEEE, 107*(8), 1474–1481. doi:10.1109/JPROC.2019.2928287

Shyam, M., & Amalasweena, M. (2023). Intellectual Design of Bomb Identification and Defusing Robot based on Logical Gesturing Mechanism. *2023 International Conference on Advances in Computing, Communication and Applied Informatics (ACCAI)*, Chennai, India. 10.1109/ACCAI58221.2023.10201034

Sisinni, E., Saifullah, A., Han, S., Jennehag, U., & Gidlund, M. (2018, November). Industrial Internet of Things: Challenges opportunities and directions. *IEEE Transactions on Industrial Informatics, 14*(11), 4724–4734. doi:10.1109/TII.2018.2852491

Solairaj, A., Sugitha, G., & Kavitha, G. (2023). Enhanced Elman Spike Neural Network Based Sentiment Analysis of Online Product Recommendation. *Applied Soft Computing, 132*, 109789. https://www.sciencedirect.com/science/article/pii/S156849462 2008389. doi:10.1016/j.asoc.2022.109789

Soori, M. & Arezoo, B. (2013). Machine learning and artificial intelligence in CNC machine tools, a review. *Sustain. Manuf. Service Econ.*

Soori, M., & Arezoo, B. (2020). Recent development in friction stir welding process: A review. *SAE International Journal of Materials and Manufacturing*, 18.

Soori, M., & Arezoo, B. (2023). Dimensional, geometrical, thermal and tool deflection errors compensation in 5-Axis CNC milling operations. *Australian Journal of Mechanical Engineering*, 1–15. doi:10.1080/14484846.2023.2195149

Soori, M., Arezoo, B., & Habibi, M. (2013). Dimensional and geometrical errors of three-axis CNC milling machines in a virtual machining system. *Computer Aided Design, 45*(11), 1306–1313. doi:10.1016/j.cad.2013.06.002

Soori, M., Arezoo, B., & Habibi, M. (2014). Virtual machining considering dimensional, geometrical and tool deflection errors in three-axis CNC milling machines. *Journal of Manufacturing Systems, 33*(4), 498–507. doi:10.1016/j.jmsy.2014.04.007

Soori, M., Arezoo, B., & Habibi, M. (2016). Tool deflection error of three-axis computer numerical control milling machines, monitoring and minimizing by a virtual machining system. *Journal of Manufacturing Science and Engineering, 138*(8), 138. doi:10.1115/1.4032393

Soori, M., Arezoo, B., & Habibi, M. (2017). Accuracy analysis of tool deflection error modelling in prediction of milled surfaces by a virtual machining system. *International Journal of Computer Applications in Technology, 55*(4), 308–321. doi:10.1504/IJCAT.2017.086015

Soori, M., Asmael, M., & Solyalı, D. (2022). Radio frequency identification (RFID) based wireless manufacturing systems, a review. *Independent Journal of Management & Production, 13*(1), 258–290. doi:10.14807/ijmp.v13i1.1497

Srinivasan, L., Selvaraj, D., & Udhaya Sankar, S. M. (2023). Leveraging Semi-Supervised Graph Learning for Enhanced Diabetic Retinopathy Detection. *SSRG International Journal of Electronics and Communication Engineering., 10*(8), 9–21. doi:10.14445/23488549/IJECE-V10I8P102

Stadnicka, D., Sęp, J., Amadio, R., Mazzei, D., Tyrovolas, M., Stylios, C., Carreras-Coch, A., Merino, J. A., Żabiński, T., & Navarro, J. (2022). Industrial needs in the fields of artificial intelligence, Internet of Things and edge computing. *Sensors (Basel), 22*(12), 4501. doi:10.3390/s22124501 PMID:35746287

Stoutenburg, E. D., Jenkins, N., & Jacobson, M. Z. (2010). Power output variations of co-located offshore wind turbines and wave energy converters in California. *Renewable Energy, 35*(12), 2781–2791. doi:10.1016/j.renene.2010.04.033

Su, H., Maji, S., Kalogerakis, E., & Learned-Miller, E. (2015). Multi-view convolutional neural networks for 3D shape recognition. *Proc. IEEE Int. Conf. Comput. Vis. (ICCV)*, (pp. 945-953). IEEE. 10.1109/ICCV.2015.114

Sundaram, S., & Zeid, A. (2023). Artificial Intelligence-Based Smart Quality Inspection for Manufacturing. *Micromachines, 14*(3), 14. doi:10.3390/mi14030570 PMID:36984977

Susto, G. A., Schirru, A., Pampuri, S., McLoone, S., & Beghi, A. (2014). Machine learning for predictive maintenance: A multiple classifier approach. *IEEE Transactions on Industrial Informatics, 11*(3), 812–820. doi:10.1109/TII.2014.2349359

Taj, I., & Jhanjhi, N. Z. (2022). Towards Industrial Revolution 5.0 and Explainable Artificial Intelligence: Challenges and Opportunities. *International Journal of Computing and Digital Systems., 12*(1), 285–310. doi:10.12785/ijcds/120124

Tao, F., Cheng, J., Qi, Q., Zhang, M., Zhang, H., & Sui, F. (2018, February). Digital twin-driven product design manufacturing and service with big data. *International Journal of Advanced Manufacturing Technology, 94*(9), 3563–3576. doi:10.1007/s00170-017-0233-1

Tekler, Z. D., & Chong, A. (2022). Occupancy prediction using deep learning approaches across multiple space types: A minimum sensing strategy. *Building and Environment, 226,* 109689. doi:10.1016/j.buildenv.2022.109689

Tien, J. M. (2020). Toward the fourth industrial revolution on real-time customization. *Journal of Systems Science and Systems Engineering, 29*(2), 127–142. doi:10.1007/s11518-019-5433-9

Torres, D. R., Alejandro, D. S. A., Roldán, Á. O., Bustos, A. H., & Luis, E. A. G. (2022). A Review of Deep Reinforcement Learning Approaches for Smart Manufacturing in Industry 4.0 and 5.0 Framework. *Applied Sciences (Basel, Switzerland), 12*(23), 12377. doi:10.3390/app122312377

Tripathi, M. A., Tripathi, R., Effendy, F., Manoharan, G., Paul, M. J., & Aarif, M. (2023, January). An In-Depth Analysis of the Role That ML and Big Data Play in Driving Digital Marketing's Paradigm Shift. In *2023 International Conference on Computer Communication and Informatics (ICCCI)* (pp. 1-6). IEEE. 10.1109/ICCCI56745.2023.10128357

Tu, Y., & Yeung, E. H. (1997). Integrated maintenance management system in a textile company. *International Journal of Advanced Manufacturing Technology, 13*(6), 453–461. doi:10.1007/BF01179041

Udhaya Sankar, S. M., Kumar, N. J., Dhinakaran, D., Kamalesh, S. S., & Abenesh, R. (2023). *Machine Learning System for Indolence Perception. 2023 International Conference on Innovative Data Communication Technologies and Application (ICIDCA)*, Uttarakhand, India. 10.1109/ICIDCA56705.2023.10099959

V. B. S., Pramod, D., & Raman, R. (2022). *Intention to use Artificial Intelligence services in Financial Investment Decisions.* 2022 International Conference on Decision Aid Sciences and Applications (DASA), Chiangrai, Thailand. 10.1109/DASA54658.2022.9765183

Van der Burg, S., Bogaardt, M. J., & Wolfert, S. (2019). Ethics of smart farming: Current questions and directions for responsible innovation towards the future. *NJAS Wageningen Journal of Life Sciences, 90*(1), 100289. doi:10.1016/j.njas.2019.01.001

Wang, F. (2023). Research on the application of artificial intelligence technology to promote the high-quality development path of manufacturing industry. *SHS Web of Conferences.* IEEE.

Wang, S., Atif Qureshi, M., Miralles-Pechuan, L., Reddy Gadekallu, T., & Liyanage, M. (2021). Explainable AI for B5G/6G: technical aspects, use cases, and research challenges. *arXiv e-prints.*

Wang, W., & Siau, K. (2018). *Ethical and moral issues with AI.*

Wang, F., Zhang, M., Wang, X., Ma, X., & Liu, J. (2020). Deep learning for edge computing applications: A state-of-the-art survey. *IEEE Access : Practical Innovations, Open Solutions, 8,* 58322–58336. doi:10.1109/ACCESS.2020.2982411

Wang, G., Chen, H., Li, Y., & Jin, M. (2012). On received-signal-strength based localization with unknown transmit power and path loss exponent. *IEEE Wireless Communications Letters*, *1*(5), 536–539. doi:10.1109/WCL.2012.072012.120428

Werens, S., & von Garrel, J. (2023). Implementation of artificial intelligence at the workplace, considering the work ability of employees. *Tatup, 32*(2), 43-9. https://www.tatup.de/index.php/tatup/article/view/7064

Wu, D., Ren, A., Zhang, W., Fan, F., Liu, P., Fu, X., & Terpenny, J. (2018). Cybersecurity for digital anufacturing. *Journal of Manufacturing Systems, 48*, 3–12. doi:10.1016/j.jmsy.2018.03.006

Wuest, T., Irgens, C., & Thoben, K.-D. (2014, October). An approach to monitoring quality in manufacturing using supervised machine learning on product state data. *Journal of Intelligent Manufacturing, 25*(5), 1167–1180. doi:10.1007/s10845-013-0761-y

Xingyu, C., & Chuntang, C. (2020, November). The Development of Machinery Manufacturing And the Application Analysis of Artificial Intelligence. *Journal of Physics: Conference Series*, *1684*(1), 012017. doi:10.1088/1742-6596/1684/1/012017

Xu, H., Chai, L., Luo, Z., & Li, S. (2022). Stock Movement Prediction via Gated Recurrent Unit Network Based on Reinforcement Learning with Incorporated Attention Mechanisms. *Neurocomputing, 467*, 214–228. https://www.sciencedirect.com/science/article/pii/S092523122 1014508. doi:10.1016/j.neucom.2021.09.072

Yadav, M., Vardhan, A., Chauhan, A. S., & Saini, S. (2023). *A Study on Creation of Industry 5.0: New Innovations using big data through artificial intelligence, Internet of Things and next-origination technology policy. Conference on Electrical, Electronics and Computer Science (SCEECS)*, Bhopal, India. 10.1109/SCEECS57921.2023.10063069

Yan, F., Liu, J., Yan, X., & Wang, G. (2022). *Application of Key Technologies of Intelligent Manufacturing in Metallurgical Industry Led by Artificial Intelligence*. 2022 International Conference on Cloud Computing, Big Data and Internet of Things (3CBIT), Wuhan, China. 10.1109/3CBIT57391.2022.00073

Yang, T., Cabani, A., & Chafouk, H. (2021). A survey of recent indoor localization scenarios and methodologies. *Sensors (Basel), 21*(23), 8086. doi:10.3390/s21238086 PMID:34884090

Yang, Z.-X., Wang, X., & Wong, P. K. (2018, December). Single and simultaneous fault diagnosis with application to a multistage gearbox: A versatile dual-ELM network approach. *IEEE Transactions on Industrial Informatics, 14*(12), 5245–5255. doi:10.1109/TII.2018.2817201

Yeredor, A. (2014) Decentralized TOA-based localization in non-synchronized wireless networks with partial. *Asymmetric Connectivity 15th international workshop on signal processing advances in wireless communications*. IEEE.

You, Z., Si, Y.-W., Zhang, D., Zeng, X., Leung, S. C. H., & Li, T. (2015). A decision-making framework for precision marketing. *Expert Systems with Applications, 42*(7), 3357–3367. doi:10.1016/j.eswa.2014.12.022

Zaccaria, V., Stenfelt, M., Aslanidou, I., & Kyprianidis, K. G. (2018, August). Fleet monitoring and diagnostics framework based on digital twin of aero-engines. *Turbo Expo Power Land Sea Air, 51128*, 10. doi:10.1115/GT2018-76414

Zhang, J., Tu, Y., & Yeung, E. H. H. (1997, July). Intelligent decision support system for equipment diagnosis and maintenance management. In Innovation in Technology Management. The Key to Global Leadership. PICMET'97 (p. 733). IEEE. doi:10.1109/PICMET.1997.653599

Zhang, P.-B., & Yang, Z.-X. (2018, January). A novel AdaBoost framework with robust threshold and structural optimization. *IEEE Transactions on Cybernetics, 48*(1), 64–76. doi:10.1109/TCYB.2016.2623900 PMID:27898387

Zhang, Q., Zhou, D., & Zeng, X. (2017). HeartID: A Multiresolution Convolutional Neural Network for ECG-Based Biometric Human Identification in Smart Health Applications. *IEEE Access : Practical Innovations, Open Solutions, 5*, 11805–11816. doi:10.1109/ACCESS.2017.2707460

Zhang, S., Zhang, S., Wang, B., & Habetler, T. G. (2020). Deep learning algorithms for bearing fault diagnostics—A comprehensive review. *IEEE Access : Practical Innovations, Open Solutions, 8*, 29857–29881. doi:10.1109/ACCESS.2020.2972859

Zhang, Z., Wang, X., Wang, X., Cui, F., & Cheng, H. (2019, March). A simulation-based approach for plant layout design and production planning. *Journal of Ambient Intelligence and Humanized Computing, 10*(3), 1217–1230. doi:10.1007/s12652-018-0687-5

Zhao, G., Si, X., & Wang, J. (2011). A novel mutual authentication scheme for Internet of Things. In: *Proceedings of 2011 International Conference on Modelling, Identification and Control*, (pp. 563–566). Shanghai, China: IEEE. 10.1109/ICMIC.2011.5973767

Zhong, R. Y., Xu, X., Klotz, E., & Newman, S. T. (2017). Intelligent manufacturing in the context of industry 4.0: A review. *Engineering (Beijing), 3*(5), 616–630. doi:10.1016/J.ENG.2017.05.015

Zhuang, D., Gan, V. J. L., Duygu Tekler, Z. D., Chong, A., Tian, S., & Shi, X. (2023). Data-driven predictive control for smart HVAC system in IoT-integrated buildings with time-series forecasting and reinforcement learning. *Applied Energy, 338*, 120936. doi:10.1016/j.apenergy.2023.120936

Zonta, T., da Costa, C. A., da Rosa Righi, R., de Lima, M. J., da Trindade, E. S., & Li, G. P. (2020, December). Predictive maintenance in the industry 4.0: A systematic literature review. *Computers & Industrial Engineering, 150*, 106889. doi:10.1016/j.cie.2020.106889

About the Contributors

D. Satishkumar is a Associate Professor in the Department of Computer Science and Engineering at Nehru Institute of Technology, Coimbatore, Tamilnadu, India-641105, where he has been since 2019. From 2019 to 2021 he served as Department Research Coordinator. From 2012 to 2016 he served as Assistant Professor of Department of Computer Science and Engineering in Nehru institute of Technology, Inc. During 2003-2007 he was a Lecturer at the KCG College of technology, Chennai, Tamilnadu, India, and in 2007-2010 he was faculty at Coimbatore Institute of Engineering and Technology, Coimbatore, Tamilnadu, India, in 2010-2012 he was faculty at Kalaignar karunanidhi Institute of Technology, Coimbatore, Tamilnadu, India. He received a B.E. from Bharathiar University in 2002, and an M.E. from the Manonmaniam sundharnar University, Tamilnadu, India. He received his Ph.D. in Computer Science and Engineering from the Anna University in 2015.

M. Sivaraja, a goal driven professional born in 1974 emerged as Gold medalist in his PG, Ph.D at Anna University and PD at SUNY Buffalo, USA under BOYS-CAST Fellowship. His dedicated and committed personality established him as the Founder Principal of N.S.N.CET, Karur during 2011 at the age of 37. His Credit includes 35 Journal publications, 95 conference papers and 24 invited talks, conducted 16 seminars and conferences, completed 11 research funded projects and honoured with many Awards.

***

Ahila A. received her BE(ECE) from National Engineering College from MS University, Tamilnadu, India in the year 1999 and completed his Masters in Engineering (Applied Electronics)from Anna University in the year of 2005 and since then actively involved in teaching and research and has 20 years of experience in Teaching. She obtained his PhD in the field of Information and Communication Engineering from Anna University in the year of 2018. At Present, she is working as an Associate Professor in Sri Sairam College of Engineering and Technology, anekal, Bangalore affiliated to Visveswaraya Technological University, her area of interest is the field of medical image processing, wireless networks and VLSI. She can be contacted at email: ahilaa.ece@gmail.com.

A. Siva Kumar is an Assistant Professor in the Department of Data Science and Business Systems, SRM Institute of Science and Technology, Kattankulathur, Chennai, India. He has 10 years 6 months of teaching experience in various reputed engineering colleges in Chennai. He received his B.Tech. degree in Information Technology from Anna University in 2010 and a Master's degree in computer science and engineering from Anna University in 2013. He received his doctorate from Anna University Chennai in 2022. He has published various research papers in international journals and conferences. His research interest includes Cloud with Blockchain, web service, Cloud Security, and Network Security.

A. V. Kalpana is an accomplished academician and researcher, currently serving as an Assistant Professor in the Department of Data Science and Business Systems at the School of Computing, SRM Institute of Science & Technology, Kattankulathur, Chennai. She completed her Bachelor degree in Computer Science and Engineering in 2004 from the University of Madras, showcasing her foundational understanding of computer science. Further enhancing her expertise, she pursued a Master degree in Computer Science and Engineering from Anna University. She holds a Ph.D. degree from Anna University, Chennai, reflecting her dedication to advancing knowledge in her field. Her research contributions are evident through numerous publications in reputable journals and international conference proceedings. Driven by a passion for knowledge, her research interests span across Machine Learning, Deep Learning, Wireless Sensor Networks, and the Internet of Things (IoT).

Lakshmi D. is presently designated as a Senior Associate Professor in the School of Computing Science and Engineering (SCSE) & Assistant Director, at the Centre for Innovation in Teaching & Learning (CITL) at VIT Bhopal. She has 17 international conference presentations, and 21 international journal papers inclusive of SCOPUS & SCI (cumulative impact factor 31). 3 SCOPUS inee book chapters. A total of 24 patents are in various states and 18 patents have been granted at both national and international levels. One Edited book with Taylor & Francis (SCOPUS Indexed). She has won two Best Paper awards at international conferences, one at the IEEE conference and another one at EAMMIS 2021. She received two awards in the year 2022. She received two awards in the year 2022. She has addressed innumerable guest lectures, acted as a session chair, and was invited as a keynote speaker at several international conferences. She has conducted FDPs that cover approximately ~80,000 plus faculty members including JNTU, TEQIP, SERB, SWAYAM, DST,

AICTE, MHRD, ATAL, ISTE, Madhya Pradesh Government-sponsored, and self-financed workshops across India on various titles.

D. Dhinakaran (M'84) was born in Chennai, Tamilnadu, India, in 1984. He is a distinguished scholar and educator with a passion for advancing the fields of Computer Science and Engineering. Dr. Dhinakaran earned his B.E degree in Computer Science and Engineering from Anna University, Chennai, in 2006, showcasing early dedication to his academic pursuits. Building upon his foundation, he pursued and obtained the M.E degree in the field of Computer Science and Engineering from Anna University, Trichy, in 2009. Demonstrating a commitment to academic excellence, Dr. Dhinakaran completed his Ph.D. in Computer Science and Engineering from Anna University, Chennai, further solidifying his expertise in the field. Currently serving as an Assistant Professor in the Department of Computer Science and Engineering at Vel Tech Rangarajan Dr. Sagunthala R&D Institute of Science and Technology, Chennai, Dr. Dhinakaran continues to inspire students and colleagues alike. His areas of specialization encompass Privacy-Preserving Data Mining, Artificial Intelligence, Internet of Things (IoT), Mobile Ad Hoc Networks (MANET), Cloud Computing, and Image Processing, reflecting his diverse and extensive contributions to cutting-edge technologies. Dr. Dhinakaran is a prolific researcher, having presented more than 30 papers at various National and International Conferences. His commitment to advancing knowledge is evident through his membership in esteemed professional societies such as IAENG (International Association of Engineers), IFERP (International Foundation for Engineering Research and Publications), and CSTA (Computer Science Teachers Association). Notably, Dr. Dhinakaran has made significant contributions to academic literature, with publications in 50 different international journals and the successful filing of two patents. His work demonstrates a keen interest in addressing contemporary challenges and pushing the boundaries of technological innovation. As an accomplished academic, researcher, and member of professional societies, Dr. Dhinakaran continues to play a pivotal role in shaping the future of Computer Science and Engineering, leaving an indelible mark on both academia and the broader technological landscape.

A. Ramathilagam is working as a Professor in the Department of Computer Science and Engineering at P.S.R. Engineering College, Sivakasi. She has 21 years of teaching experience. She completed her Ph.D. in Information and Communication Engineering from Anna University in the year 2018. She obtained her M.E Computer Science, Engineering in the year 2004, B.E Computer Science, and Engineering in the year 1999 from Arulmigu Kalasalingam College of Engineering, Krishnankovil. She has authored many book chapters in the reputed publishers like springer, CRC Press etc. She has published 15 papers in reputed international/national journals and presented 20 papers in National and International conferences. Her research interests are Computer Network, Security, Cloud Computing, Big Data Analytics, Machine Learning and Data Science. She is the reviewer of Journal of Communication systems (Wiley publishers), Journal of Super Computing (Springer), and Journal of Intelligent System (Wiley publishers). She has published seven patents. She has received Rs.3 Lakhs from AICTE for organizing STTP. She also holds professional membership in ACM and Life Member in ISTE.

Selvaraj Damodaran is working as a professor in the Department of Electronics and Communication Engineering at Panimalar Engineering College. He obtained his B.E Degree from Madras University, ME degree from Anna University and Ph.D. from Sathyabama University. He has a vast teaching experience of 21 years. He has published more than 75 papers at various refereed journals and conferences. His area of interest is in the field of signal processing, image processing, Biomedical Instrumentation, wireless communication, Artificial Intelligence, Machine learning, Data Mining and Antenna Design. He is a member of IEEE and life member of IETE

K. Arthi, working as Associate Professor in SRM University, has more than 13 years of teaching and research experience. She has earned her M.E. and Ph.D. in Computer Science Engineering from College of Engineering, Anna University, Chennai in 2005 and 2015 respectively. Specializing in IoT and Machine Learning, she has published more than 25 articles in reputed journals and holds two patents in her area of specialization. She has professional membership in ACM and ISTE, through which she has organised multiple events.

K. R. Senthilkumar working as a Librarian in Sri Krishna Arts and Science College, Coimbatore. His most notable contributions to the field of E- Library and the Development of Library Web page. His research interests span both bibliometrics and Web 2.0. Much of his work has been on improving the understanding, design, and performance of Information systems, mainly through the application of E- Library, Survey, and Compare evaluation. In the Information Science arena, he has worked on TN Public Online Library . He has explored the presence and implications of self-similarity and heavy-tailed distributions in Open Source Journals. He has also investigated the implications of Web workloads for the design of scalable and no cost-effective Web Pages. In addition, he has made numerous contributions to research papers like Journals, Conference and Book Chapters

Geetha Manoharan is currently working in Telangana as an assistant professor at SR University. She is the university-level PhD programme coordinator and has also been given the additional responsibility of In Charge Director of Publications and

Patents under the Research Division at SR University. Under her tutelage, students are inspired to reach their full potential in all areas of their education and beyond through experiential learning. It creates an atmosphere conducive to the growth of students into independent thinkers and avid readers. She has more than ten years of experience across the board in the business world, academia, and the academy. She has a keen interest in the study of organisational behaviour and management. More than forty articles and books have been published in scholarly venues such as UGC-refereed, SCOPUS, Web of Science, and Springer. Over the past six-plus years, she has participated in varied research and student exchange programmes at both the national and international levels. A total of five of her collaborative innovations in this area have already been published and patented. Emotional intelligence, self-efficacy, and work-life balance are among her specialties. She organises programmes for academic organisations. She belongs to several professional organisations, including the CMA and the CPC. The TIPSGLOBAL Institute of Coimbatore has recognised her twice (in 2017 and 2018) for her outstanding academic performance.

Malathi Murugesan was born at Namakkal, India. She completed her B.E in Electronics and Communication Engineering from Government College of Engineering, Salem, M.E in Communication Systems from Kumaraguru College of Technology, Coimbatore and Ph.D in Annamalai University, Chidhambaram. Currently she is working as an Associate Professor and Head in the Department of Electronics and Communication Engineering in E.G.S Pillay Engineering College, Nagapattinam and has published papers in many International and National journals and patents. She is having 19 years of teaching experience and her research interest include Image Processing, Antennas, Communication Systems and Wireless networks.

Priyanka N. is currently working as an assistant professor at the Vellore Institute of Technology, Vellore. She did her Ph.D. at Anna University, Chennai. She did her Master's degree in software engineering at Anna University, Chennai. Her areas of interest include software-defined networks, next-generation wireless networks, the Internet of Things, and network security.

N. Suresh Kumar, Principal and Professor in ECE Department of Velammal College of Engineering and Technology, Madurai, obtained his B.E from Thiagarajar College of Engineering, Madurai, M.E from AlagappaChettiar College of Engineering, Karaikudi and Ph.D from Madurai Kamaraj University. He has more than 33 years of Teaching and Research Experience. He has a significant contribution in carrying out several research projects in EMI/EMC, he has published and presented in many papers in journals and conferences, He is a life member of IEEE, ISTE, IETE and IE.

P. Rajeswari, Associate Professor of ECE Department of Velammal College of Engineering & Technology, Madurai, obtained her B.E., degree from Madurai Kamaraj University, Madurai and M.E. degree from Anna University, Chennai. She has more than 18 years of Teaching experience. Completed Ph.D. in Anna University, Chennai in EMI/EMC. She has published and presented many research papers in journals and international conferences. Her area of research includes EMI/EMC and Wireless communication. She has a significant contribution in carrying out several research projects in Electromagnetic interference and Compatibility, She is a life member of ISTE, IETE and Society of EMC Engineers (India).

P. Vijayakumar, is currently working as an Assistant Professor (SG) at Department of Aeronautical Engineering, Nehru Institute of Technology (Autonomous), Coimbatore. He served totally 10+ years in teaching field at various colleges. He obtained his doctorate degree in Thermal Science and published more than 15+ articles in various reputed journals, published patents.

Hosnna Princye received her BE(E&I) from Sapthagiri College of Engineering from Periyar University, Tamilnadu, India in the year 2002 and completed her Masters in Engineering(Applied Electronics) from Anna University in the year of 2004 and since then actively involved in teaching and research and has Fourteen years of experience in Teaching. She obtained his PhD in the field of Information and Communication Engineering from Anna University in the year of 2018. At Present, she is working as an Associate Professor in Sri Sairam College of Engineering and Technology, anekal, Bangalore affiliated to Visveswaraya Technological University, her area of interest is the field of medical image processing, signal processing and VLSI.

R. Renugadevi completed her Ph.D in 2022 from Anna University, Chennai. She has published many papers in International journals and Conferences. She is currently working as Associate professor in R.M.K Engineering College. She has more than 16 years of teaching experience in engineering colleges. Her interest includes Machine learning, Internet of Things, Wireless Sensor Networks and Cloud computing. She is life member of IAENG and ISTE.

S. Edwin Raja, Assistant Professor, Department of Computer Science and Engineering in Vel Tech Rangarajan Dr.Sagunthala R&D Institute of Science and Technology, Chennai. He has completed his Ph.D. in Information and Communication Engineering from Anna University-chennai in the year 2021. He obtained his M.Tech. (Network Engineering) in the year 2010 from Kalasalingam University, Krishnan kovil and B.E. (Computer Science and Engineering) in the year 2006 from

C.S.I Institute of Technology, Thovalai. He has more than thirteen plus years of teaching experience and his area of interests are Cyber Security, Big Data Analytics, Machine Learning & Computer Networks. He has published more than 4 research articles in various journals and conferences. Also, he has published 3 patents, acted as reviewer in reputed journals.

S. Nalini currently working as an Assistant Professor in the Department of Computing Technologies, School of Computing, SRM Institute of Science and Technology, Kattankulathur, Chennai. She awarded Ph.D in Anna University, Chennai. She earned a Silver Medal for completing her M.E. (Computer Science and Engineering) at Vinayaka Missions University, Salem. She has published more papers in referred journals and international conference proceedings. She has supervised more than 30 undergraduate and 10 post graduate students. She has acted as co-investigator in many funding agencies like CSIR, DST, DRDO and organized many workshops, symposiums and technical events for the benefit of students, faculty members and research scholars. Her field of specialization includes Machine learning, Deep learning, Quantum Computing and Quantum Machine Learning.

Satheesh Kumar S. is working as an Assistant Professor (SG) in the Department of Aeronautical Engineering at Nehru Institute of Technology. He is an Entrepreneur heading the Market Metro Group, Coimbatore. He graduated in Mechanical Engineering at PSNA College of Engineering and Technology, Dindigul, Tamilnadu, India. He secured his Masters in Thermal Engineering at Government College of Technology, Coimbatore, Tamilnadu, India. He is Pursuing Ph.D. in the field of Tribology at Anna University, Chennai, India. He has got Industrial exposure and he is in the teaching profession for more than 12 years. He has organized International Conferences and Science Expos. He has published seven papers in International Journals. He has presented a number of papers in National and International Conferences. His main area of interest includes Thermodynamics and Renewable Energy.

S. Vaishnavi received her Doctor of Philosophy from Anna University, Chennai. She received the Gold medalist in Master Degree from RMD Engineering College, Chennai. She is currently working as Assistant Professor Senior Grade in Manipal Academy of Higher Education, Bangalore. Her research area is Internet of Things and Security. She has published 2 articles in peer reviewed International journals presented 4 papers in International Conferences.

Udhaya Sankar S. M. is currently working as Professor & Head, Department of CSE (Cyber Security), RMK College of Engineering and Technology (RMKCET), Tiruvallur, India. He has completed his Ph.D. degree in the Faculty of

Information and Communication Engineering from Anna University, Chennai in 2018. He received his Master degree in Computer Science and Engineering from Velammal Engineering College, Anna University, Chennai in 2008 and Bachelor's degree in Computer Science and Engineering from Sethu Institute of Technology, Madurai Kamaraj University, Madurai in 2000. He has nearly 20 years of experience in teaching at under graduate and graduate level. He has attended several conferences, workshops and published few papers in international/National Journals and Conferences. He is a Member of ACM,ISTE, Internet Society and IAENG. His research interest includes Network Security, Wireless Network, Ad Hoc Network, Information Security & IoT.

J. Shobana received Ph.D. in Computer Science and Engineering. Working as an Assistant Professor in the department of Computer Science and Engineering, SRM Institute of Science and Technology. She completed her masters M.E in Madha Engineering College. She has been serving the Education Profession for the past 17 years. Her area of interest is Text Mining,Natural Language Processing, Artificial Intelligence and Machine Learning.

T. Chandrasekar, a distinguished professional with a Ph.D. in Management Studies from Anna University, Chennai, currently leads as the Head of Business Administration at Kalasalingam Academy of Research and Education, India. With a Six Sigma Black Belt certification from MSME, he brings over 18 years of teaching experience. Author of three impactful textbooks, he has 16 publications in Scopus indexed journals, holds a patent, and serves as a Reasoner for MoUs with MSME. His expertise spans Operation Management, Strategic Management, Human Resource Management, Marketing Management, Lean Six Sigma, Control System, and Digital Signal Processing. Formerly the Controller of Examinations at an autonomous institution, he exhibits administrative powers and a commitment to academic excellence.

T. Cynthia Anbuselvi obtained her B.E. degree from Kamaraj college of engineering and technology, M.E. degree from Thiagarajar college of Engineering, and PhD (Full time) from Anna University in 2009, 2011 and 2020 respectively. She is currently working as Assistant Professor in S.E.A College of Engineering and Technology, Bangalore. She has published papers in reputed SCI Indexed journals with high impact factor. She has published papers in the national and international conferences. Reviewer in SCI Indexed journals. Her research interest includes Wireless Communication, Machine Learing, Cognitive Radio Networks and Smart Grid.

Index

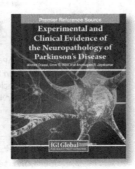

Printed in the United States
by Baker & Taylor Publisher Services